Gabriel Lippmann's Colour Photography: Science, Media, Museums

Edited by
Hanin Hannouch

Routledge
Taylor & Francis Group
LONDON AND NEW YORK

The publication of this book is made possible by a grant from Preus Museum: Norway's National Museum for Photography (Horten) and Photo Elysée (Lausanne)

First published in 2022 by Amsterdam University Press Ltd.

Published 2025 by Routledge
4 Park Square, Milton Park, Abingdon, Oxon OX14 4RN
605 Third Avenue, New York, NY 10158

Routledge is an imprint of the Taylor & Francis Group, an informa business

ISBN: 9789463728553 (hbk)
ISBN: 9781041179962 (pbk)
ISBN: 9781003696292 (ebk)
NUR 670

Cover illustration: Richard Gustav Neuhauss (1855-1915). Garten im Grosslichterfelde. Bootstr.17 [no original title]. Lippmann process. 6,5 cm × 8,4 cm. Glass plate with wedge prism. Undated. Albert Narath Collection. Preus Museum.

Cover design: Coördesign, Leiden

DOI: 10.5117/9789463728553

Gabriel Lippmann's Colour Photography

Table of contents

Gabriel Lippmann's Colour Photography: A Critical Introduction

Hanin Hannouch

Abstract
This introduction, in lieu of a streamlined biography of physicist Gabriel Lippmann (1845–1921), complicates photography historian Georges Potonniée's three canonical predicates about Lippmann's interferential colour photography. These predicates, which have shaped most studies about him so far, are: Lippmann's association with France, his position as the sole inventor of interferential colour photography, and the myth of the "non-use" of this medium. After presenting the state of the art, the introduction offers readers interdisciplinary methods and considerations that embrace the complexity of the interferential image, positioning it at the interplay between science, media, and museums.

Keywords: Gabriel Lippmann, interferential colour photography, Wilhelm Zenker, media history, history of photography, history of science

> Who made any money out of colour photography by the interference method? No one. Nevertheless, would anyone dare to challenge the right of [Gabriel] Lippmann to be called an inventor? (Potonniée 1936, 126).

In his seminal *Histoire de la découverte de la photographie* (1925, translated in 1936 by Edward Epstean), photography historian Georges Potonniée mentions physicist and photographer Gabriel Lippmann (1845–1921) in passing as he charts the claims of both Louis-Jacques-Mandé Daguerre (1787–1851) and Joseph-Nicéphore Niépce (1765–1833) to the first photograph. The commercial success of the daguerreotype long overshadowed the scientific contribution of Daguerre's collaborator Niépce. To stake a larger claim than what has been attributed to Niépce in the history of photography, Potonniée bolsters his argument by referring to Gabriel Lippmann whose own input, though

Hannouch, H. (ed.), *Gabriel Lippmann's Colour Photography: Science, Media, Museums.*
Taylor & Francis Group, 2022
DOI: 10.5117/9789463728553_INTRO

significant to colour photography, has been consigned to oblivion since it proved to be financially unviable. Yet, for Potonniée, its commercial failure does not diminish the scientific achievement underpinning it, nor does it, as this book contends, justify its marginalisation in the history of photography.

Lippmann disclosed a colour photograph on glass featuring the spectrum of light on February 2, 1891 at the French Académie des sciences in Paris. The photographic process Lippmann used is called "interferential colour photography," since the colours visible in the image are the result of light waves "interfering" with their own reflection. This interference principle allows us to see rainbow colours in soap bubbles or in seashell nacre. Lippmann had thus succeeded in recording these optical standing waves in a single-layer, ultra-fine-grain emulsion made from silver iodide and/or silver bromide in albumen. His work differs from the application of pigments or dyestuff unto its photographic surface also known as hand-colouring, and from the reliance on three-colour photography whose separation processes split light into three primary colours which can then be reconstructed through projection, the use of viewers or printing (Hannouch 2022).

The year 2021 marks the centenary of Lippmann's death. *Gabriel Lippmann's Colour Photography: Science, Media, Museums* uses this as an opportunity to reflect on Lippmann's legacy and reconsider his colourful endeavour. This book seeks to enrich the debates about his life, his work, and his international reception in order to enable scholars to approach him from a variety of vantage points. It hopes to furnish readers with interdisciplinary methods and considerations to navigate colour photography; ones that embrace the complexity of its image, thus operating at the interplay between science, media, and museums. These three forces, which shaped the emergence of Lippmann photography, likewise constitute this book's three sections: in the first section, the authors provide nuanced answers and a clear orientation to elemental questions surrounding the science of colour, light waves, and photography. In the second section, writers complicate Lippmann photography's position in media history, delve into the aesthetic appreciation for the colour image, and pluralise its ties to cultural history and to neighbouring technologies. The third part of this volume surveys the renewed enthusiasm and material care for Lippmann plates, both in museums and exhibition spaces today, as well as among contemporary analogue photographers. Readers of *Gabriel Lippmann's Colour Photography: Science, Media, Museums* will immerse themselves in colour photography as its own field of inquiry. They will explore the questions contouring it, pertaining to the nature of scientific knowledge, its interdependence on other media, and the impact politics and culture have had on it. Moreover, archivists, conservators, curators, and photographers tending to interferential colour photographs in their museum or personal collections will gain knowledge on how to address their materiality, preservation, and exhibition.

Researching Colour Photography

This introduction begins with a state of the art meant to help readers situate Gabriel Lippmann and colour photography in the history of photography as such. In lieu of a streamlined biography of the scientist, this introduction also investigates Potonniée's three predicates about Lippmann, which are embedded in his rhetorical question cited above to help the audience orient themselves in the topic's various historical, cultural, and medial entanglements: Lippmann's association with France, his position as the sole inventor of the interferential colour process, and the myth of the "non-use" of his colour photography. The Lippmann process was internationally discussed for almost two decades (from 1891 until around 1910), Lippmann even won the Nobel Prize for Physics in 1908 against quantum physicist Max Planck (1858–1947), and the aesthetic quality of the photographs he and practitioners of his method produced was well-received. Nevertheless, there is no book-length history of interferential photography or a published biography of Lippmann so far. Moreover, researching his life and work today is arduous since writers have to grapple with the historiographic vacuum around colour photography at the turn of the last century. In fact, throughout the nineteenth century, the absence of colour was constantly bemoaned as a "problem" across a steady stream of publications, such as Ducos de Hauron's 1869 *Les Couleurs en photographie: solution du problem*, Paul Liesegang's 1884 edited volume *Die Heliochromie : das Problem des Photographirens in natürlichen Farben*, and Hermann Krone's 1893 *Über das Problem, in natürlichen Farben zu photographiren*. With this in mind, it is surprising how little scholarly attention photographers' solutions to this predicament have received during the emergence of the grand narratives about the medium in the 1920s (Brunet 2017). The current literature on photographic histories continues to exclude colour, except for the autochrome. The "autochrome turn," which is the increased investigation of the Lumière brothers' colour process over the last two decades, has bestowed upon it the honorary and exceptional title of the "first" commercial colour technology (Bjorli and Jakobsen 2020, Fuchs 2017, Lavédrine 2009, Boulouch 2008, Amad 2007, Boulouch 1995, Wood 1993). While this approach has fruitfully directed research to colour photographs produced before or during World War I, it has rarely led to the study of the multitude of colour media that were coeval to the autochrome. It has also ignored the numerous debates about colour processes in the nineteenth century that shaped public sensitivities to and expectations of colour images, such as interferential colour photography.

The fragmentary nature of the secondary literature about this photographic process and its uneven weight in favour of the history of science and the under-representation of photographic or media history compounds the difficulties scholars face. Beyond a handful of writers, engagements with the Lippmann process and the scientist whose name it bears are as sporadic as they are significant. They have emanated more from science historians and practitioners and less from photography theorists and artists, arguably because interferential colour photography stems from the French scientific milieu. Daniel Jon Mitchell's unpublished doctoral thesis "Gabriel Lippmann's Approach to Late-Nineteenth Century French Physics" (2010) sketches Lippmann's intellectual biography beyond his investigation of colour, including Lipmmann's study of electrical standardisation. In doing so, the dissertation is unique in its scope, being the only research about the scientist's life. Klaus Hentschel's 2002 book *Mapping the Spectrum: Techniques of Visual Representation in Research and Teaching* positions Lippmann's success in colourfully recording electric and solar arc spectra within spectroscopy's broader imaging techniques and their interpretation. Also, Pierre Connes's 1987 foundational article "Silver Salts and Standing Waves: The History of Interference Colour Photography" traces the circulation of knowledge about standing waves across Europe, with a focus on Lippmann's theoretical formulation.

The subsequent use of Lippmann photography's emulsion for holograms has garnered interest for Lippmann among holography enthusiasts. Historian of science and artist Susan Gamble's doctoral dissertation "The Hologram and Its Antecedents 1891–1965: The Illusory History of a Three-Dimensional Illusion" (2004) surveys Lippmann photography's emergence and reception and its relationship to Dennis Gabor's (1900–1979) discovery of the hologram in 1948 and to other scientists. Gamble's creative work as part of the artist duo Wenyon & Gamble combines art, optics, and colour. Also, Sean Johnston's book *Holographic Visions: A History of New Science* (2006), which explores holography as a distinct field of inquiry, its past and its present, shows Soviet physicist Yuri Denisyuck's (1927–2006) reliance on Lippmann's conceptualisation of three-dimensional images for the technology he produced.

Among contemporary optical engineers and scientists, Hans I. Bjelkhagen is a key figure in the revival of Lippmann's method. The multiple primary sources about it, which he made available online for free, are tremendously helpful. So, too, are his studies of its relationship to holography in articles such as "Lippmann Photography" (1999) and "Silver Halide Emulsions for Lippmann Photography and Holography" (1996), and in his book *Ultra-Realistic Imaging: Advanced Techniques in Analogue and Digital Colour Holography* (2013). William R. Alschuler's international, decades-long survey of over four hundred Lippmann plates in private and public collections, as well as his studies of their materiality, of process history, and of their connection to holography, combines his multifaceted and interdisciplinary scholarship with connoisseurship, clarifying both the theory and the physicality of Lippmann plates. Must-read articles by Alschuler include "Physical and Visual State of 100–year-old Lippman Color Photographs" (1998) and "Lippmann Photography and the Glory of Frozen Light: Eternal Photographic Color Real and False" (2003). Jean-Marc Fournier likewise delved into interferential images' past and holographic connection by analysing their microstructure, the results of which he published in articles such as "An Investigation on Lippmann Photographs: Materials, Processes, and Color Rendition" (1994).

Despite the keenness of the scientific community about Lippmann, contemporary photography historians have paid little attention to him. A notable exception is Kim Timby's monograph *3D and Animated Lenticular Photography* (2015), focusing on Lippmann's preoccupation with lenticular photography modelled on insect eyes, which emerged out of his colour research "in the hopes of creating images as realistic as a view out a window" (Timby 2015, 13).

To integrate Lippmann further in the history of (colour) photography, unpacking Potonniée's undertones are key, starting with his association of Lippmann to France, a relationship that Potonniée takes for granted.

Gabriel Lippmann and France

Potonniée includes Lippmann in a conversation about Daguerre and Niépce, two photographers whose productions helped position France as the giver of the medium to humankind (Brunet 2011, 2017). Yet, Lippmann was born in Luxembourg, where his family had been living for generations (Wagener 2013, 40). Due to antisemitism in Luxembourg, the Lippmanns struggled with naturalisation and voting rights (Wagener 2016, 158–9) and their efforts to settle there were thwarted. Gabriel's father, Isaïe Lippmann, was excluded from the electoral roll to vote in his community of Hollerich in 1845 (Wagener 2013, 39), despite his best efforts, because he was Jewish.

Therefore, the Lippmanns moved to France, where the introduction of the Code Napoléon allowed them to remain, and they took their leather glove manufacturing company with them. France, however, was hardly free from discrimination. Lippmann's disclosure of the colour spectrum at the Académie des sciences in Paris in 1891, and his subsequent research about interferential colour photography, unfolded in light of a major political scandal in the Third Republic. Starting in 1894, the Dreyfus affair tore apart the fabric of French society. The scandal lasted for over a decade, during which, due to antisemitism and a spectacular miscarriage of justice, Captain Alfred Dreyfus (1859–1935) was sentenced to life imprisonment for treason, having allegedly divulged French military secrets to the German Empire. This event polarised the French public at the time, and several scientists in Paris positioned themselves in relation to the Dreyfus affair. Lippmann considered himself a *dreyfusard* and asserted Dreyfus's innocence (Noronha-DiVanna 2010, 45), despite the continuous rise of the French far right and its xenophobic, misogynistic, and antisemitic overtones.

Also, since Lippmann had been elected member of the Académie des sciences in 1886, he presented the research of non-member scholars, such as Marie Skłodowska and Pierre Curie's first paper on radioactivity, titled "Rayons émis par les composés de l'uranium et du thorium" (1898), thus beginning a long cooperation with Skłodowska. Lippmann amiably worked with her, was her doctoral advisor, and was part of the defence committee for her thesis (which is why his exclusion of her in a petition nominating Pierre Curie and Henri Becquerel for the 1903 Nobel Prize in Physics is disappointing) (Lestel 2007, 112). This cooperation did little to better their social positions. Skłodowska was wrongly considered Jewish, and she was vilified for being a homewrecker due to her clandestine love affair with physicist Paul Langevin (1872–1946) after Pierre Curie's death, which had detrimentally reached the press. Lippmann refused to publicly condemn her private life, and along with mathematician Henri Poincaré (1854–1912) and other scientists, he openly supported her ambition of being elected into the Académie des sciences, which rejected her nomination in 1911.

In his support for Skłodowska, combined with his status as a *dreyfusard*, Lippmann became the target of *La France profonde*, a protofascist set of ideas (Weissmann 2010, 4) underscoring the importance of rural life away from the centrality of Paris. Monarchist journalist and editor-in-chief of the nationalist newspaper *L'Action Française* Léon Daudet (1867–1942) championed this conservative movement (Weissmann 2018, 166). He repeatedly attacked both Skłodowska and Lippmann in writing, referring to Lippmann as "the Jew of colour photography" (Daudet 1911), among numerous insults. Yet, Lippmann's

Nobel Prize win in 1908 against Planck was heralded in France as a national and nationalist victory against the German Empire, while newspapers in Luxembourg appropriated his academic success and hailed his local origins (Pier and Massard 1997). Not much critical reflection on the reasons for his troubles in France or on his family's departure from Luxembourg (Wagener 2013, 40) took place in light of his win. Following Lippmann's death a mere decade after these events, scientist and photochemist Daniel Berthelot (1865–1927) published a lengthy obituary of him in the conservative journal *La Revue des deux mondes*, which had previously been "prudently reserved" about the Dreyfus affair. In it, Berthelot refers to the "sentiment of grandeur of loss" that French science experienced with Lippmann's demise (Berthelot 1922, 19). He indicates the religion of Lippmann's parents as "Israelite" (Berthelot 1922, 43) and that he was interested in the "the ironic and malevolent gaze of the foreigner" Albert Einstein's (1879–1955) relativity theory. Unsurprisingly, Berthelot silences the extent to which Lippmann's stance on Dreyfus and on Marie Skłodowska fuelled the hateful discrimination they both faced.

Read in this light, Potonniée's brief evocation of Lippmann in the same breath as two other major French photographers is significant, because it obfuscates the historical conditions under which Lippmann and his family grappled with antisemitism both in Luxembourg and in France. It also flattens Lippmann's ambiguous position: he was the *French* Nobel Prize winner seen from the outside, but the Jewish scientist seen from within French society. Potonniée also absorbs him into a French photographic memory in which he occupies the uncertain place of a local inventor whose genius nobody doubted, but whose work nobody implemented.

For scholars today, bringing interferential photography out of its marginalisation in the history of colour photography has to begin by historicising the scientist and by placing his research within broader cultural, political, and aesthetic trajectories.

Gabriel Lippmann as Sole Inventor

Photography's foundational dates – 1891, the year of Lippmann's presentation at the Académie des sciences, being one of them – are the crux of the hierarchy of knowledge around which practitioners and historians articulate both the medium's prehistory and its "general" and often nationalistic historiographies. Potonniée's assertion that nobody doubted Lippmann as the sole inventor of the colour process turns his history of photography written from a French vantage point into an overarching and universal narrative. Lippmann's success propelled him into the disputed terrain of the first colour photograph between the French and the German Empires. It led authors on both banks of the Rhine to produce compelling accounts of national success and to domesticate larger epistemological and cultural divergences between themselves as exclusive Western European contenders. The variety of terminologies used in reference to interferential photography, such as "direct colour photography," "the Lippmann process," and "the Zenker-Lippmann process," reveal this much. They all describe the same photographic process and its complicated relationship to standing light waves. They differ, however, in relation to the positionality of the history of colour photography, as in the stance from which historians understood and located its technical genesis, both geographically and epistemologically. Doubt about Lippmann as the sole inventor was palpable, and nowhere was his reception more heated or contentious than in the German Empire.

Gabriel Lippmann had numerous biographical ties to his neighbouring country. His multilingualism (he spoke French, German, and English) opened up several avenues for him within local and international academic exchange, which Daniel Jon Mitchell extensively explored. After being admitted to the Parisian École Normale Supérieur in 1868, Lippmann worked with Pierre-Augustin Bertin (1818–1884), who tasked him with the translation and analysis of German and English scientific publications, which were widely discussed among department members. Lippmann's initial interest in electrical phenomena and his inquiry into the discharge of condensers emerged out of Bertin's preoccupation with electricity and galvanism (Mitchell 2010b, 37–38). Yet, the eruption of the Franco-Prussian War (1870–71) and the ensuing violence during the Paris Commune drew Lippmann away from the *agrégation*, to which he became inadmissible in 1872. This qualification could have secured him a junior laboratory position, later a doctorate, and could have facilitated his career in French academia. By that time, the Ministry of Public Instruction considered the rise of German laboratories and universities a possible threat to French science, whose marginalisation within the international research landscape was deemed a major culprit in France's military defeat (Mitchell 2010b, 39). Therefore, the ministry commissioned numerous French scientists to report on the state of research in disciplines such as chemistry, physiology, anatomy, and pathology practised across the border. At its behest in 1872, and to avoid becoming a secondary school teacher without the *agrégation*, Lippmann could continue his research about electricity at his laboratory of choice in the German Empire, provided that he represented French science, whose rehabilitation facing its neighbouring country became "a patriotic duty of government associated with the extraction of revenge" (Mitchell 2010b, 40). At the University of Heidelberg, Lippmann worked at Wilhelm Kühne's (1837–1900) physiology laboratory and Gustav Kirchhoff's (1824–87) physics laboratory. With Kirchhoff's encouragement, he passed his examination for the title of Doctor of Philosophy with distinction in 1875, with a thesis on the change in the surface tension of a liquid when electrically charged, titled "Relations entre les phénomènes électriques et capillaires" (Fechete 2016, 1). Both Hermann von Helmholtz (1821–94) and Kühne served, respectively, as doctoral and academic advisors to Lippmann. He wished to continue reporting on German institutions (Mitchell 2010b, 86–87) and moved to Berlin to work alongside Helmholtz, whose theory of the electrical double layer was widely discussed and whose notion of three-colour vision was instrumental to three-colour photography, but was recalled back to Paris. There, Lippmann successively held the chairs of mathematical physics (1883) and of physics (1886), presiding over the prestigious Laboratoire des Recherches Physiques de la Sorbonne. Hence, Lippmann was known in some scientific circles in the German Empire and was himself familiar with some of the ongoing research there. This set the ground for the antagonistic reception of his colour process across the Rhine, especially the claim to priority by German scientist Wilhelm Zenker (1829–99).

The supposed theoretical precedent set by Wilhelm Zenker long haunted interferential colour photography's historiography. It can be exemplified by this question: Did Gabriel Lippmann invent colour photography in 1891 (as Potonniée asserted), or was Lippmann merely proving the theories that Zenker initially expressed in 1868?

Lippmann's colourful 1891 session took place right after physicist Marie Alfred Cornu (1841–1902) presented Otto Wiener's (1862–1927) research about the orientation of the polarisation plane of light rays in relation to their propagation direction at the Academy (Mitchell 2010b, 190), based on Wiener's initial article "Stehenden Lichtwellen und die Schwingungsrichtung polarisirten Lichtes" of 1889. In his text, Wiener briefly cites scientist Wilhelm Zenker's self-published (1829–90) *Lehrbuch der Photochromie – Photographie der Natürlichen Farben* of 1868, a mention which Cornu reiterates. In *Lehrbuch*, Zenker uses standing waves to explain the genesis of colour in the images produced by Becquerel, Sir John Herschel, and Johann Seebeck. The book's reception was virtually non-existent among scientists (Connes 1987, 152), and it only began in the German-speaking photographic circles starting in 1891 – after Lippmann disclosed his interferential method. Wiener's reference to Zenker brought him back from photographic oblivion, and the latter published an article, "Die Entstehung der Farben in der Photochromie" about the wave theory in the 1891 edition of Josef Maria Eder's *Jahrbuch für Photographie und Reproduktionstechnik*. In it, Zenker posits that a "general interest in the photographic reproduction of colour" (Zenker 1891, 294–95) can be sensed, and that while it is "possible" that accidents lead to progress, it is "particularly advantageous" (*von besonderem Vortheil*) that one inquires deeper into pre-existing scholarship, since it offers guidance for one's own research. The author explained the emergence of colour in Becquerel's plates, cited Wiener's work, and made no mention of Lippmann at all, if indeed the article was published after February 2, 1891. In parallel to Zenker, just four days after Lippmann's presentation in Paris, his colleague and promoter physicist Alphonse Berget (1860–1933) publicised his discovery as entirely autonomous of German or British (Lord Rayleigh's) scholarship, underlining its exclusive French origin. He reaffirmed this stance in his 1901 treatise, writing that "it has nothing in common with the photochromatic essays attempted until this day. It is so ingenious, so elegant in its genius simplicity that it is certainly a scientific oeuvre more beautiful than the actual result" (Berget 1901, VII). In doing so, he rebutted the hegemonic narrative in the German Empire, where Lippmann was thought to have merely confirmed a pre-existing theory through a novel research method: colour photography. Furthermore, as early as April 10, 1891, at the Physikalische Gesellschaft zu Berlin, which hosted regular meetings about new colour technologies and theories of vision, and which was still presided over by Hermann von Helmholtz (Schreier, Frankeunter, and Fiedler 1995, 41–42), Hermann Wilhelm Vogel (1834–98) presented the interference theory of light and colour photographs of spectra made by Lippmann. Vogel, who was the first professor of photochemistry at the Technical University of Berlin and who was known for his sensitisation of the orthochromatic plate, by which he added "aniline dyes to his emulsions to stretch its sensitivity well into the green, and eventually he extended his films to orange sensitivity" (Wilder 2009b, 166), was an authority on photographic matters. He added fuel to the fire and solidified the German claim to priority over Lippmann by anchoring the latter's work in Zenker's *Lehrbuch der Photochromie* from 1868 and by refuting any British claim to Lord Rayleigh's priority (Vogel 1892, 34–35). Also, in a survey of the year 1892's progress in photography and reproduction technologies in Eder's *Jahrbuch*, Lippmann's colour photography and Wiener's inquiry were considered as confirmations (*bestätigen*) of Zenker's theory, stressing how the silver salts in the emulsion are stored in both irregular layers as Lippmann predicated as well as in regular ones as Zenker "stipulated" (*verlangt*) (Anonymous 1892,

331). In his 1893 article "*Ueber die Enstehung der Farben im Lippmann'schen Spectrum,*" Zenker himself began referring to the "Lippmann-Zenker'sche Theorie" (Zenker 1893, 115), hyphenating his name to that of the French physicist whose images he deemed a mere method (*Methode*) validating his own concepts (Zenker 1893, 114). Anthropologist and doctor of medicine Richard Neuhauss, who was a key Lippmann photographer in Berlin, also emphasised how the Lippmann process, in making images fixable, actually *applies* Zenker's notions (Neuhauss 1894, 296). With this much recognition, in 1895, Zenker became an honorary member of the Deutsche Gesellschaft von Freunden der Photographie in Berlin, which considered him the founder (*Begrunder*) of photography in natural colours, as announced in the journal *Photographische Rundschau* that year. *Lehrbuch der Photochromie* was republished posthumously in 1900, and Neuhauss reviewed it along the German nationalist line, heralding it as the work in which Lippmann's foundational theory is expressed *avant la lettre* (Neuhauss 1900, 148). Nevertheless, around that time in Paris, A. Berthier's introduction to his *Manuel de Photochromie Interferentielle* subtly but clearly maintains Lippmann as "the creator himself" (*le createur lui-même*) (Berthier 1895, 2) of the photographic technology.

Nobody attempted to explain the two-decade gap between *Lehrbuch*'s release in the German Empire and the 1891 announcement in Paris or why Zenker never tried colour photography himself to prove his own theory and gain broader recognition for his work at the time of its initial publication. The "Zenker before Lippmann" historiographic paradigm, which began as early as 1891 and continued until ca. 1910, thus countered its coeval "Lippmann as sole inventor" narrative in France, broached by Potonniée.

The relationship between Lippmann and Zenker, whether real or imagined, along with the fact that the presentation of Lippmann's colour spectrum in 1891 sits somewhat awkwardly in the panorama of his work at the time, has led to various claims to priority in France, along with contemporary speculations about colour photography's relation to Lippmann's previous research , which Alschuler's chapter in this book surveys. The note Lippmann read to the Academy that year was accompanied by little detail. It also contained no references to other scholars and no clarification about prior experiments, and nothing in Lippmann's lectures at La Sorbonne alludes to any investigation of standing waves before 1891 (Connes 1987, 157). Therefore, the date which scholars consider to be the starting point of Lippmann's research is equally pivotal for grasping his relationship to Zenker. Alschuler states in his chapter that Lippmann's 1889 attempts to expand the sensitivity of orthochromatic emulsions and his previous work on three-colour photography informed his research on the interferential image. He also agrees with historian of science Pierre Connes that the emergence of Lippmann's optical standing wave theory (and, by extension, his photographic practice) can be linked to his previous research on acoustic waves (Connes 1987, 151), thus rebutting the possibility that he relied on Zenker's work since their starting points were different. Connes insists that Lippmann maintained the starting date of his research to be 1886 (Connes 1987, 153), while Alschuler's text points to Lippmann lodging the start of his research even earlier than that – in 1879 – thus strengthening the argument that his work on acoustics predates his possible knowledge of Wiener (and, by extension, the possibility of him knowing Zenker via Wiener, in both cases). Also, as Klaus Hentschel points out in his contribution to this volume, Lippmann's overall mastery of physics gave him enough empirical and theoretical knowledge to undertake the research he saw fit without needing to rely on a scientist as obscure as Zenker. This marks "the acoustic to optical wave" trajectory as a plausible explanation.

The unclarity about the origins of interferential colour photography persisted during the first years after its presentation in 1891. Daniel Jon Mitchell (2010b) asserts that Lippmann suppressed the extent of his chemical research during and before interferential photography's disclosure in order to highlight its theoretical elegance, for which he won the Nobel Prize for Physics. And if Lippmann's starting point was purely theoretical, as he declared, then it is surprising that he only revealed the photographic process's Fourier transform; as in the mathematical rendition of the theory underpinning it, three years after the image's revelation (Mitchell 2010b, 326). This marks the "chemistry VS physics" contention. Yet, as Alschuler claims, "any chemistry with sufficient resolution to resolve the interference pattern to a 10th of the wavelength will succeed in recording it in satisfactory detail" (W. R. Alschuler, 2021, personal communication, March 2021). This means that any silver halide–based photographic material (be it albumen, gelatin, or collodion, etc.) will manage to resolve light – provided it is fine and consistent enough. For this reason, Lippmann did not need to elaborate on his chemistry because all emulsions worked. Hence, the interplay between chemistry and physics in interferential photography and the establishment of disciplinary boundaries between these two fields should fuel future considerations of Lippmann photography.

Potonniée, in his brief but crucial mention of Lippmann, suppresses these contentions to Lippmann being the sole scientist behind interferential colour photography, although his colleagues at the time were clearly aware that Zenker set a theoretical precedent to Lippmann's disclosure, one powerful enough to alter the geographic location of the medium's emergence. Zenker risked dimming the lustre of interferential photography as a French achievement by inadvertently aligning Lippmann with the German research tradition on optical waves. And in the German Empire, it did just that, becoming a drawn-out chauvinistic battle for the first colour photograph and of the precedence a theory about colour photography can take over the image.

Nevertheless, disputes about Lippmann's merits and limitations have unfolded at the expense of the multiple ways in which colour photography and the wave theory of light came into being in various parts of the world. Chapters in this volume evoke colour photography's numerous practitioners in the German Empire, such as Richard Neuhauss, Hermann Krone, and Hans Lehmann; in Spain, through the photography of science by Santiago Ramón y Cajal; in Austria, through the writings of Edward Steichen and Josef-Maria Eder; and in the United States, with the work of Herbert and Frederic Ives. Knowing this, this book acknowledges its focus on the Global North. It recognises that the handful of countries it mentions cannot account for the international fascination with the medium, and it asserts that remaining within this geographic epistemology can only propel research about the history of this neglected medium up to a point. Readers are encouraged to pursue the fundamental disparity at the heart of *all* "beginnings" as historiographic constructions in order to dispel the chimera of their (Northern) unifying origin. Future scholars are called upon to transregionally examine colour photography's "first" images and to activate future discussions about its manifold outsets and numerous histories beyond Western Europe and the United States. In this book, Alschuler opens up fascinating avenues for this kind of work: Lippmann was read and reviewed in Argentina as early as 1893, which means that interferential photography was known there. Also, in the Russian Empire, physicist Ivan Filippovich Usagin (1855–1919) published "Practical Techniques of Photographing by Lippmann's Method Used for Making Photos of Solids and Vapours Spectrums" in the 1903 edition of the journal *News of Russian Association of Photography Amateurs*. He also demonstrated a colour image of the light spectrum at the world fair in Paris in 1900, which marks his reception there (Sogreshilin 1966, 36). Additional scholarship is needed to extend the history of interferential colour photography to these countries and to many more.

Moreover, expanding the state of the art about Lippmann has to account for another parameter that shapes historical inquiry and writing: gender. "In order to create a richer, truer history of photography, we need to identify the 'masculine' bias that distorted photography's early history" (Hudgins 2020,

1). This book is aware that it mentions exclusively White male scientists who practised interferential photography but still contends that their history is not universal and must be rebalanced by other genders. The various chapters presented here merely hope to provide enough questions about a historically marginalised practice for future scholarship to overcome the narrative of the "great male scientist," especially since the spread of photography throughout the nineteenth century "coincided with feminist challenges to prevailing gender relations" (Latimer and Riches 2006). As authors Pauline Martin and Carole Sandrin mention, little is known about Laurence Cherbuliez (1857–1934), Lippmann's laboratory assistant and spouse, beyond these two roles she assumed. Still, there is enough to indicate that she played a more substantial role in his scientific and photographic practice ca. 1900 than previously assumed, which future scholarship should be attuned to. Also, women were a crucial part of Lippmann's audience at the Royal Society in London. Their appreciation of the material quality of his photographs; whose brilliance and their "jewel-like quality" persisted as an observation category throughout the twentieth century (Gamble 2004, 39) and needs further scrutiny. In Berlin, the Lette-Verein founded in 1866 played a key role in women's professionalisation in the photographic industry ca. 1900. Since "technical developments, new processes from abroad and requirements from industry and science shaped the curricula, teaching methods and the time required for the various branches of education" (Vitten 2018), and given the three interferential colour photographs made by Richard Neuhauss which I located in the Verein's collection, it is highly probable that women learned interferential photography and were more than just passive recipients, so their contributions must be mapped as well. That this research about Lippmann photography and women has not yet been undertaken only means that the absence of evidence is not evidence of absence, and there is still a learning curve ahead.

Beside repositioning Lippmann within the political history of France, deconstructing Eurocentric approaches to him, and using gender to criticise and renew the history of colour photography, assumptions about colour also require an update. The uselessness of interferential colour photography reiterated by Potonniée is a key notion precluding further inquiries.

The Myth of the "Non-Use" of Interferential Photography

A long-reiterated myth about the Lippmann process that Potonniée notes is its purely theoretical nature. Scholars on both sides of the Atlantic have maintained that, as a "direct" colour photography, the Lippmann process succeeded in transcribing the shades of light, but due to its technical difficulty, it remained unused. For this reason, it was quickly upended by "indirect" methods, such as three-colour photography and the autochrome, methods that reconstitute the perception of colour based on the recombination of its three primary hues. Thus, historians have often conflated the commercial viability of a medium with the various manners in which its practitioners implemented it. In doing so, they have inadvertently blurred the line between the actual and unique uses of interferential colour photography and the historiographic constructions that led to the omission of these practices.

This book is sensitive to the implementations that interferential photography *did* have, instead of lingering on the ones it did not, thereby connecting it with the various scientific, medial, and institutional frameworks that gave rise to it and without which it could not have thrived as long as it did. It addresses the mechanism of the exclusion of the Lippmann process to help rewrite its history by criticising its dominant understanding as a theoretical exercise at the expense of the materiality of its images. The understanding of interferential photography as an abstraction rather than as a material object has several reasons. First, it emerged due to the overall unreliability of the process, resulting in the failure of many photographs to show any colour to the point where the belief in its experimental viability stood on shaky grounds. Second, the Lippmann plate, as a direct positive on glass, is irreproducible which means that opportunities for the spectators to connect with the physical object hinged on the one photograph and were thus scarce. Third, its visibility is such that the surface reflection, which is the unwanted light reflected from the top surface of the emulsion, interferes with viewing the image – even at the best angle. The best angle and its range are restricted by the angle sensitivity of the interference process. The surface reflection issue was dealt with by cementing a wedge prism on the image. When beholding a Lippmann plate, however, even with a prism, there are special conditions under which the colours may be seen at their best, and viewers must take their time to find the right angle from which its shades will reveal themselves. This means that even when an interferential colour photograph is a successful experiment, seeing it remains a task all on its own. Hence, these material conditions underpinning the Lippmann plate have widened the gap between it as object and as colour theory. This distinction was visible during the early as well as modern formation of the history of colour photography, and has become a standard narrative, one which this book deconstructs.

British photography historians wrote in 1900 that "unfortunately there seems at present but *faint hope of carrying out in practice* that which appears *simple in theory*, but M. Lippmann appears to hope almost against hope" (Bolas, Tallent, and Senior 1900, 59, my emphasis). Beaumont Newhall's *Photography 1839–1937* (1937) picked up this rift, dedicating one paragraph to Lippmann's "direct" photography, deeming it "remarkably true to nature" and its image "marvelously brilliant," although it was "never a practical technique" (Newhall 1937, 82). In the German Empire, Erich Stenger's *Geschichte der Photographie* (1929) and his *Die Photographie in Kultur und Technik* (1938) restate interferential photography's "difficult implementation" – (*schwierige Ausführbarkeit*) (Stenger 1929, 31 and Stenger 1938, 154) – and stopping at that.

In a more modern canonical narrative about colour photography, *Colour Photography: The First Hundred Years 1840–1940*, the author Brian Coe starts his chapter with the "hope – even expectation – that soon a way would be found to record the colours of Nature directly, and in a manner as simple as that of recording light and shade in a monochrome photograph" (Coe 1978, 20). The chapter is then concluded with the brick wall of the "long-hoped-for fully practical method of direct colour photography [that] has never been found; instead, successful colour photographs have been produced by indirect methods of analysing and reproducing colour" (Coe 1978, 25). Moreover, as François Brunet has pointed out, when Potonniée employed the term "photograph" in his *Histoire*, he did so in reference primarily to a monochrome image on paper that is neither in colour nor on glass (Brunet 2017, 182–83). From this vantage point, he is indifferent to the Lippmann plate, because his focus was on a different kind of photographic materiality altogether. Potonniée only paid attention to glass slides used in projection when writing the history of cinema and of modern visual communication, briefly mentioning interferential photography in his subsequent books.[1]

Over time, the existence of the interferential colour photograph as material object became almost impossible to imagine. It turned into a mere horizon, an expectation, and a theory, which was quickly thwarted by other technologies and then forgotten. Yet, the physical photograph did exist and continues to do so, thereby negating the purely conceptual nature of the medium on a fundamental level: several photographers produced Lippmann plates – and did not merely theorise about interference – and these photographs are part of museum and private collections. This book is an invitation to reconnect with the Lippmann plate as a material object, one that is also a conduit for various forms of knowledge which have structured

1 Over a decade after *Histoire*, in *Cent ans de photographie: 1839–1939*, Lippmann re-emerges as an "illustrious scientist" and is accompanied by a short description of his colour photographic process (Potonniée 1940, 85).

its historical implementations and continue to configure its current uses, implementations that the authors centre on.

They dispel the myth of the "non-use" of interferential photography by exploring its "symbiotic relationship" with science (Wilder 2009b, 163), its use in commercial entertainment through projection sessions ca. 1900, and today, its importance as a museum object and as a medium of artistic expression by contemporary practitioners exploring its creative dimension.

The first part of this volume positions the Lippmann plate as both a scientific photograph representing physical phenomena (such as the wave nature of light) and as the object of scientific inquiry (Wilder 2009a, 12). William R. Alschuler's chapter introduces the history of the Lippmann process, as in the Lippmann plate and the knowledge about light waves embedded in it, as revealed by Lippmann himself in France, as well as in the context of its international reception among scientists and photographers. Furthermore, as Klaus Hentschel asserts, interferential colour photography operates at the crossroads of the rise and institutionalisation of scientific photography at the turn of the last century as well as spectrographers' visual practice of documenting and mapping the solar spectrum. Lazaros C. Triarhou and Manuel del Cerro investigate the inquiry of Spanish scientist Santiago Ramón y Cajal into the genesis of interferential colour. They present how Cajal transmitted practices from histology and microscopy to colour photography and the ties that bind these two fields. Susan Gamble studies Lippmann photography and the scientific theories underpinning it as the object of transatlantic competition and of textual translation from French to English by Frederick and Herbert Ives. In doing so, she sheds light on all that was lost in translation during this transmission, thus highlighting the misunderstanding of Lippmann in the United States.

Beyond its centrality to scientific knowledge, the second part of this book aims to help readers straddle the technical and discursive complexity of the Lippmann process by grounding it in a matrix of media, art, and cultural history. Just as interferential colour photography is marginalised by photography historians, contemporary historians of vision have fared no better. Jens Schröter's article in this volume criticises historian of vision Jonathan Crary's "rhetoric of ruptures," which posits a shift between the seventeenth century's geometrical optics and the eighteenth century's physiological optics, accounting for the disappearance of the camera obscura and the increasing importance of stereoscopic vision. Given Crary's focus on physiological optics around 1820, he excludes several photographic media. Schröter favours an approach he terms "layering replacing succession," as in the sedimentation of various layers of optical knowledge in media over time, rather than considering how one indefinitely replaces the other. Concretely, Lippmann photography did not vanish with the rise of three-colour technologies or the autochrome; it merely acquired another place in the landscape of media. Lippmann photography's position among media should interest photography historians attuned to the "synchronic plurality and [the] heterogeneity of regimes of vision" in order to overtake the teleological history of colour photography. My article highlights how anthropologist Richard Neuhauss's use of interferential photography was connected to and entangled with an assortment of media that were foundational to its genesis in an imperial context: taxidermy allowed Neuhauss to work consistently with the same test object – a stuffed parrot. The use of various optical media, including projection, helped the Lippmann plate's high-resolution colours reach individuals as well as collective spectators ca. 1900. And, thanks to the hype around it, Lippmann photography rivalled Hermann Wilhelm Vogel's three-colour printing method, all of which consolidated its presence along other technologies. Moreover, as Elizabeth Cronin points out, interferential photographs were exhibited in world fairs and made headlines in the photographic literature. They were seen as jewel-like objects and admired just as much for their beauty and their brilliance as for the complicated science behind them. The aesthetic framing of Lippmann plates proves that they were more than the results of "difficult science". Rolf Sachsse's article reflects upon the cultural history of Lippmann photography and the motivations for its exclusion from the iconic exhibition "Farbe im Foto" in Germany. Sachsse also connects the scientist with his surroundings, and most importantly with his family situation by proposing that his use of liquid mercury in the photographic process could have stemmed from his family business in leather manufacturing, thereby linking colour photography to craftsmanship.

The third section of this book presents interferential photography as used today by scientists, museums, and artists who together reconfigure and expand the limit of its implementation. Jens Gold's article points out how researching the materiality of colour plates made by Richard Neuhauss and Hans Lehmann is central to Preus Museum: Norway's Museum for Photography. Gold positions the collection as worthy of conservation care and of protection from degradation to make sure that, through the preservation of their material, knowledge is safeguarded for future generations. As Pauline Martin and Carole Sandrin propose, the extensive collection of interferential plates made by Gabriel Lippmann at the Musée de L'Elysée in Lausanne (Switzerland) challenges and enlarges the very concept of the museum and the future of its interdisciplinarity in research and in cooperation. Also, Hans Bjelkhagen not only explores its past but also highlights its steady attraction for contemporary photographers and holographers, which has been increasing since the 1980s. The colour image's allure has led them to reconfigure its emulsion and update the process, ascertaining its continuity in the second half of twentieth century. Last, my interview with contemporary Lippmann photographer Filipe Alves underscores the joys and frustrations of practising this kind of colour photography today – both as a hands-on practice and as a way to empirically connect to its history.

Therefore, readers of *Gabriel Lippmann's Colour Photography: Science, Media, Museums* are invited to take part of a wave of interest in interferential photographs as objects of scientific knowledge, as an anchor in the sea of media, culture, and art, and as material museum objects that are appreciated by scientists, conservators, and curators. All of these are functions Potonniée did not predict.

Acknowledgements

I thank Darran P. M. Green, William R. Alschuler, and Mark Jacobs for the inspirational discussions we had. I also thank Daniel Jon Mitchell for sharing his PhD thesis with me.

Works Cited

"Die Forschritte der Photographie und der Reproduktionstechnik in den Jahren 1890 und 1891." 1892. *Jahrbuch für Photographie und Reproduktionstechnik* 6: 287–487.

Alschuler, William R. 1998. "On the physical and visual state of 100 year old Lippmann color photographs." In *Sixth International Symposium on Display Holography, 21–25 July 1997, Lake Forest, Illinois*, edited by T. H. Jeong and H. I. Bjelkhagen, 84–94. Bellingham: SPIE.

Alschuler, William R. 2002. "Lippmann Photography and the Glory of Frozen Light: Eternal Photographic Color Real and False. " In *Proceedings of the 31st Applied Imagery Pattern Recognition Workshop, Washington, DC, October 16–18, 2002*, edited by David Schaefer and IEEE Computer Society. doi:10.1109/AIPR.2002.1182248.

Amad, Paula. 2010. *Counter-archive film, the everyday, and Albert Kahn's Archives de la Planète*. New York: Columbia University Press.

Berthelot, Daniel. 1922. "La vie d'un savant: Gabriel Lippmann." Revue Des Deux Mondes (1829–1971) 10, no. 1: 19–46.

Berthier, A. 1895. *Manuel de Photochromie Interférentielle. Procédés de reproduction Directe des Couleurs*. Paris: Gauthier-Villars et Fils Imprimeurs-Libraire.

Bjelkhagen, H. I. 1996. "Silver halide emulsions for Lippmann photography and holography." *SPIE Milestone Series* 130: 102.

Bjelkhagen, Hans I. 1999. "Lippmann Photography: Reviving an Early Colour Process." *History of Photography* 23, no. 3: 274–80.

Bjelkhagen, Hans I., and David Brotherton-Ratcliffe. 2013. *Ultra-Realistic Imaging Advanced Techniques in Analogue and Digital Colour Holography*. Boca Raton: CRC Press, Taylor & Francis Group.

Bjorli, Trond, and Kjetil Jakobsen. 2020. *Cosmopolitics of the Camera: Albert Kahn's Archives of the Planet*. Bristol : Intellect Books.

Bolas, Thomas, Alexander A. K. Tallent, and Edgar Senior. 1900. *A Handbook of Photography in Colors*. New York: E. & H.T. Anthony & Co.

Boulouch, Nathalie, and José Vovelle. 1995. *La photographie autochrome en France (1904–1931)*. Lille: Atelier national de reproduction des thèses.

Brunet, François. 2011. "Nationalities and universalism in the early historiography of photography: (1843–1857)." *History of Photography* 35: 98–110.

Brunet, François. 2017. *La photographie: histoire et contre-histoire*. Paris: Presses Universitaire de France.

Coe, Brian. 1978. *Colour photography: the first hundred years, 1840–1940*. London: Ash & Grant.

Connes, P. 1987. "Silver Salts and Standing Waves: The History of Interference Colour Photography." *Journal of Optics* 18, no. 4: 147–66.

Fechete, Ioana. 2016. "Jonas Ferdinand Gabriel Lippmann: The pioneer of color photography or primus inter pares." *C. R. Chimie* 19: 903–8.

Fournier, Jean-Marc. 1994. "An investigation on Lippmann photographs: materials, processes, and color rendition." In *Practical Holography VII*, edited by S. A. Benton. Proceedings SPIE International Symposium on Electronic Imaging: Science and Technology 2176: 144–52.

Fuchs, Caroline. 2017. *Das Autochrom in Großbritannien: Revolution der Farbfotografie*. Berlin: De Gruyter.

Gamble, Susan A. 2005. *The Hologram and its Antecedents 1891–1965: The Illusory History of a Three-Dimensional Illusion*. Camrbidge: University of Cambridge Press.

Goldsmith, Barbara. 2008. *Obsessive Genius The Inner World of Marie Curie*. New York: Paw Prints.

Hentschel, Klaus. 2002a. *Mapping the Spectrum. Techniques of Visual Representation in Research and Teaching*. Oxford: Oxford University Press.

Hudgins, Nicole. 2020. *The Gender of Photography: How Masculine and Feminine Values Shaped the History of Nineteenth-Century Photography*. Abingdon and New York: Routledge.

Johnston, Sean F. 2009. *Holographic visions: a History of New Science*. Oxford: Oxford University Press.

Latimer, Tirza, and Harriet Riches. 2006. *Women and Photography*. Oxford: Oxford University Press.

Lavédrine, Bertrand. 2009. *Photographs of the Past: Process and Preservation*. Los Angeles: Getty Conservation Institute.

Lestel, Laurence. 2008. *Itinéraires de chimistes, 1857–2007: 150 ans de chimie en France avec les présidents de la SFC*. Les Ulis: EDP sciences.

Mitchell, Daniel Jon. 2010a. *Gabriel Lippmann's Approach to Late-Nineteenth Century French Physics*. PhD diss., Oxford University.

Mitchell, Daniel Jon. 2010b. "Reflecting Nature: Chemistry and Comprehensibility in Gabriel Lippmann's 'Physical' Method of Photographing Colours", *Notes and Records of the Royal Society of London* 64 (4), 319–337.

Neuhauss, Richard. 1894b. "Die Photographie in natürlichen Farben." *Photographische Rundschau* 8, no. 10: 295–302; 11: 327–29; 12: 359–64.

Neuhauss, Richard. 1900. "Bücherschau". *Photographische Rundschau*. 14: 148.

Newhall, Beaumont. 1937. *Photography: 1839–1937*. New York: The Museum of Modern Art. Exhibition catalogue.

Noronha-DiVanna, Isabel. 2010. *Writing History in the Third Republic*. Newcastle upon Tyne: Cambridge Scholars.

Pier, J.P, and J. A. Massard (eds). 1997. *Gabriel Lippmann, Commémoration par la Section des Sciences Naturelles, Physiaues et Mathématiques de l'Institut Grand Ducal de Luxembourg du 150e Anniversaire du Savant né au Luxembourg Lauréat du Prix Nobel en 1908*, Luxembourg: Jean-Paul Pier & Jos. A. Massard.

Potonniée, Georges. 1925. *Histoire de la Découverte de la Photographie*. Paris: Montel.

Schreier, Wolfgang, Martin Frankeunter, and Annett Fiedler. 1995. "Geschichte der Physikalischen Gesellschaft zu Berlin 1845–1900". *Physik Journal*. 51 (1): F-9–F-59.

Schröter, Jens. 2014. *3D: History, Theory and Aesthetics of the Transplane Image*. London: Bloomsbury Academic.

Sogreshilin, V. I. 1966. *Usagin Ivan Filippovich*. Moskva: Izd-vo Prosveshchenie.

Stenger, Erich. 1929. *Geschichte der Photographie*. Berlin: VDI Verlag.

Stenger, Erich. 1938. *Die Photographie in Kultur und Technik Ihre Geschichte während 100 Jahren*. Leipzig: E.A Seemann.

Timby, Kim. 2015. *3D and Animated Lenticular Photography*. Berlin: De Gruyter.

Vitten, Anne. 2018. "Professionalisierung und Vernetzung. Berufsfotografinnen im 19. und 20. Jahrhundert". *Fotogeschichte* 149. http://www.fotogeschichte.info/bisher-erschienen/hefte-126–149/149/anne-vitten-berufsfotografinnen/ Accessed March 30, 2021.

Vogel, H.W. 1892. "Ueber Lippmann's Photographien des Spektrums in natürlichen Farben". *Verhandlungen der Physikalischen Gesellschaft zu Berlin im Jahre 1891.*10: 33–35.

Wagener, Renée. 2016. "Verspätete Verbürgerlichung. Politische Partizipation Luxemburger Juden im 19. Jahrhundert." In *Identitätsbildung und Partizipation im 19. und 20. Jahrhundert: Luxemburg im europäischen Kontext*, edited by Thorsten Fuchshuber, Université du Luxembourg, Jean-Paul Lehners, Renée Wagener, Sonja Kmec, Norbert Franz, and Michel Pauly, 147–74. Frankfurt: Peter Lang Publishing Group.

Wagener, Renée. 2016. "Gabriel Lippmann, ein Moderner Wissenschaftler." *Ons Stad* 113: 38–40.

Weissmann, Gerald. 2010. "Radium, Telomeres, and Ribosomes: Glass Ceilings Break in Stockholm". *The FASEB Journal: Official Publication of the Federation of American Societies for Experimental Biology.* 139 (1): 1–5.

Wilder, Kelley. 2009a. *Photography and Science.* London: Reaktion Books.

Wilder, Kelley. 2009b. "Photography and the art of science", *Visual Studies* 24 (2), 163–168.

Wood, John. 1993. *The Art of the Autochrome the Birth of Color Photography.* Iowa: University of Iowa Press.

Zenker, Wilhelm. 1891. "Die Entstehung der Farben in der Photochromie." *Jahrbuch für Photographie und Reproduktionstechnik* 5: 294–303.

Zenker, Wilhelm. 1893. "Ueber die Entstehung der Farben im Lippmann'schen Spectrum." *Jahrbuch für Photographie und Reproductionstechnik* 7: 114–121.

About the Author

Dr. **Hanin Hannouch** is curator for analog and digital media at the Weltmuseum Wien (Vienna). She has guest-curated the exhibition "Slow Color Photography" about interferential colour photography at Preus Museum: Norway's National Museum for Photography and is the guest-editor of the journal PhotoResearcher Nr. 37 "Three-Colour Photography around 1900: Technologies, Expeditions, Empires". She was post-doctoral researcher at the Kunsthistorisches Institut in Florenz, Max-Planck Institut and at the Ethnologisches Museum in Berlin. She is currently writing her monograph about the history of colour photography in Imperial Germany.

Part I

The Science of Colour Photography and the Colour Photography of Science

1. How Lippmann Froze Light and Passed It around for Others to Taste

William R. Alschuler

Abstract

This essay examines how Gabriel Lippmann decided to pursue the definitive demonstration of the wave nature of light, his conceptualising of the scheme he adopted, and the difficulties he encountered in reaching that goal. It then examines what he foresaw as the possible applications of his invention of a permanent form of colour photography and how it was understood, received and questioned. This questioning pertains both to its theoretical basis, practical aspects, modifications as well as to its applications by other scientists, professional and amateur photographers, photographic manufacturers, and the general public, both in France and around the world.

Keywords: wave theory of light, Gabriel Lippmann, interferential colour photography, history of photography, history of science

Introduction

Lippmann was home-educated until relatively late, arriving at the École normale supérieure at an age five years older than his classmates. He showed early promise in doing scientific research, and knowing Lippmann's facility with languages (he spoke, read, and wrote French, English, and German), a member of the physics faculty, Pierre Auguste Bertin (1818–1884), asked him in 1868 to keep current with physics journals in Europe and to summarise new developments, in some cases critically, and publish them in the *Annales de Chemie et de Physique*, which was run by Bertin (Appell 1923).

In 1874, as he graduated from the Sorbonne, Gabriel Lippmann failed to take his examinations to receive his Aggregation, the licence to teach in

Hannouch, H. (ed.), *Gabriel Lippmann's Colour Photography: Science, Media, Museums.*
Taylor & Francis Group, 2022
DOI: 10.5117/9789463728553_CH01

the French public school system. His dedication to physics research and his general experimental abilities were, however, recognised, and the government awarded him a visiting research post to Germany, from which he was to eventually report back to the French government on the state of German science. Although Germany had defeated France in a war just two years before, he seems to have received a warm welcome. He worked in several physics laboratories and most likely reported back as required by the grant (though I have found no copy of this report), because he was awarded a second stay and ended up in Kirchhoff's laboratory in Heidelberg in 1874, and he received access to the lab – even in Kirchhoff's absence. While there, he took a suggestion by German physicist Georg Hermann Quincke (1834–1924) to investigate what was known somewhat widely as the "mercury beating heart" phenomenon. This refers to the demonstration that a drop of mercury, sitting in a dish of sulphuric acid with a drop of oxidiser added, when contacted with an iron nail, will begin to oscillate, becoming nearly spherical, losing contact with the nail and then flattening out and thus recontacting the nail. As it oscillates, it can have various vibration modes and sometimes has a circular base and sometimes a polygonal one. The oscillation continues until chemical reactions exhaust the solution. This was clearly an electrochemical effect. This effect is called *electrocapillarity* and depends on the stop-and-go flow of electrons to the droplet surface alternately oxidising and then reducing its surface, which cycles its surface tension. Lippmann decided to investigate it thoroughly and generated a doctoral thesis, for which he received a degree from the Rupert-Carolla University in 1874. In the thesis, he proposed a chemical reaction pathway and gave a set of equations that described it and the oscillatory shape changes that were produced in experiment. His explanation was widely accepted, and he returned to Paris to complete a doctoral thesis there on the same subject, which he received in 1875.

In the process of completing his thesis, he made two ingenious inventions that utilised the effect, which he termed the *electrocapillarity effect*. The first was an electrocapillary motor, one of the strangest motors ever invented. It ran by utilising the rise and fall of mercury encapsulated in capillary glass tubing clusters, submerged in sulphuric acid, and capped with pistons. Lippmann had a demonstration model constructed as a two-cycle engine, and it ran at one hundred cycles per minute. It has had no practical applications that I am aware of. The second invention was an electrocapillary volt-ammeter, with a physical arrangement generically similar to a mercury barometer. Again using electrocapillarity, he achieved a sensitivity to voltages ten thousand times better than those available from existing equipment. He derived all the equations for this equipment and built working models.

The publication of this result and the later commercial production of the meters by others based on his design set a new standard for electrical precision and earned Lippmann a considerable reputation. Étienne-Jules Marey (1830–1904), a physiologist well known for his stop-motion photography, asked Lippmann to collaborate on measuring the currents and voltages of various muscles as they flexed (Picard 1931), and Augustus Desiré Waller (1856–1922) collaborated with Lippmann on registering the weak electric currents of the human beating heart, and they succeeded in making the first known electrocardiograms. Willem Einthoven (1860–1927) later briefly used the same equipment to measure electrocardiograms.

This electrocapillarity work, and also perhaps familiarity with his father's former leather gloves business (mercury was used in conditioning the leather), left Lippmann experienced with the properties and handling of mercury, which would be of use in his invention of interferential colour photography.

Pre-professorship

In 1875, after getting his Sorbonne doctorate, Lippmann followed a reasonably usual trajectory, becoming first an assistant to Professor Jules Jamin (1818–86), who was running the Sorbonne physics experimental laboratory. While doing this, he was still living with his parents in an apartment in Paris and apparently took some Sorbonne equipment home to carry out experiments on his own time. Then, in 1878, he became a maître des conférences (lecturer) (Appell 1923, 109).

It appears that, during this period in the late 1870s, pursuing his duties to organise classroom demonstrations for his professor, Lippmann wished to devise an experiment that would definitively demonstrate the wave nature of light. He had made use of demonstrations that showed the wave nature of sound by setting up a single-frequency sound source facing a wall and using it as a reflector, creating a standing wave as a result. The wave's presence was revealed as he moved a candle flame along the sound path, into regions of wave maxima where the flame wavered, and to minima where the sound waves cancelled and where it was stable, which were separated by half wavelengths of the sound.

The principle of creating standing waves must have been powerful for him, as it is the technique he adapted to make his colour photographic process work to demonstrate that light is waves. He later used this audio demonstration as an analogy to his photo process at public lectures he gave on his process.

Professorship

When a faculty position came open at the Sorbonne in 1883 for a chair in Probability Calculations and Mathematical Physics, which had been occupied by Charles Auguste Briot (1817–1882), Lippmann won it, even though he was better described as an experimental physicist than a mathematical one. He was encouraged by many mathematics professors to apply for the position (Picard 1931). A major factor was, no doubt, an admiring endorsement of his mathematisation of electrocapillarity, as demonstrated in his thesis, and his later work on piezoelectricity. Sorbonne professor Charles Hermite (1822–1901), the great geometer is quoted in part:

> [...] in applying the differential equations which were established tools for study of diverse phenomena he showed that you could not only analyse but also make predictions from scratch. In this way Lippmann announced that a crystal would lengthen if subjected to an electric field and calculated in advance the numerical values for this elongation under specified conditions. These results were later verified by Pierre and Jacques Curie. This is perhaps the first time in the area of electricity that calculation preceded experiment. This short discussion shows that Mr. Lippmann has attached his name to work of the first importance, concerning the biggest and most important questions of physics in our time. (Appell 1923, 109–110, my translation)[1]

In 1886, when Jamin died, Lippmann applied for Jamin's chair in Experimental Physics and his directorship of the Laboratory of Experimental Physics, along with increased space and funding for it. He got what he asked for.

Pre-interferential Colour Photographic Work

Lippmann worked and published mostly in the area of electrocapillarity for many years, and in a range of other physics areas, but he started

1 "en appliquant les équations différentielles qu"il établi a l'étude des divers phénomènes, il montre que l'on peut non seulement les analyser, mais en prévois le nouveaux. Ainsi M. Lippmann a annonce qu'un cristal [...] doit s'allonger lorsqu'en le soumet a l'influence électrique et il a indique d'avance la valeur numérique de cet élongement dans des conditions determinées. Or ces résultats du calcul ont été vérifié plus tard par MM. P. et J. Curie. C'est peut-être le première fois qu'en électricité le calcul ait devance expérience [...] Cet expose succinct montre que M. Lippmann a attaché son nom a des travaux de premier ordre, concernent les plus hautes et plus importantes questions de la physique a notre époque." (Appell 1923, 109–110)

publishing on photography a few years before his discovery of interference colour photography. In a short memoir (Lippmann 1889), he noted that black-and-white emulsions of the time were greatly sensitive to blue light and ultraviolet light, but still only slightly sensitive to longer wavelengths, though acknowledging that improvements in green and yellow had been made by photochemist Hermann Wilhelm Vogel's (1834–1898) innovations in sensitising dyes. Lippmann undertook to fix this problem for rendering proper black-and-white tonalities by sequentially exposing one of these orthochromatic plates using a planned sequence of on-the-lens filters while keeping the camera still and shooting a subject at rest (the latter presumably to avoid blurring and ghost images). His sequence was a short exposure with a blue filter to capture the blue. Then he replaced it with a green filter that blocked the blue and gave a forty-second exposure. After that, he replaced the green with a red filter that blocked blue and green and exposed for one thousand seconds. He found that this procedure gave tones that reproduced visual sensation. He still had to deal with this non-panchromatic sensitivity problem in his early experiments on his interference process, and he applied the lessons learned for this paper to his early experimental trials; in fact, his work in this area may have been motivated by his long-running project to demonstrate his interference process.

Colourful Success

Lippmann realised that, to record light waves interferentially, he would have to find a method to freeze light and that he could do that by creating standing waves, whose maxima and minima could be recorded in a photographic emulsion. There were several difficulties with this idea: some purely practical and some theoretical. On the purely practical side, light waves are very small. The emulsion would have to have good enough resolution to faithfully resolve maxima and minima separated by a half wavelength of the colour to be recorded. Since visible light has wavelengths of between about 400 nm (blue) and 700 nm (red), or 0.4 to 0.7 microns, the resolution of the emulsion would have to be better than five thousand lines per millimetre to record the shortest waves. That was and still is extremely high resolution and extremely fine grain. For comparison, conventional film has had resolutions of 100–300 lpmm. The only emulsions with nearly the required resolution currently commercially available are made for holography. In fact, Lippmann images are a type of hologram (which are made with lasers as light sources), but made with ordinary light, which record colour but cannot record depth. A consequence of this extreme resolution requirement is that the emulsion will be slow, as sensitivity is inversely proportional to grain size. Experiments I carried out show that today's commercial holographic emulsions, which I have used in my Lippmann photography experiments, have an American Standards Association (ASA) equivalence of one or less. A second practical consideration was that, because he wanted the light to pass through the emulsion, bounce back off a reflector, and interfere, the emulsion had to be not only fine-grained but transparent. To achieve this, it had to be absorption- and scattering-free, not be anti-halation backed (a layer that suppresses reflection that is conventional in ordinary emulsions), and the ratio of silver halide to gelatin had to be on the low side compared to ordinary concentrations.

He announced his success at a meeting of the Académie des sciences on February 2, 1891, which was published in the proceedings and was less than a page in length, giving almost no details (Lippmann 1891a, 274–75). Lippmann begins:

> I proposed to myself to obtain the image of a spectrum with its colours, in such a way that it would remain fixed and could remain exposed to daylight indefinitely and without alteration.
> I have been able to resolve this problem while using sensitising substances, developers, and fixers currently in use in Photography, and modifying simply the physical conditions of the experiment. The essential conditions for obtaining colours in Photography number two: 1st the continuity of the sensitive layer; 2nd the presence of a reflecting surface in contact with this layer [...] The ordinary emulsions in use today are found to be excluded. A continuous layer is transparent ordinarily with a slight blue opalescence. I have used as support albumin, collodion and gelatin; as sensitisers silver iodide and bromide; all of these combinations gave good results [...] The visible colours on the exposure are thus of the same nature as those of soap bubbles. They are just more pure and more brilliant [...] This follows since they form in the depth of the sensitive layer [from] a large number of superposed thin layers: around 200, if the sensitive layer is 1/20 mm thick. For the same reason the purity of the colour increases as the number of reflecting layers grows. (my translation)[2]

2 "Je me suis proposer d'obtenir sur un plaque photographique l'image du spectre avec ses couleurs, de telle façon que cette image demeurât désormais *fixée* et put rester exposée indéfiniment au grand jour sans d'altérer. J'ai pu résoudre ce problème en opérant avec les substances sensibles, les développateurs et les fixatifs courants en Photographie, et en modifiant simplement

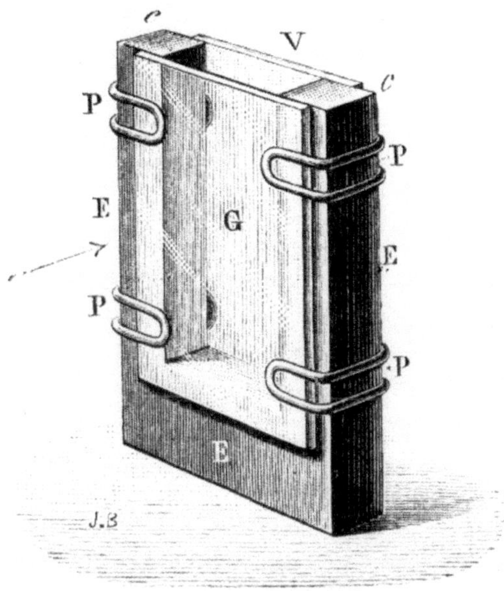

Fig. 1 A drawing of Lippmann's earliest mercury plateholder G is the sensitised plate, emulsion facing away from the light (the arrow). The cavity will be filled with mercury. (From Berget A. 1901. Photographie des couleurs par la method interferentielle de M. G. Lippmann. Gauthiers-Villars, Paris, p. 4)

It is notable that this first notice by Lippmann was immediately followed in the journal by a longer notice by Edmund Becquerel, who used it to discuss in detail why Lippmann's process was distinguished from the one Becquerel published in 1848. In spite of this, others have occasionally claimed they were fundamentally the same process, both being based on interference (see below).

Lippmann published a similarly short note in the Revue d'Astronomie Populaire (Lippmann 1891b).

As stated above, Lippmann first succeeded using silver halides in albumen as his medium (shortly followed by collodion and gelatin), coated on glass. He used a carbon arc light, made by Cancé, as his light source and spread

les conditions physiques de expérience. Les conditions essentielles pour obtenir les couleurs en Photographie sont au nombre de deux: 1° continuité de la couche sensible; 2° présence d'une surface réfléchissante addosée a cette couche [...] L'emploi des grossières émulsions usitées aujourd'hui se trouve par la exclu. Une couche continue est transparent, sauf ordinairement une légère opalescence bleue. J'ai employé comme support l'albumine, le collodion et la gélatin; comme matières sensibles, l'iodure et le bromure d'argent; toutes ces combinaisons donnent de bons résultats [...] Les couleurs visible sur le cliché sont ainsi de même nature que celles des bulles de savon, Elles sont seulement plus pure et plus brillantes [...] Cela tient a ce qu'il se forme dans l'épaisseur de la couche sensible un très grand nombre de lames minces superpose: environ 200, si la couche a, par exemple 1/20 de millimètre. Pour les mêmes raisons, la couleur réfléchie est d'autant plus pure que le nombre de couches réfléchissantes augmente." (Lippmann 1891, 4)

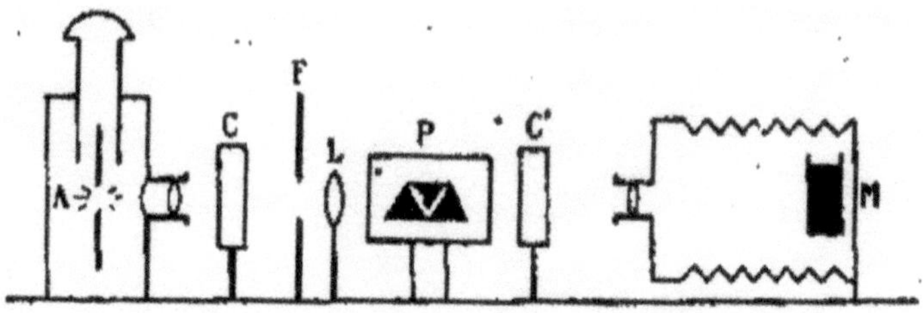

Fig. 2 Lippmann's original optical setup for recording spectra from an arc lamp. A is the arc, C is a heat-absorbing cell of water, L a collimator, P the prism, C the cell with intercheagable filter and M the plateholder. (From Niewenglowski, G.-H. 1909. Traite Pratique de Photographie des Couleurs. Garnier Frères, Paris. Pg. 93)

the light into a spectrum with a direct vision Amici prism, from which the light entered a glass trough filled with a succession of dyes serving as colour filters for a series of exposures on the same schedule as in Lippmann's 1889 exposure compensation scheme, and then into a conventional camera with a special plate holder (see Fig. 2). The plate holder was handmade with a set of clips to seal the coated glass plate, at its perimeter, to a spacer frame, with emulsion facing away from the camera lens, and another glass plate clipped to the opposite face of the spacer to form the rear wall. The frame was just a rectangular annulus and was thick enough to create a cavity about 3 mm deep, open at the top, into which Lippmann poured mercury to form a mirror in perfect contact with the emulsion (see Fig. 1). He attached the plate and poured the mercury in the dark, placed it in a box at the focus of a front lens, then did his exposure, poured the mercury back out, cleaned the plate of any residual mercury (it has a slight tendency to stick), and then developed it. The exposure times were one or two minutes for the violet, twenty minutes for the green, and two hours for the red, because he still had the problem of non-panchromaticity of sensitisation (Berget 1891a, 38).

He chose a carbon arc as his light source because it provided more intense illumination than he could easily get from the sun, a more continuous source, but which would have required large mirrors and tracking to achieve his needed exposures. He chose the spectrum as his first subject for the obvious reason, one true for any new colour process, that he wanted to demonstrate colour. In addition, any gaps would indicate sensitisation problems, and for his process, spreading the light out in effect made the image a band of pure colours at each point along its dispersion, and the interference of pure colours was easy to explain. Finally, if the colour faded out towards the blue, it would indicate that the grain size was not small enough to resolve the increasingly fine interference pattern there. He was able to improve the sensitisation of his plates considerably within a few months, perhaps with the help of the

Lumière brothers, to the point that he could expose a complete spectrum in one exposure without the colour filters in under six minutes (Berget 1891a, 39). By December 1893, he could do a single panchromatic spectrum with an exposure of thirty seconds (Lippmann 1893, 11).

In thinking back over his work, Lippmann remembered it had taken him twelve years to succeed in making an emulsion that satisfied both the resolution and transparency requirements. (Lippmann 1897 and 1898). This means Lippmann began working on this problem in 1879, a year after he became *maître des conferences*. That it took him so long to get the desired results is somewhat curious, as a recipe for such emulsions had been published by German chemist and photographer Eduard Liesegang in the 1880s (H. Bjelkhagen, personal communication, 1998) and Lippmann was very conversant with European journals (see above).

Initial Public Reception

Lippmann never wrote a book about his invention, but Alphonse Berget, his student and research lab assistant, who managed demonstrations of the technique at Lippmann's lectures on it, published a short book about it within the year (Berget 1891b). Berget describes an early experiment that demonstrated that the process can work with a version of mixed as well as pure colours: Lippmann inserted two filters side by side, one green and one blue, in the light path of his original set-up (Fig 3). They were recorded successfully. When examined through a spectroscope, each filter displayed the whole visible spectrum, but in differing proportions (Berget 1891b, 55). In his conclusion, Berget states:

> One can say that now Photography has taken the last step that remained to be accomplished: the general solution to the only problem that still remained to be resolved and discovered and it is definitive because it is scientific and rational." (my translation)[3]

Though Lippmann usually published his work in the French journal of Académie des sciences and sometimes in other French journals, as well as in German journals, and occasionally in British ones, the announcement

3 "On pu dire que maintenant la Photographie a franchi la dernière étape qui lui restât a fournir: la solution générale du seul problème qui fut encore a résoudre est trouvée et elle est définitive parce qu'elle est scientifique et rationnelle." (Berget 1891/1901, 55)

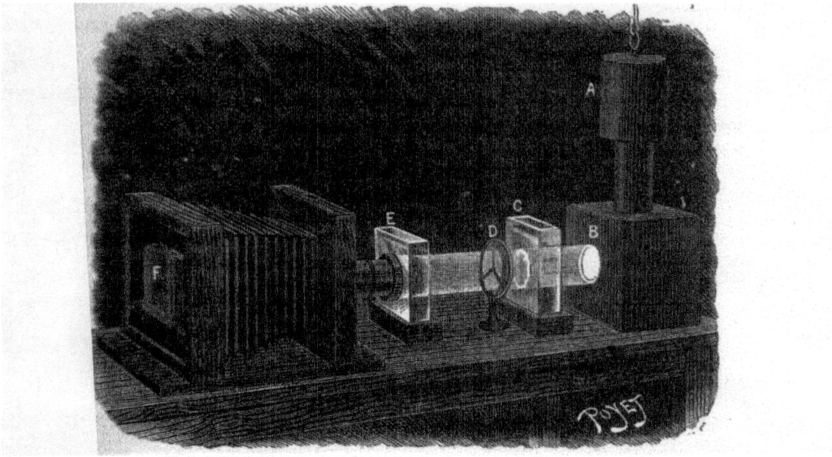

Fig. 3. Lippmann's early setup for recording mixed colours. A is the carbon arc, C is the heat-absorbing cell. The colour filter has been replaced by a two-colour piece of stained glass D. The lamp emits 800 candles. (From Wallon, E. 1891. "La Photographie des couleurs. Experiences de M.G. Lippmann." Photo-Gazette. 25 Fevrier, 1891. Pg.51)

in the *Comptes Rendus* was quickly picked up elsewhere, including in an anonymous article in a completely non-scientific journal *Les annales politiques et litteraires* (!) (*La Decovèrte de M. Lippmann* 1891a, 109). The article explained the equipment Lippmann used and a tiny bit about the technique, and it noted that Lippmann showed a spectrum and that "objects, landscapes stained glass, flowers, etc., with their exact colors" were in the future. It discussed the idea of thin film interference, interference fringes and Newton's rings. It placed the new process in the context of predecessors: Becquerel, Niépce de Saint-Victor, Poitevin, and Ducos du Hauron. It also mentions the limited colour successes of Campbell in the United States and Gaedicke in Berlin. Tissandier (1891, 161) says Lippmann presented numerous plates of the solar spectrum at the meeting on February 2. This is a bit surprising, as the first spectra were of an arc light. There are two illustrations: one showing the initial handmade plateholder, held together by clips (Fig.1), and the other a sectional diagram showing the glass plate, the emulsion, and the mercury in contact with that and dotted in lines portraying the interference maxima planes, a half wavelength apart. Tissandier notes that portraits and other real objects are problematic as subjects due to the lengthy exposure times required. In the widely read journal *l'Illustration*, Guerronan (1891, 190–91) gives a somewhat lengthier description of the process. He begins by stating that physicists consider light to be vibrations in the elastic medium they call the ether. Vibrations of molecules as the waves travel enter the conversation. Then he gives details of interference theory, the wavelengths of visible colours, the equipment, and thin film interference, with illustrations.

In the popular French magazine *Cosmos*, Maumene (1891, 379) comments that Lippmann searched for a way to record colour that was not primarily photochemical, as were the processes of prior attempts, but a purely "physical" way. He discusses interference in soap bubble films in detail, noting that the thickness of the bubbles typically varies as time passes and each thickness displays a different colour, using this analogy to explain how Lippmann's process records the various colours. He also refers to light as a wave in a medium. He says: "Light is due to very regular undulations of a fluid called the ether (the ether of poets), analogous to water, but much more subtle" (my translation).[4] He also considers the practical problems of searching for a substitute for the mercury mirror, noting it has to be non-reactive to the emulsion, highly polished, easy to remove without damaging the emulsion, and in perfect contact with it during the exposure. This last point is a key requirement that will be further discussed.

It is not mentioned in the earliest accounts of Lippmann's discovery that he cemented a thin wedge prism to the finished emulsion with about a ten-degree wedge angle, the purpose of which was to send the reflection of the viewing light out of the field of view so that it did not intermingle with and kill the brilliance of the image. The cement was Canadian balsam, which had a refractive index of about 1.5 that matched both the prism and the emulsion, preventing additional internal reflections. This technique has been attributed to Wiener at a later date, but I have seen what I believe are very early images by Lippmann that have the prisms.

News of the discovery spread rapidly, even to the United States. *The New York Times* has a brief account in February 1891, just after the announcement in Paris, titled: "NATURE HOLDS IT LOCKED, Attempts to Solve the Secret of colour Photography, Some Measure of Success has been Accomplished, but the Results Most Desired Have Never been Attained. It begins with a summary of preceding attempts." It then pronounces:

> As yet he appears to have succeeded in reproducing the primary colours. The plates he shows are of the solar spectrum. All the colours of the rainbow are reproduced, including the infrared and the ultraviolet, which are invisible to human eyes. He thinks that he will be able to reproduce composite hues [...] But this he has not yet attempted.

4 "La Lumière est due aux ondulations très régulières d'un fluide appelé éther (l'éther des poètes), analogue a l'eau, mais beaucoup plus courtes que celles d'eau." (Maumene 1891, 379)

The mention of infrared and UV is surprising, as they would likely have been out of reach of his emulsion sensitisers at the time. Also, even if the UV could be recorded, its shorter wavelength would require an even greater resolution than the extremely high values Lippmann achieved. The author goes on to say that any worthwhile photography process would generate a negative capable of making duplicates, which the Lippmann process does not. This fact persisted. A few of Lippmann's images were shown in 1891 at the National Conference of Amateur Photographers in New York (Brown 1994, 29).

Priority

Becquerel had always maintained that his process, a colour version of a daguerreotype, was based on photochemistry. Becquerel said (Becquerel, 1891, 275–76):

> I wish to note all the differences that exist between the entirely physical process that Mr. Lippmann has described for photographically recording the colours of light, and the photochemical process that I discovered in 1848 for obtaining coloured images of luminous spectra and then images of objects in their correct colours; it's with the use of the same chemical substance, the sub-chloride of silver, formed on the surfaces of silver layers [...]
> If you submit these photographic colour images to the solvents of silver chloride such as ammonia or sodium hyposulfite, the nuances disappear and, if luminous rays have exercised their action, there remains on the surface of layers of silver a slight trace formed by a thin layer of metallic silver, that, if it is a bit humid, manifests feeble tints, complementary to those that had existed in the same places. These effects, which it is difficult to explain a priori, show perhaps that the depths of the deposited layers play a role in the production of the phenomena of colouration.[5]

5 "Je désire faire remarquer toute la différence qui existe entre le procédé entièrement physique que vient d'exposé M. Lippmann pour reproduire photographiquement les couleurs de la Lumière, et le procédé photochimique qu j'ai découvert en 1848 pour obtenir les images colorées du spectre lumineux ainsi que les images des objets avec leur couleurs propres; c'est a l'aide d'une même substance chimique, le sous-chlorure d'argent, forme a la surface de lames de argent [...] .Lorsqu'on soumet les images photographiques ainsi colorées a l'action réductrice d'un des dissolvants du chlorure d'argent, tells que l'ammoniaque ou hyposulfite de soude,les nuances colorées disparaissent et, la ou les rayons lumineux ont exerce leur action, il reste a

Becquerel could also point to the results of Sir John Herschel's experimentation; in 1840, Herschel pointed out that, among the silver halides, silver chloride, if painted onto paper, could (temporarily) record some of the colours of the spectrum, and no mirror was involved in that, so the colour was not interferential. Becquerel produced his media by first submerging a silver plate in a solution of chlorinated water and sending an electric current through the solution to the plate, which formed one pole of the circuit. This gradually produced a thin film of silver chloride on the surface that could be monitored for its colour. Becquerel found that when he stopped at the right thin film colour, the plate would then record colour. However, no way of fixing the image could be found. Only keeping it in the dark preserved it reasonably well. It could be pulled out to be viewed briefly in subdued light. By many contemporary accounts, the colour was reasonably good. However, E. J. Wall wrote, as recorded originally in an early (1910s) edition of his *Wall's Dictionary of Photography*: "The colours so obtained are only first approximations, however, and it requires the eye of faith to detect more than a muddy purple, a blue, and an orange in the record of a spectrum made in this way" (Wall 1942, 137).

I have looked at a number of examples of Becquerel plates. Several were in the Musée Niépce at Chalon sur Saone, where I saw them in 1999. There was not much colour. The predominant colour was a kind of aubergine. There was a hint of red in some places and a bit of orange on plates labelled as of the solar spectrum. A similar situation was found with a Becquerel plate of the solar spectrum at the Science Museum in London. I saw that in 2002. It was apparently a gift from Becquerel to the renowned physicist Sir John Herschel and was accompanied by viewing notes by Herschel and his son William Herschel, which were added to over a period of almost fifty years (up to 1897), in which the colours (red and orange, no blue or green) were said to be unchanged each time. Again, the predominant colour I saw on the plate was aubergine. All these examples had been stored in the dark between viewings. In the United States, Levi Hill produced "Hillotypes," coloured versions of daguerreotypes in a process that had about thirty steps and involved numerous dangerous chemicals (see Alschuler 2007a, 315–318) for further discussion[6]. In all the above cases, I also saw some of the

la surface des lames d'argent une légère trace formée par une lame mince d'argent métallique, qui, lorsqu'elle est encore humide, manifeste de faibles teintes, complémentaires de celles qui existaient auparavant aux mêmes places. Ces effets, dont il et difficile de se rendre compte *a priori*, montrent que peut-être les épaisseurs des couches disposes jouent un rôle dans le production des phénomènes de coloration." (Becquerel 1891, 275–76).

6 There are two older papers of interest: Jourdain (1900) and Connes (1987).

same thin film general colour that Becquerel noted above. I have another example of a somehow related process, a daguerreotype made, like with Becquerel's photograph using silver chloride, produced and given to me by the daguerreotypist John Hurlock in 2005. It shows a clear yellow wall on the image of a house and a blue sky.

In spite of Becquerel's disavowal that his process was interferometric, others maintained that it was and offered in support the idea that his colours were unfixable due to the fixer, sodium hyposulfite, collapsing the supposed interference layers. Berget even offered this idea, which is surprising considering that Lippmann did not (Berget 1891b, 52).

Several aspects of these Becquerel-type images appear to mark them as of a different nature than Lippmann's. First, they do not show any colour shift when the viewing angle or illumination angle used is shifted. In interference images and thin film interference in general, of which Lippmann-type images are a special case, the greater the angle, the bluer the image gets. Second, Becquerel succeeded using only silver chloride, and the other halides failed. This recalls Sir John Herschel's partial success at recording photographic colour using silver chloride on paper back in the 1840s, which also completely failed with the other silver halides. Lippmann succeeded with every emulsion combination, sensitising dye, and photosensitive molecule he tried. Third, Lippmann-type images can be sectioned vertically through the emulsion, and when examined under a microscope, they will display the layer cake structure of interference planes.

Three other people were offered by some as possibly having priority over Lippmann. The first was Otto Wiener, who in 1890, had published a beautiful experiment that demonstrated that light waves could be photographically recorded (Wiener 1890). His arrangement was simple. He aimed a beam of monochromatic light through the short face of a right-angled prism that was bonded on its hypotenuse to a photo plate by the index-matching fluid benzol (benzene) to dampen interfacial surface reflections (benzol's n=1.5, like the prism glass, the photo plate glass, and the emulsion), and the photo plate surface was similarly bonded to a silvered mirror with a layer of benzol (see the figures from Wiener 1890 231–32.) The light bouncing off the mirror interfered with the light arriving and created a set of standing waves that Wiener was able to record on the emulsion, revealed by vertical section under microscopic viewing (see his last figures, Wiener 1890, 244). The layers were canted forty-five degrees to the emulsion surface due to the geometry of the beams. Wiener made no attempt to record images by this technique. His interest was to show experimentally the presence of light waves. He also wanted to measure their wavelengths, and assuming that they were electromagnetic in nature, whether the maxima he recorded were due to the electric part or the magnetic part, following on earlier attempts by Hertz. It should be noted that Lippmann never tried to investigate the electromagnetic properties of light, only their wave nature.

The second person various people pointed to as deserving of credit for Lippmann's process was Wilhelm Zenker, who published the idea that colour could be imaged by photographically recording standing waves of light. In his 1868 *Lehrbuch der Photochromie*, Zenker proposes that reflection of light off a metal mirror or the emulsion or air interface could give rise to standing waves, and thus record colour in the photo emulsion, at least for a pure colour, though he did not understand the coherence length issue. He further proposed that the colour visible in Becquerel's process images was of this nature (see also Jones 1974, 370–71). Lippmann never acknowledged this, and Becquerel himself disagreed with this idea, though Wiener agreed (Wiener 1895 and 1898). I examine this controversy in detail in Alschuler (2007b).

The other person sometimes pointed to as a precursor was Lord Rayleigh. He made suggestions, somewhat less explicit than Zenker's, and did not carry out experiments to prove them (Jourdain 1900).

Criticism of Lippmann's Process

Initially, some accounts of the process were sceptical, in part, perhaps, because of prior disappointments with other colour processes. There was an anonymous note in the *British Journal of Photography* that is rather critical of the process results and explanation (Professor Lippmann's Experiments on the Production of Coloured Films 1891, 178–179):

> Occasionally the colours obtained on the film will to some extent recall or resemble those spectral tints which acted on the same parts of the film; but more often this will not be the case. In other words, colours may be obtained, but generally they will be the *wrong colours*.
> Doubtful as we are as to whether Dr. Lippmann has actually photographed even the spectral colours, a consideration of his theory is barely incumbent upon us at present [...] we now come to the remarkable assumption that the light waves are not impulses traveling in straight lines, but that they work in and out, snake fashion [...] If all the above assumptions were correct, we might see the possibility of photographing the spectrum, or any homogeneous light, by Lippmann's mode of procedure, but not

> photographing natural objects, which seldom reflect homogeneous [monochromatic] light.

So this observer denied the wave theory of light in favour of the particle theory, in addition to doubting everything else about the results! The first part above is a bit puzzling in light of the fact that an early spectrum by Lippmann still exists, in private hands (Trewinard Collection), which I have seen, and it is excellent colour. I have also seen another, unlabelled, but virtually identical, in the collection at the Musée de l'Elysée in Lausanne. These both look almost the same as a modern carbon arc spectrum, digitally scanned (Ingalidakis 2020).

One of the most critical of those who reacted to Lippmann's process was the American Frederic Ives. He is most famous for pioneering four-colour lithography, but had a horse in the photography race because he, too, was interested in and an innovator of colour photography. He commercialised a camera and viewing system based on colour separation, which he called the Kromskop, in 1897. Another version, called the Photochromoscope, was published in 1894 (Ives 1894). His camera was fitted with internal filters and mirrors that split light from twin lenses into three pairs of beams – red, green, and blue – that recorded on three separate black-and-white emulsions on glass plates. These served as negatives to generate six positive black-and-white images and were placed back in the camera behind their filters for viewing. This system produced stereo colour images.

Ives had earlier remarked in Philadelphia that some of the Lumière Lippmanns were beautiful (Ives 1893a). However, Ives spoke at length to the British Photographic Society, as published in a paper in the *Photographic Journal* (Ives 1893b). He asserted that the photographs he had seen (perhaps including some at the Columbian Exposition of that year in Chicago?) suffered grave defects, such as metallic, changeable, and unnatural colours, in some cases colours not confined to outlines of the coloured objects, and with dark whites. He also said that Berget (Lippmann's assistant) and others ignored these defects in their presentations and publications. He goes on to say that Lumière had produced good images by the process. He next discusses the production of whites in the images:

> We do not know whether the "whites" in these photographs are due to the reflection of light from interference laminae within the film, in accordance with Lippmann's theory, or whether they are nothing more than the normal reflection from the surface of the glass and gelatin film. Lippmann's theory, as I understand it, calls for a different series of laminae

within the film for every wavelength of light, amounting, where white light acts, to over 30,000 laminae within a space no thicker than a single wavelength of light. (Ives 1893b, 124–27)

Here, Ives states a large but arbitrarily defined number of wavelengths that fill the visible spectrum, and he also misunderstands Lippmann's published theoretical explanations and the close analogy of light wave interference to sound. In the latter, a superimposition of multiple frequencies leads, not to separated tones, but to a blend, more or less pulse-like depending on the number of frequencies present This superimposition does not suffer any such crowding problem such as Ives asserts[7].

Ives claims that the discrepancies he sees between the expected density of whites versus those in blacks seen in transmission versus those expected, based on Lippmann's theory, invalidates Lippmann's theory and that the differences depend on reversing the image after its initial development. This again misunderstands the functioning of interference fringes in reconstructing the images. The most exposed areas can be dense or dark, like an ordinary negative, reversed, or transparent, depending on how the emulsion is processed, and in each case, the image will be reconstructed with proper values, as is true for holograms.

Ives (Ives 1893b, 124–27) mentions that Valenta (at the Graphische Bundes-Lehr- und Versuchsanstalt in Vienna) noted to him that the Lumière images were mounted with thin wedge prisms cemented to the emulsion with Canadian balsam, an early mention of this practice, which greatly improves the colour saturation by rejecting the front surface reflection out of the field of view. He notes that breathing on the (naked) emulsion shifts the colours and can make red appear at what was the blue end of a spectrum, a feature he calls curious. That feature was used repeatedly by Lippmann in his lectures to demonstrate the presence of interference layers, with their swelling caused by the emulsion imbibing humidity, as a way to confirm their nature. The president of the British Photographic Society added at the conclusion of the talk that he himself had done Lippmann photos of the solar spectrum and that he had seen variations in the positions of the reds and blues due to variations in exposure time. (It is true that Lippmann image results are sensitive to exposure time, perhaps more so than ordinary colour photos.) He also doubted that Lippmann images could produce pure colours, especially reds, which he saw as mostly orange. This is odd and reflects an incomplete understanding of the theory, since the purer

7 See the associated Lippmann 1894 Fourier analysis in this article on page 50.

the colour, the more well-defined and contrasty the interference fringes and the stronger the resulting colours, other things being equal. Likely, his observations resulted from poor sensitisation in the red, which was still a general problem for photography at the time.

Further on Lippmann image's whites, Edward Steichen is quoted as saying in a 1908 letter to Alfred Stieglitz:

> Professor Lippmann has shown me slides of still-life subjects by projection, that were as perfect in color as in an ordinary glass-positive in the rendering of the image in monochrome. The rendering of white tones was astonishing, and a slide made by one of the Lumiere brothers, at a time when they were trying to make the process commercially possible, a slide of a girl in a plaid dress on a brilliant sunlit lawn, was simply dazzling and one would have to go to a good Renoir to find its equal in color luminosity. (Quoted in Newhall, 1982, 272).

I have myself seen Lippmann images in the collection at the Palais de la Decouvèrte (since returned to the Faculté des Sciences in Paris) that show great whites, one summer and one winter image, made at the Bois de Boulogne. As noted above, Ives had a hand in the game in the form of his own efforts to commercialise another colour process, but there is also a curious irony to Frederic Ives's criticism. His son, Herbert Ives, who became a noted pioneer of television, published a paper that combines colour separation negatives (like his father's process), a three-colour screen, and the final recording made on a Lippmann emulsion. Saving the three negatives allows for the making of copies, and Ives claims good results (Ives 1907, 103–7). He did his doctoral thesis on Lippmann photography (!), which was published the next year, when Lippmann won the Nobel Prize (Ives 1908).

In 1892, in a book on colour photography, Brandt (1892, 65), remarks that a plate of a spectrum, using the Lippmann process, presented by the Lumière brothers to the Conservatoire des Arts et Métiers, was left exposed to light all day long, without the least alteration. This was said to brilliantly solve the problem of colour photography, though perhaps not yet in the most practical manner.

Further photochemical and emulsion success

Lippmann (1892a) reports that improvements of his dye sensitisers allowed him to produce full colour on silver bromide in albumin using Azaline plates produced by Vogel sensitised with cyanine, in exposures of five to thirty seconds, without interposition of colour filters to provide different exposures for different parts of the spectrum, though he does not specify the source of light or its intensity or that the sensitisers may have been provided by the Lumières. Their colours seen in reflection are intense enough that you can see their colour complements when you look through them in transmission. He asserts that mixed colours will record just as well as pure single colours and offers four examples of such success: a four-colour stained-glass window, oranges on a red plate, drapery, and a multicoloured parrot. Later that year, he adds an image of a green holly branch with red fruit, shown by his assistant, Alphonse Berget, at a meeting of the Photo-Club de Paris, where Berget showed the images and explained the process. The images were greatly admired by all (Berget 1892).

Lippmann further demonstrates that his process works on any emulsion and photochemistry that has the sensitivity and resolution to record the interference fringes by making some images on bichromated albumen and bichromated gelatin with brilliant results (Lippmann 1892b).

In 1893, an article in an Argentinian journal reported Lippmann's announcement of 1891. It reports that the colours are permanent and can be recorded with various combinations of photochemistry and emulsion. It includes a graphic showing a section through the plateholder (Fotografia de los colores, 1893). In 1893, Lippmann sent several images to the Columbian Exhibition in Chicago, as did the Lumière brothers. Their images, which they arranged to have projected, as opposed to Lippmann's, which were passed to viewers, included a bouquet, a landscape, a cottage, place cards, and a Japanese screen (Brown 1994, 30). St. Florent showed images with another alternate photochemistry using iron salts in gelatin, and Wall describes this and that complimentary colours can be seen in transmission of well-exposed Lippmann plates in his report on Lippmann images in 1894 (Wall 1894).

No Mercury

The use of mercury in the process was not seen as a great convenience, and almost from the beginning someone experimented with its omission. An anonymous author (*British Journal of Photography,* 1891c) wrote: "Interference colours can be obtained on the plate whether the reflecting surface of mercury is used or not, but more readily and quickly when the mercury is used." This seems to pre-date others who claimed priority for making

Lippmann images without mercury. These included Krone (Krone 1892) and later Rothé (Rothé 1905a, 1905b).

Lippmann's Complete Theory of How Information Is Recorded and Images Formed: 1894

Lippmann's first publication states briefly the idea of his process, and the case of pure colour light wave interference was easily understood and accepted by almost everyone who knew anything about wave interference. Many people had demonstrated light wave interference in various forms, from Newton's "rings" in the 1600s to work by Young and Fresnel in the early and middle 1800s. It is curious that Lippmann did not publish his detailed theory until 1894. He had already demonstrated a version of mixed colours in late 1891. Perhaps it took him that long to work out its details.

Lippmann seems to have understood the need to have the mirror in perfect contact with the emulsion for interference to take place and create recordable standing waves, but never mentions the ideas of coherence or of coherence length, defined as the *difference* in distance of travel over which two waves of slightly different frequency from the same source can still interfere, and its relation to that need. The coherence length of white light is about 1000 nm (1 micron), so no interference is recorded if the there is a gap between mirror and emulsion of that size or larger. Nor anywhere does he use any of Maxwell's theory of light as electromagnetic waves to underpin his theory. This appears to be a conservative decision on his part, one based on his not needing it to provide his explanations of the phenomena. This avoidance is also consistent with the remarkable fact that he does not mention Maxwell's theory anywhere in his Optics course at the Sorbonne, which he taught with nearly unchanging content over many years. In his 1899 introduction to his course on Acoustics and Optics, he stated that the internal nature of light and electricity is unknown, in contrast to that of sound.

However, he must have been aware of the possible relevance of Maxwell's theory to his process, as he delivered a paper later in 1891 on behalf of M. Labatut to the Académie des sciences (Labatut 1891, 128) that discusses this point, at least implicitly. Labatut notes:

> During exposure this external face [of the emulsion], in contact with the mercury mirror, is a *vibrational node*; the first maximum is found a distance ¼ wavelength inside, the second ¾ wavelength, the third 5/4 wavelengths, etc. Let us admit, for a moment that the planes of

photographic reduction are situated at the vibration maxima. [My translation and my emphasis.][8]

That the surface of the mirror is a wavefront node makes sense only because the wave, which is partly electric, must reach zero amplitude at its contact with the mercury, because the latter is an electrical conductor. If there is no mercury (or any metallic mirror in perfect contact), then the interface of the emulsion is with air which is a dielectric, a much weakened partial reflectance results (~4 per cent) and the wavefront is a maximum at the surface. If the light is one pure colour, this is not too severe a result, as there are many other layers a half wavelength apart in the depth of the emulsion, but if it is ordinary mixed colours, approximately half of the blended first (and only) maximum is lost, and since that is where the wavefront information is stored, it greatly diminishes the colour rendition and brilliance.

Lippmann (1894) explains why there is only one major maximum for mixed colours and that all the single-frequency components that make it up will be recovered when the recorded pattern is illuminated. He based this finding on applying Fourier's mathematical discovery (published in 1822) that almost any continuous or discontinuous function of a variable could be approximated to any level of precision by a properly chosen set of sines and cosines, with their frequencies and coefficients chosen by a method he disclosed. Lippmann analogised this finding for light to the more familiar example of sound, with pure tones and mixed tones that can be made up of pure tones. Indeed, Fourier's analysis has had wide applications for finding distinct frequencies present in mixed signals, for detecting signals in noise, and for other purposes in many fields, and it continues to be in wide use to the present day. Lippmann's use of Fourier analysis to explain his results stands as a beautiful example of its application.

Lack of Patent

Lippmann was occasionally asked why he did not patent his process. He replied that he had devised the process while he was an employee of the state and saw no reason why he should profit from it. This was generally seen as

8 "Pendant la pose, cette face externe, en contact avec le miroir de mercure, est un nœud de vibration; le premier ventre s'en trouve distant de $\lambda/4$ du rouge, le deuxième de $3\lambda/4$, le troisième de $5\lambda/4$, etc. Admettons, pour un instant, que les plans de réduction photographique soient situes aux ventres de vibration." (Labatut 1891, 128)

noble, but it would have been possible to patent it and then transfer the patent to the Sorbonne, which could then have licensed it to an industrial developer. That would have given a developer added security and likely would have justified increased efforts and money to improve and market the process. Specifically, the Lumière brothers might have extended their attempts to commercialise the process beyond the three to four years they devoted to it before trying the process they called chroma (trademarked "χρωμα"), a wash-off dye-transfer process that they never fully commercialised, and then going on to invent autochromes, a random dot colour screen process, which then became, for decades, the most successful colour photographic process.

1895

The Société Française de Photographie sponsored a competition in the production of colour photographs by Lippmann's method. The object of the competition, says *Le Moniteur* and as translated by the *British Journal of Photography*, is to popularise the Lippmann method and enrich society with the work of its members:

> We hope both ends will be attained, but remembering Mr. Fred Ives's statement, that half a dozen examples shown in the country a year ago or so were the result of some thousands of trials, we are not sanguine on this point. ('Competition for Lippmann Photographs' 1895, 23)

Clearly, there was awareness that the process had not achieved widespread use. It is also about this time that the Lumières gave up on the process and started working on their "χρωμα" process.

A plateholder was designed by Contamine with the mercury reservoir internal to the frame so that tubing and a separate reservoir could be dispensed with. The mercury was installed from its reservoir to fill the cavity behind the plate by inverting the plateholder. After exposure, the holder was reinverted to restore the mercury. This is described and illustrated in Dillaye (1895). It does not seem to have been commercialised.

1897

Lippmann delivered a long talk to the British Photographic Society in 1897. During this talk, he looked back and mentioned that "for twelve years my investigations were impeded on account of the difficulty which I experienced in the preparation of transparent film." This contradicts a statement made several years earlier by his assistant Berget that it took only three years to develop the process (Berget 1894, 187). In the same talk, he mentions trying to make copies of his images by contact copying, without success, and this task has never been completed as far as I know. Lippmann also suggested that one might lift the completed image emulsion off its glass plate and transfer it to a photo paper, but there is no evidence that this idea was executed. In the same conceptual area, he noted that if the emulsion were stretched laterally, it would thin out, compressing the interference layers, which would shift the image colour towards the blue (Lippmann 1897 and 1898). Lippmann's process was described in a book on colour photography published in Italian (Bonacini 1897).

1898

In a series of notes, author G. R. Baker discusses the newly announced prizes funded by Nobel, their amounts, and considerations for awards. He suggests that colour photography is a field ripe for a prize. He also discusses a talk given by Lippmann before the Royal Photographic Society, in which he notes that an aphengescope can be used to project an enlarged image from the plate and that this image could be projected onto a new plate to make a duplicate at the same or enlarged scale (Baker 1898, 2). The Society's editor W. Gamble mentions that British supplier of photographic equipment Penrose is making plateholders intended for Lippmann photography (Gamble 1898, 167).

Others Try the Process

The Lumières expressed early interest in the process and contacted Lippmann, who in a note to them in August 1891, requested their help in improving gelatin-based emulsions. They responded and it is likely that they made improvements to the emulsion sensitisers, which cut exposure times significantly and were likely on the plates used and shown by Lippmann in spring 1892. (Rittaud-Hutinet et al. 1995, 4–5). With these improved emulsions, it became possible to make portraits with exposure times of about thirty seconds, not too demanding, but about as long as it took to make a daguerreotype. In 1893, the Lumières presented the first photographic colour portrait, a picture of one of their daughters seated at a table with her head laid down on one arm. The original of this has not been found but was reproduced (in black and white) in Eder's *Geschichte der Photographie* (1905, 447).

Whole books were published dedicated to Lippmann's process. These included Valenta (1894), Neuhauss (1898), Lehmann (1906), and a substantial part of Cajal (1912). One of the best illustrated and most practical is a long chapter in Niewenglowski (1909).

Lippmann-Type Images at "Amateur" Exhibitions

Various people who were neither scientists nor professional photographers tried Lippmann's process and in some cases put their images on public display. Perhaps the largest number of these were in England. For example, in the catalogue for the 45th Annual Exhibition of the Royal Photographic Society in 1900, we find listed: item 356, "Section through the picture surface of the red spectral zone of a spectrum prepared after Lippmann's interference method of colour photography, enlarged 4000 diameters:1897" by Neuhauss; item 583, "A Parrot" by Neuhauss; 584, "A Solar Spectrum by Neuhauss"; item 612, "Versailles and Flower Beds" by Lippmann (offered by Fawcett); and "Experimental Results with the Lippmann Process upon a Coloured Window," by Fawcett; and then images by other processes, such as items 554 and 555, by the Joly-McDonough process (RGV line screen process), and item 557, a stereo image negative and print by the Sanger Shepherd colour process. Other Royal Photographic Society shows had Lippmann-type images by philosopher Edward Douglas Fawcett (1866 – 1960) and Edgar Senior (birthdate and death date unknown). Senior was a professor of art and photography at a polytechnic in London and co-author of a book on colour photography, authoring the chapter on Lippmann photography (Bolas et al. 1900).

1908: The Nobel Prize

In 1908, various modest developments took place in or about the Lippmann process. There was even a book published on photography for young people that mentioned the process and describes its theory (Jenks 1908). In the *Comptes Rendu*, a report was published on Lehman's recent paper on Lippmann photography, reporting a special plateholder, a plate viewer, and a projector (a version of the aphengescope) offered by Zeiss (*Le Procedé Lippmann Quelques Ameliorations* 1908, 34–35). Later in the same publication, Lippmann announces his new three-dimensional process, "Photographies Intégrals" (Lippmann 1908a). Cajal published several papers on Lippmann's process. These include processing of the emulsions and a careful study of the internal cross-sectional structure of the interference layers produced under various illuminations. He shows these in beautiful microphotographs. This should be no surprise, as Cajal was a superb microscopist who won the Nobel Prize in Physiology or Medicine in 1912 for his microscopic characterisation of the human nervous system (Cajal 1908).

Lippmann spoke in public about his colour process on various occasions throughout the 1890s and early 1900s in both England and France, though he focused on other projects in physics for most of this time. His name was put forward by French scientists for the Nobel Prize after its creation in 1901 for several years until 1908. That year, the Prize in Physics was controversial within the selection committees. Major sentiment on the prize committee was to award it to Max Planck for his innovations at the beginning of quantum theory, but no agreement could be reached because there was felt to be insufficient evidence for his theory and its impact, beyond resolving the problem of the "ultraviolet catastrophe" (the disagreement between theory of and experiments on the black body spectrum). The committee turned to Lippmann's invention as a sort of compromise, based on the fact that it had demonstrated a breakthrough method of recording permanent photographic colour (in spite of its failure to commercialise), that this method was founded in physics, and that it came with a well worked out mathematical theory for its explanation. At his Prize acceptance speech, Lippmann offered the narrative that the images he showed had been taken on vacation and easily developed and processed on the mantel of his hotel room.

The award made the news in France and internationally for a brief time, and possibly led to new or reissued equipment for the Lippmann process in the form of plate holders, plates, viewers, and projectors ("Le Procedé Lippmann Quelques Ameliorations" 1908). This did not lead to any major new work at process improvement or significant commercialisation. There were some publications that looked back and discussed the process anew[9].

9 See: Dunn, Carlton E. 1936. *Natural color processes*. Boston: American photographic Publishing Company and Friedman, Josef S. 1944. *History of Color Photography*. Boston: American Photographic Publishing Company.

Reasons for Lack of Commercial Success and Conclusion

Lippmann succeeded brilliantly in confirming the wave nature of light, which was a goal from a very early stage of his career, by inventing and demonstrating his photographic process and giving a full mathematical explanation of the resulting images' properties. His achievement was finally recognised by his receipt of the Prize in Physics. His achievement is now directly connected to the theory and practice underlying the current practice of holography. However, the process never was successfully commercialised.

There is likely no single reason why the Lippmann process never achieved commercial success. It had two important advantages over all other processes, including all analogue photographic and digital colour processes, from its time up to the present day: its extremely high resolution and its image permanence. It may someday be possible to make digital processes of equal resolution. If so, they would evade the resolution or sensitivity trade-off of all photographic materials. It also may seem at first blush that digital processes are permanent ("things on the internet last forever"), but storage media are subject to degradation and storage formats are subject to change, requiring the maintenance of software and data transfer media to ensure long-term access. Lippmann images just need to be kept from damage and need a good diffuse viewing light source for "interpretation." Lippmann photography also has the comparative advantage of working with any of the relatively simple black-and-white photochemistries and needing no colour dyes to form the image colours, which contribute to its image permanence. In fact, it will work with any photosensitisers and emulsions with sufficient resolution and transparency.

The reasons for failure are most likely 1) the difficulty of achieving reliable imaging due to conditions of exposure and other variables, cited above with respect to the Lumières' attempts, which was too high an economic investment barrier due to the failure of Lippmann to patent his process, and which led to their passing on to the "χρωμα" and the autochrome processes and 2) the long exposure times, which prevented it from becoming a universal amateur snapshot process, in spite of its incredible resolution. The fact that it gave only positive originals, not easily copied like a negative, was a disadvantage but one that would not otherwise have ruled out its success, as demonstrated by the success of the autochromes, which shared that characteristic. Also in the late 1890s, other colour processes, some colour separation, and some coloured screens, such as Kromscope, Omnicolor,

Dioptochromes, Padgett, Joly, and Agfacolor, entered the market, and they quickly became more or less snapshot processes.

It was not picked up for news photography for the above reasons and also because of the requirement for carting mercury around and the weight of the plates involved, which by the 1890s, began to be superseded by film marketed by Kodak and others (though if anyone had cared to try, it would have been possible to create Lippmann emulsions on film).

Perhaps there is a commercial future for the Lippmann process in two applications, both of which have recently begun to be explored: first for secure, uncopyable images (see, for example, Bjelkhagen [1999]) and, second, for archival documentation of coloured objects, whose colours fade. It otherwise appears to have a future among photographers with a taste for historic processes, as Alves and other have gathered together in a Facebook group to share their passion for the medium[10].

Works Cited

"Fotografia de los colores." 1893. *Anales dela Sociedad Cientifica Argentina, vXXXVI, segundo semester*: 45–46.

"La Decoverte de M. Lippmann. La Photographie des couleurs." 1891. *Les annals politiques et litteraires.*

"Le Procedé Lippmann Quelques Ameliorations." 1908. *Comptes rendus hebdomadaires des séances de l'academie des sciences. La Photographie des Couleurs. Nouvelle Serie* 2: 33–36.

"Professor Lippmann's Experiments on the Production of Coloured Films." 1891. *British Journal of Photography* (March): 178–79.

"The Illustrated Catalogue." 1900. *The Photographic Journal* XXV, 1: 42–43.

Alschuler, William R. 2007a. "Color Theory and Practice 1800–1860." In *Encyclopedia of 19th Century Photography*, edited by John Hannavy. 315–18. New York: Routledge.

Alschuler, William R. 2007b. "The Source and Nature of Inherent Color in Early Photographic Processes." *The Photo Historian, Journal of the Historical Group of the Royal Photographic Society* 152: 9–19.

Appell, P. 1923. Notice Nécrologique, Annales de l'Ecole Normale Superieure.

Baker, G. R. 1898. *British Journal of Photography, Supplement* (January): 2.

10 See Alves and others on the Lippmann Facebook group at: https://www.facebook.com/groups/1422495841341822/permalink/2857389027852489/

Becquerel, H. 1891. *"Photographie – Observations de M. Edm. Becquerel sur la Communication de M. Lippmann au sujet de la reproduction photographique des coleurs." Comptes Rendus de l'Academie des Sciences* 112: 275

Berget, Alphonse. 1891a. "La Photographie des Couleurs." *Revue Rose* XLVIII (July): 38

Berget, A. 1891b. *La photographie des couleurs para la methode interférentieele de M. Lippmann.* Paris: Gauthier-Villars.

Berget, Alphonse. 1892. Bulletin de Photo-club de Paris. (2): 196–7

Berget, Alphonse. 1894. "Gabriel Lippmann," edited by C. W. Ward. *The Photogram* 1, no. 8: 186–88.

Berget, 1901, *Photographie des Couleurs par la Method Interferentielle de M. Lippmann. 2eme ed.* Paris:Gauthier-Villars.

Bjelkhagen, H. Private Communication, 19--) WRT Liesegang emulsion.

Bjelkhagen, Hans I. 1999. "New optical security device based on one-hundred-year-old photographic technique." *Optical Engineering* 38, no. 1: 55–61.

Bolas, Thomas, Alexander A. K. Tallent, and Edgar Senior. 1900. *A Handbook of Photography in Colours.* New York: E. & H.T. Anthony & Co.

Bonacini, Carlo. 1897. *La Fotografia dei colori, trattato teorico-pratico del dottor Carlo Bonacini.* Milan: U. Hoepli.

Brandt, Ch. 1892. *La Photographie des couleurs, état actuel de la question, suivi des procédés de reproductions photographiques en couleurs ... par Ch. Brandt.* Paris: C. Mendel.

British Journal of Photography. 1891. March 20: 178–9.

British Journal of Photography. "Competition for Lippmann Photographs." 1895. (11):23.

Brown, Julie K. 1994. *Contesting Images: Photography and the World's Columbian Exhibition.* Tucson: University of Arizona Press.

Cajal, Ramón y Santiago, Primer Centenario, concesión del Premio Nobel Don Santiago Ramón y Cajal. Madrid: Ministerio de Educación y Ciencia, 2006.

Cajal, S. R. 1908. "La Photographie des Couleurs." *Comptes Rendus de l'Académie des Sciences. Nouvelle Serie* 1 (April): 110–12.

Connes, P. 1987. "Silver Salts and Standing Waves: The History of Interference Colour Photography." *Journal of Optics* 18, no. 4: 147–66.

Dillaye, Frédéric. 1895. *Les Nouveautés photographiques : complément annuel à La théorie, la pratique et l'art en photographie / par Frédéric Dillaye.* Paris: Librairie illustrée. http://gallica.bnf.fr/ark:/12148/bpt6k54928532.

Dunn, Carlton E. 1936. *Natural Color Processes.* Boston: American photographic American photographic Publishing Company.

Eder, Josef Maria. 1905. *Geschichte der Photographie mit 148 Abb. u. 12 Taf.* Halle (Saale): Knapp.

Friedman, Josef S. 1944. *History of Color Photography*. Boston: American Photographic Publishing Company: 23–24.

Gamble, M. 1898. "Colour Processes." *The Photographic Journal*. (28) February. 167.

Garrett, A. E. 1911. *The Advance of Photography: its History and Modern Applications*. London: Kegan Paul, Trench, Trübner & Co: 231–239

Guerronan, A. 1891. "Photographie des Couleurs." *L'Illustration* (2505) February 28.

Ingalidakis. 2020. https://ingalidakis.com/spectroscope/amici.html#colorphotos Accessed 11/17/20.

Ives, Frederic Eugene. 1893a. "The Lumiere Lippmann Color Photographs." A Communication to the Photographic Society of Philadelphia, Nov 8, 1893" Journal of the Photographic Society of Philadelphia I (March 1894) cited in Brown, 1994, *op cit*, 135.

Ives, Frederic Eugene. 1893b. "The Lumiere-Valenta-Lippmann Colour Photographs." *The Photographic Journal, Photographic Society of Great Britain* 22: 236.

Ives, Frederic Eugene. 1894. *Handbook to the Photochromoscope: With Chapters on the Nature of Light, and Theory of Color, by Some of the First Authorities*. London: Simpkin.

Ives, Frederic Eugene. 1907. "Three-color Interference Pictures." *Physical Review* 24: 103–7.

Ives, Herbert E. 1908. "An Experimental Study of the Lippmann colour Photograph." *The Astrophysical Journal* 27: 325–52.

Jenks, Tudor. 1908. *Electricity for Young People*. New York: Frederick A. Stokes Company.

Johnson, G. Lindsay. 1909. *Photographic Optics and Colour Photography: Including the Camera, Kinematograph, Optical Lantern, and the Theory and Practice of Image Formation*. New York: D. Van Nostrand Co: 216–218.

Jones, B. E.1974. In *Encyclopedia of Photography*. Edited by Jones, B. E. and Sobieszek, R. New York: Arno Press.

Jourdain, P. E. B. 1900. "Notes on the Development of Interference Photochromy." *British Journal of Photography* 47: 612, 613, 693, 694.

Konig, E. and E. J. Wall. 1906. *Natural-color Photography*. [Iliffe & Sons]: London: 12–15.

Krone, Herman. 1892b. "Über Farbenphotogramme von Spectren." *Annalen der Physik und Chemie* 3, no. 46: 426–30.

Labatut, M. 1891. "L'absorption et la photographie des couleurs." *Comptes Rendus de l'Académie des Sciences*. 1891, 2eme semester. 63, no. 3: 126–129.

Lehmann, Hans. 1906. *Beiträge zur Theorie und Praxis der direkten Farbenphotographie mittels stehender Lichtwellen nach Lippmanns Methode*. Freiburg: Trömer.

Lippmann, Gabriel. 1891a. "La Photographie des Couleurs." *Comptes Rendus des scéances hebdomadaires de l'Académie des Sciences* 112, no. 5: 274–75.

Lippmann, G. L. 1891b. "Photographie des Couleurs." *Revue d'Astronomie Populaire.*
Lippmann, G. 1892a. "Sur la Photographie des couleurs: deuxieme note." *Comptes Rendus de l'Académie des Sciences* CXIV, no. 17: 124–25.
Lippmann, Gabriel. 1892b. "Photographies colorées du spectre, sur albumine et sur gélatine bichromatées." *Comptes Rendus Hebdomadaires des Séances de l'Académie des Sciences* 115: 575.
Lippmann, G. L. 1894. "Sur la théorie de la photographie des couleurs simples et composées par la méthod interferentielle." *Comptes rendus des Scéances Hebdomadaires de l'Académie des Sciences* 118: 92–100.
Lippmann, G. L. 1908a. "Épreuves Reversibles, photographies integrals." *Comptes Rendus de l'Académie des Sciences. La Photographie des Couleurs. Nouvelle Serie* 5 (May): 121–24.
Lippmann, G.L 1893. *La Photographie en Couleurs.* Conferences Publique sur la Photographie. Session 3. Paris: Gauthier-Villars.
Lippmann, G. L. 1889. "Sur l'obtention de photographies en valeurs justes par l'emploi de verres colorés." *Comptes Rendus de l'Académie des Sciences* 108: 871–73.
Lippmann, G. L. 1897. "Color Photography." *The Photographic Journal*: 124.
Lippmann, G. L. 1898. *British Journal of Photography Supplement* (January): 3.
Maumene, E. 1891. "Photogrpahie des couleurs." *Cosmos* 319 (March). 370–378.
Newhall, Beaumont. 1982. *The History of Photography.* New York: Museum of Modern Art.
Niewenglowski, Gaston-Henri. 1909. *Traité de photographie des couleurs.* Paris: Garnier frères: 57–115.
Picard, E. 1931. *La Vie et l'Ouvre dc Gabriel Lippmann.* Academie des Sciences. Paris: Gauthier-Villars.
Ramón y Cajal, Santiago, and Gerardo F. Kurtz. 1994. *La Fotografía de los colores: fundamentos científicos y reglas prácticas.* Madrid: Clan.
Rittaud-Hutinet, J., *et al*, 1995. *Letters Auguste and Louis Lumiere*, Faber and Faber, London: 4–5.
Rothé, M. E. 1905a. "Nouvelles Recherches sur la Photographie des Couleurs." *Congres pour l'avancement des Sciences, session de Cherbourg* (August): 511–22.
Rothé, M. E. 1905b. "La Photographie interferentielle des Couleurs." *Bulletin de la Société Dauphinoise d'Amateurs Photographes* 33 (April): 7–23.
Shulman, Arnold Roy. 1970. *Optical Data Processing.* New York: Wiley: 581–582.
Story, Alfred Thomas. 1915. *The Story of Photography.* New York: Forgotten Books: 141–142.
Stroke, George W. 1969. *Introduction to Coherent Optics & [and] Holography.* New York: Academic Press: 200–201.
Tissandier, 1891. "La Photographie des Couleurs". *La Nature.* (924) February 14.
Valenta, Eduard. 1894. *Die Photographie in natürlichen Farben unter besonderer Berücksichtigung des Lippman'schen Verfahrens.* Halle (Saale): Knapp.

Wall, E. J. 1941. *Wall's Dictionary of photography and reference book for amateur and professional photographers*, edited by Arthur Lindsay McRae Sowerby, F. J. Mortimer, and E. J. Wall. 16th edition. London: Published for the Amateur Photographer by the Fountain Pres.

Wall, E. J. 1894. *The Photographic Journal, Photographic Society of Great Britain* (June) 26: 290–299.

Wiener, Otto. 1890. "Stehende Lichtwellen und die Schwingungsrichtung polarisirten Lichtes." *Annalen Der Physik* 276, no. 6: 203–43.

Wiener, Otto. 1895. "Farbenphotographie durch Korperfarben und Mechanische Farbenanpasungein der Nature," *Annalen Der Physik und Chemie.* 55: 225–49 and translated in Wiener, Otto. 1898. *Color Photography by Means of Body Colors, and Mechanical Color Adaptation in Nature*. Smithsonian Inst. Annual Rept. for 1896. Washington: Smithsonian Inst.

Zenker, Wilhelm. 1868. *Lehrbuch der Photochromie: (Photographie in natürlichen Farben) : nach den wichtigen Entdeckungen von E. Becquerel, Nièpce de St. Victor, Poitevin u. A. : nebst einer physikalischen Erklärung des Entstehens der Farben.* Berlin: self-published. Reprinted Friedrich Vieweg und Sohn, Braunschweig, 1900.

Lippmann in Books on Color Photography:

Bolas, et al, 1900. Bolas, T., Tallent, A. K., and Senior, E. 1900. *A Handbook of Photography in Colors*. Marion & Co: 315-343 extensive recipes.

Garrett, A. E., 1911. *The Advance of Photography.* Kegan Paul, Trench Trubner & Co. London: 231–239 (a great discussion that includes H. Ives work of floating mirror)

Johnson, G. L. 1909. *Photographic Optics and Colour Photography.* Ward and Co. London: 216–218 (has a section offered by Senior)

Konig, E. and Wall, E .J. 1906. *Natural Color Photography.* Iliffe & Sons, London. pgs. 12–15.

Niewenglowski, G.-H. 1909. *Traite Pratique de Photographie des Couleurs.* Garnier Freres, Paris: 57–115

Story, A. T. 1915. *The Story of Photography.* D. Appleton and Co., New York, pgs. 141–142.

Later mentions of Lippmann with process recipes and discussion (Traces of Lippmann's process after the advent of Kodachrome)

Dunn, C. E. 1936. *Natural Color Photography.* American Photographic Publishing Co. Boston: 190–191.

Friedman, J.S. 1944. *History of Color Photography.* American Photographic Publishing Company, Boston: 23–24.

More Modern References to the Lippmann Process in Texts on Holography

Shulman, A. R. 1970. *Optical Data Processing.* John Wiley and Sons, New York: 581–582.

Stroke, G. 1969. *An Introduction to Coherent Optics and Holography.* Academic Press, New York: 200–201.

About the Author

William R. Alschuler earned a doctorate in astronomy from University of California, Santa Cruz and has been at the School of Critical Studies at the California Institute of the Arts since 1992. He has previously published articles on holography and Lippmann photography, has compiled a catalogue raisonné of Lippmann photographs, and contributed ten articles to the *Encyclopedia of 19th Century Photography* (2007). walschuler@hotmail.com

2. Lippmann's Interferential Colour Photography: A Juncture between the Histories of Spectroscopy and Photography

Klaus Hentschel

Abstract

This contribution to the histories of science and photography locates Lippmann photography at the crossroads between two trajectories: the early modern pursuit of improved and precise documentation and mapping of the solar spectrum by novel experimental and technical methods, on one hand, and the rise and institutionalisation of scientific photography at the turn of the nineteenth into the twentieth centuries, on the other. The award of the Nobel Prize in Physics to Gabriel Lippmann in 1908 for developing his process of interferometric colour photography expresses this juncture between scientific and technical inquiry. Even though Lippmann photography was practiced for only about fifteen years by very few scientifically oriented photographers, it was nonetheless an important predecessor to the later holograph.

Keywords: interferential colour photography, spectroscopy, interferometry, history of science, history of photography

Since the time of Sir Isaac Newton (1643–1727), when the spectrum first became a research object in its own right and shook off its bad reputation as a disturbing, even frightful phenomenon, the recording of the coloured spectrum in its full variegated glory proved to be a protracted challenge (Hentschel 2002). Various techniques were tested out and used, among them woodcut, copper engraving, and lithography, and, since 1839, also

Hannouch, H. (ed.), *Gabriel Lippmann's Colour Photography: Science, Media, Museums.*
Taylor & Francis Group, 2022
DOI 10.5117/9789463728553_CH02

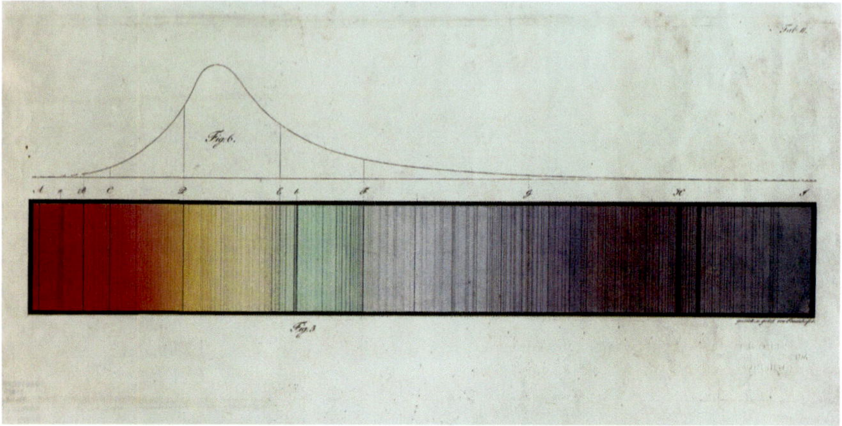

Fig. 4. Fraunhofer's handcoloured plate from 1815. Deutsches Museum (Munich).

photography. In the early nineteenth century, hand-colouring of engravings produced some of the most impressive results (such as here in Fig. 4).

However, hand-colouring depended on the skill of the artist in capturing the subtle gradations of tone. Such high-quality plates were consequently rare and not reproducible at the time. Chromolithography, which was widely practised since the 1830s, took a substantial step forwards, but the generation of colours by superpositions of layers of various printed primary colours (up to six) remained insufficient. Daguerreotype, talbotype, and other early forms of photography all had the disadvantage of being recordings in black and white. Furthermore, early photography had severe problems with colour sensitivity, especially in problematic colour ranges such as orange or green. The solar spectrum actually served as the preferred test object for every new photographic process. The resulting exposure directly displays the relative sensitivity to the different components of the continuous spectrum. Any broader gaps in the bright continuous spectrum became immediately visible. Thus, the main goal of spectroscopists and photographers alike was to obtain a faithful representation of the whole spectrum in colour.

Motivated by Jean-Baptiste Biot (1774–1862), who had been instrumental in making the daguerreotype process publicly known, Edmond Becquerel (1820–91), the nineteen-year-old son of one of Biot's colleagues in the Parisian Academy, took up this photochemical analysis of prismatically decomposed solar light, also examining various artificial light sources in 1839 (Lehmann 1906, 1–2, Becquerel 1868, Hentschel 2002, 193ff, 205 ff.). Becquerel's initial idea was to cover chemically sensitised paper with a tinted plate of glass to see how the different absorptions of various colours influenced the chemical reactions. He noticed that, after a few minutes exposure to the sun through red

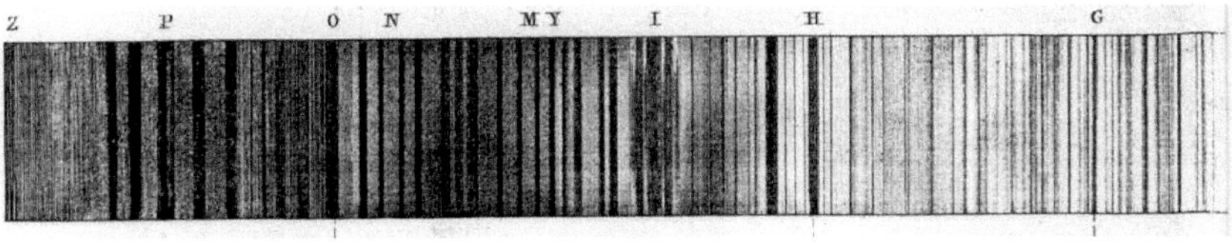

Fig. 5. E. Becquerel's plate showing the solar spectrum with dark lines beyond the violet of the visible spectrum, 1842. Copper engraving by Dulos on the basis of Becquerel's photography. From Becquerel 1842a fig. 2 (note: the line labels are not identical to the ones later adopted).

and yellow glass, no action was observable on the surface, whereas it was soon blackened through blue glass. When he first exposed the surface to violet rays for a short time and then placed the paper under red glass, the chemical action continued nearly as intensively as it did under blue glass (Bequerel 1839, 702f.).

Three years later, Edmond Becquerel advanced to the position of assistant at the Paris Muséum d'histoire naturelle, where his father had occupied the chair for physics since 1837. In June 1842, Becquerel junior utilised both photographic processes by Daguerre and Talbot for the first permanent recordings of a chemical spectrum extending substantially beyond the violet end of the optical spectrum (Becquerel 1842).

Becquerel – as well as other pioneers of spectrum photography such as John William Draper – systematically compared his spectrum photographs against Fraunhofer's visually compiled map of the solar spectrum (Hentschel 1999, 2002b). The clear coincidence of the strongest among Fraunhofer's dark lines confirmed that photography faithfully recorded all the visible lines. Moreover, it revealed that daguerreotypes disclose parts of the spectrum not visible to the eye. It became apparent that the chemical action of the spectrum on a photographic plate was *minimal* precisely in those areas of the spectrum where the eye gauged a *maximum* intensity. These early photographic recordings of the solar spectrum were all still black and white.

In the mid-1840s, Edmond Becquerel succeeded in making a few colour photographs of the solar spectrum. He called his process "héliochromie," since the sun seemed to be inscribing its own coloured light. The novelty was that he coated a well-polished silver plate with a subchloride (Ag_2Cl) gelatin emulsion. To obtain a homogeneously even thickness of the photosensitive layer, Becquerel preferred a galvanic process for its production. Another improvement in the results was gained by exposing this plate to diffuse light for a short time and then projecting the solar spectrum onto the intricately sensitised plate for just a few minutes: "again it takes on a bluish tint, distinctly registered within prismatic blue, discolours slowly toward yellow, and turns into pink within red" (Becquerel 1848b, 452,

Fig. 6a. Edmond Becquerel. "Spectres solaires". 1848. Musée Nicéphore Niépce, Ville de Chalon-sur-Saône.

448ff.).[1] One of the best specimens of Becquerel's colour photographs of the spectrum is actually preserved inside an ornamental box in the archives of the Conservatoire national des arts et métiers in Paris.[2]

Becquerel's plates retained the colour only "as long as subsequent light action is avoided" (Eder 1945, 665, Zenker 1868b, 19, 106ff.)[3] and all preserved plates have substantially lost their intensity and definition. Becquerel's plates fade quickly under strong illumination, so they had to be stored in complete darkness and could only be exhibited for relatively brief times under feeble

1 Further details in Hentschel (2002a, 205f.).

2 Bellone and Fellot (1981, 38f.) include a colour reproduction of the spectrum photograph and its box, labelled "image photographique du spectre solaire obtenue directement avec ses couleurs par M. Edmond Becquerel (1848)." The Musée Nicephore Niepce displays their four exemplars of Becquerel's "image photographique du spectre solaire obtenue directement avec ses couleurs" under weak illumination. See *Spectres solaires image photographique du spectre solaire obtenue directement avec ses couleurs par E. Becquerel* (1848) © musée Nicéphore Niépce, ville de Chalon-sur-Saône.

3 These authors report on the great mistrust that Becquerel's results faced, which were theoretically incomprehensible at that time.

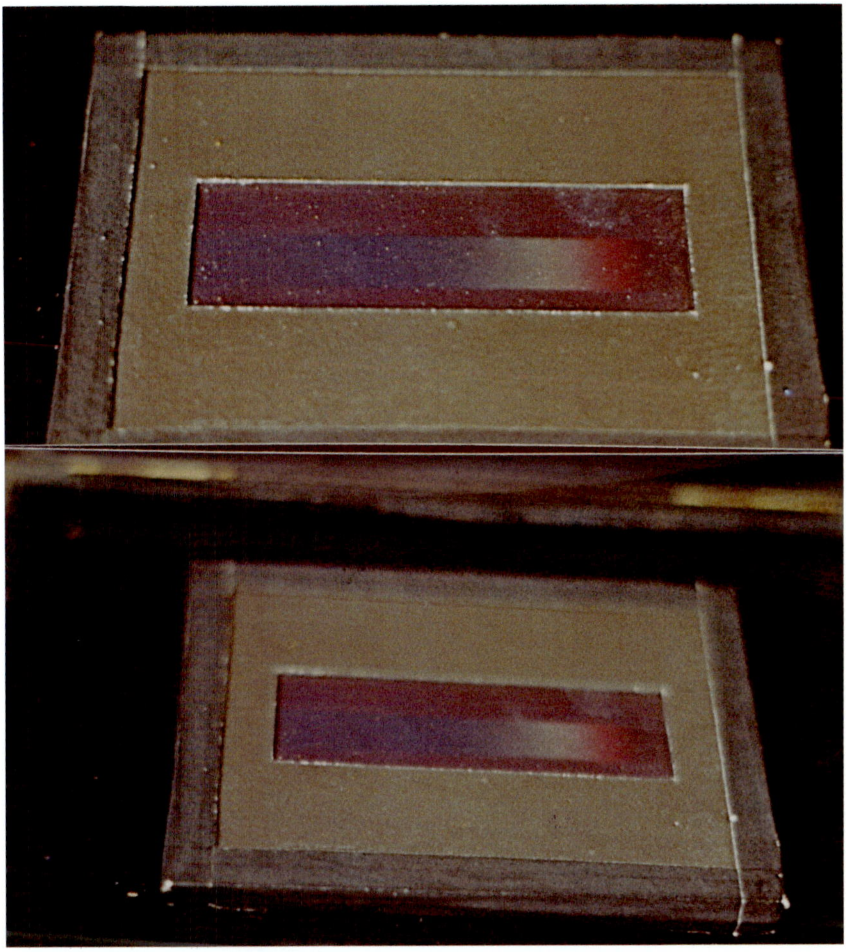

Fig. 6b. Alexandre Edmond Becquerel. 1848. Héliochromie. "Spectre solaire". 1848 © Musée des arts et métiers, Paris / photo Studio Cnam (Inv. 09728).

lighting.[4] Another problem besides conservation was that, at the time, no one could point to any theoretically successful avenue towards obtaining a permanent colour photograph. All one could do was experimentally probe the immense range of possibilities in chemical and physical preparations of silver salts. After nearly forty years of such exploratory work, with no sign of

4 Bjelkhagen (1999) reports that, during the 1855 world exhibition, one of Becquerel's colour photographs was shown in a pitch-dark tent. For this reason, all existing photos of Becquerel's original plates are of poor quality, since they can only be taken in a rush under insufficient lighting conditions.

a tractable way, William Jerome Harrison (1845–1909), himself a dedicated photographer and one of the first historians of photography, put it thus:

> There is probably only one thing which it is safe to predict about the problem of obtaining permanent photographs which shall represent objects in their natural colours, and that is that the discovery, if it is ever made, will not be the result of an accident. The question must be studied and the conditions mastered before the attempt can be made with even the least chance of success. (Harrison 1887, 117)

Various contemporary ad hoc hypotheses about different molecular reconfigurations of the silver salts explain the results were just vague post-facto rationalisations without any predictive force, many of them also easily refutable (Becquerel 1848b, 457, 1849, Wiener, 1895).[5] For many physicists at the time, it was tempting to compare the colours of Becquerel's plates with the colours appearing on the surface of a pearl, soap bubbles, or oily film on water. These Newtonian rings were known to be effects of interference – one of the keystones of the wave theory of light that had been revived by Thomas Young and Augustin Fresnel in the early nineteenth century (Buchwald 1989). However, whereas Newtonian rings change their colours with thickness of the oil, soap, or pearl layer, Becquerel's colours were immune against changes in the thickness of the photographic emulsion. A different cause for the emergence of colours had to be found.

The first feasible explanation of Becquerel's colours was provided in 1868 by the versatile scientist Wilhelm Zenker (1829–99), who was active in the fields of anatomy and physiology as well as astronomy and astrophysics, often working on their applications of photography. Zenker's ingenious idea was thus to interpret Becquerel's colours as the result of optical "standing waves" generated inside the sensitive layer of Becquerel's plates because of constructive interference. These standing waves, generated by the stable superposition of incoming and reflected waves, impress their periodic structure onto the gelatin film. At the zones of maximum intensity, interference would thus cause an intense reduction of silver chloride to silver precipitate. The stable suspension in gelatin sets the reduced silver in layers at regular intervals, soon known as "Zenkersche Blättchen," henceforth translated

5 The latest research by de Seauve et al. 2020 has shown that the colours obtained by Becquerel were due to the unequal presence of metallic silver nanoparticles. Differences in the distribution of these nanoparticles, not the interferometric Zenker layers, produce the colour in Becquerel's plates.

as *Zenker layers* (Zenker 1868, 1900, 116ff, 134ff, Valenta, 1894, Neuhauss 1898, Lehmann 1909a, 786). According to Zenker, they consist of metallic silver reduced chemically by the energy of the incoming light. They are concentrated at all zones of constructive interference, which meant that they succeed each other at regular intervals of half the wavelength of the light to which the layers had been exposed. Otto Wiener (1862–1927), then still a lecturer of experimental physics in Strasbourg, confirmed the existence of such optical standing waves microscopically in 1889/90 in thin film of silver chloride suspended in collodion (Wiener 1890). Although Wiener's experiments were not yet tied into the discussion about colour photography, they empirically supported Zenker's hypothesis. Thus reinforced, Zenker reiterated this explanation of the colours in Edmond Becquerel's plates in an article for Josef Maria Eder's *Jahrbuch für Photographie und Reproductionstechnik* in 1891 (Zenker 1891). The latest research by de Seauve et al. (2020) has shown that there are no such periodic patterns and that Zenker's and Wiener's explanation was therefore false for Becquerel's plates. But we shall see that it was exactly right for the Lippmann process, to which we now turn.

Lippmann's Photochromie

In 1891, after many years of frustrated efforts to use interferometric methods for photographic purposes, Gabriel Lippmann (1845–1921) was able to present his first results (Lippmann 1891, 1892a). He transformed Zenker's and Wiener's earlier fundamental considerations and experiments into a practicable, albeit still exotic, photographic technique. Lippmann was born in Luxembourg and studied at the École Normale Supérieure in Paris. Between 1872 and 1874, he spent time in Germany and studied physics, chemistry, mathematics, and spectroscopy (inter alia with Kirchhoff), submitting his PhD dissertation on electrocapillary effects to the university of Heidelberg in 1874, which got the predicate "summa cum laude" (Pier and Massard 1997, 7 Wagener 2016). As member of a government commission to study methods of teaching science in Germany, he also visited Hermann von Helmholtz in Berlin. In 1883, he became a member of the Paris Académie des sciences and professor of physics at the Sorbonne, where he also taught acoustics and optics after swapping his chair for mathematical physics for a chair in experimental physics (succeeding Jamin in 1886). He thus had no need for any obscure German textbook as motivation to study the wave theory of light. He developed his own interference theory of photographic colour production as early as

1886 – four years before Wiener published his experimental verification of his standing wave theory (Lippmann 1908/09, Bjelkhagen 1999).

Before I discuss Lippmann's technique and procedure, let me briefly say why these papers about interferometric techniques in photography were so important for its history. They stand for the growing intrusion of techniques from the experimental sciences, especially from physical optics and photochemistry, into the interdisciplinary research area of photography. Zenker's and Wiener's explanations of Becquerel's and later also Lippmann's colour photography necessarily presuppose the wave model of light, since they assume constructive interference of light to explain the layered changes within the photosensitive emulsion, which in turn explains why colours are seen when looking at plates that only used to yield contrast images. Interferometry had already been applied by the "master of light" Albert A. Michelson (1852–1931) in many other fields of physical optics, geodesy, metrology, and physics more generally (Michelson 1903, Staley 2003). It was shown that interferometry also plays a part in the much more lowbrow field of photography. Complex 'high theory' could henceforth be applied to it, and more theoretically inclined 'scientific photographers' hoped to thereby raise the status of their field and perhaps turn it into a veritable scientific discipline alongside photochemistry and physical optics (Zenker 1891, Krone 1894, Lehmann 1906, Ramón y Cajal 1907, 213).

On February 2, 1891, Lippmann presented his first report on his recently developed version of "photochromie" to the Parisian Académie des Sciences (Lippman 1891). In the Academy session one week earlier, Lippmann's colleague Marie Alfred Cornu (1841–1902) presented Wiener's experiments, demonstrating the existence of periodic layers due to interference of light in photosensitive fine emulsions (Cornu 1891). This must have made matters urgent for Lippmann, who then used the next Academy meeting to present his own results. Like Wiener, Lippmann also exploited the interference of incoming and reflected light but with the clear intent of recording colours in the fine-grained sensitised emulsion. Initially, albumen containing 0.5–0.6 per cent potassium bromide was sensitised in a silver bath, then washed off with a cyanine solution to make its spectral sensitivity as uniform as possible, and allowed to dry. The trick in Lippmann's procedure is to have the minute silver bromide grains suspended in a layer of transparent albumen (i.e., egg white) or gelatin that is thin enough to allow light to pass through the whole layer and bounce back again, allowing interference to occur between the incoming and reflected light. To get phase shift-free reflection, Lippmann exchanged Becquerel's polished silver mirror against mercury. Quicksilver – owing to its strong adhesion – does not leave any air gaps between the light-sensitive emulsion and the reflector, which would have spoiled the effect, but since mercury is fluid at room temperature, Lippmann needed a specially designed cassette (Neuhauss 1898, Lehmann 1906, Mitchell 2010, 322–24). Only where constructive interference occurs do the silver bromide grains suspended in the gelatin reduce to silver. This clever procedure allowed for the first time to fixate the exposed colour photograph – something that had not been possible with Becquerel's plates in 1848. Lippmann's initial photographs were still somewhat defective because of variations in the sensitivity of the film, but he himself, as well as the Lumière brothers, among others, soon improved the technique. Until 1894, a few more of his brief progress reports on recipes and procedures followed (Lippmann 1892, 1894, Lumière 1893).

In the end, Lippmann succeeded in recording solar and electric arc spectra in brilliant colours from the blue to the red (Fig. 4) and even beyond into the ultrared and ultraviolet regions. The colours on the best Lippmann plates were intense, and reflections from metallic surfaces also looked natural. However, failures abounded even for experimentally gifted experts. On average, only a quarter of their trials led to success, and sometimes not a single plate worked for months on end.

Explanation of How It Works

Lippmann himself was the first to complement his report about his successful experiments with a clear theoretical explanation quite similar to Zenker's theory of Becquerel's photographs, without quoting Zenker or Wiener explicitly, however (Lippmann 1891, Valenta 1894, 21ff, 30ff, Wiener 1899, Zenker 1891, 1893, Donath 1906, 23ff, Lehmann 1906).

Each colour produces constructive and destructive interference at different levels of the sensitive emulsion, depending on its wavelength. But only the zones of constructive interference at regularly spaced intervals will react chemically – only there is the sensitising silver bromide reduced to precipitate silver. Thus, Wiener's standing light waves create regular patterns inside the sensitised emulsion, layered patterns with interstices equal to half the wavelength of the incoming light and hence different for the different colours of the light (Zenker 1868, Rayleigh 1887, 156–58, Wiener 1890, Valenta 1894, Donath 1906, 23ff., and Lehmann 1906). The exposed and developed plate is then viewed in white light under the same angle of incidence as during the recording. The same standing waves pass through the plate at the spots where each colour had originally hit the sensitised emulsion.

Fig. 7. Cutout from a Lippmann plate of the solar spectrum from 1891, original size 67 x 52 mm. From the collection of Peter & Ruth Herzog. printed on the cover of M. Pagneux et al.: Photographie - Der Lange Weg zur Farbe, DU - Zeitschrift für Kultur 708 (2000); cf. also Bellone & Fellot 1981: 92ff and Triarhou & Cerro 2008: 5.

All the other colours are impeded and absorbed (at least in principle – see below for deviations from this elegant theory). Therefore, when white light falls onto the "interference photograph" – as Lippmann's "photochromie" was thus dubbed – only those wavelengths exactly matching the distances between the layers of constructive interference will pass through the layer, be reflected back again off the plate interface, and ultimately become visible. The observer sees exactly those colours that had fallen onto the sensitised plate earlier on at exactly the same spots. Optical interference of incoming and reflected light thus creates wonderfully vivid colour photographs. Lippmann's images were more intense than Newtonian rings or other interference phenomena observed on an oil-coated water surface, because not one, but several layers of constructive interference filter the white light. In fact, Lippmann estimated having obtained approximately two hundred such interference-generated silver layers in a sample of 1/20 mm thickness, and he suspected that the intensity of colour would steadily increase with the number of such layers (Lippmann 1891, 275, Valenta 1894, 23, Neuhauss 1898, Donath 1906, 34).

In 1891, when Lippmann's experimental success was first announced – not only in academies and scientific journals but also in newspapers all over the world – the existence of optical standing waves had just been demonstrated by Wiener (1890). For some of the more theoretically inclined experts in physical optics, who were only familiar with the publications, Lippmann's procedure actually appeared to be but one more – albeit elegant – demonstration of the correctness of Zenker's and Wiener's theory (Zenker 1893, 1900, Wiener 1895, Schütt 1896, and Lehmann 1906, 3, 1908, 280). Lippmann himself and other experts, too, were later eager to stress the link between theory and experiment (Lippmann 1894, Krone 1894).

Verification of this theoretically pleasing explanation had to wait a few years. The first experimenter able to provide this evidence was Richard Neuhauss (1855–1915), a physician in Berlin-Lichterfelde (Neuhauss 1898, Ramón y Cajal 1907, Lehmann 1908, and Ives 1908) who was prone to employ photography in all kinds of fields and became editor of the *Photographische Rundschau* in 1894. Between 1894 and 1908, he produced roughly 2,500 Lippmann plates, including the famous photograph of a parrot, along with landscapes and other motifs.[6] Aside from brilliantly coloured objects like his

6 Neuhauss's intense work on colour photography and various recipes and exposure times for his sensitised gelatin plates are found in his laboratory journals for the year 1894 to 1900, recently published in transcription by Hanne Holm-Johnsen, curator at the Preus museum in Horten, Norway, where the Neuhauss journal and some of his Lippmann photographs are

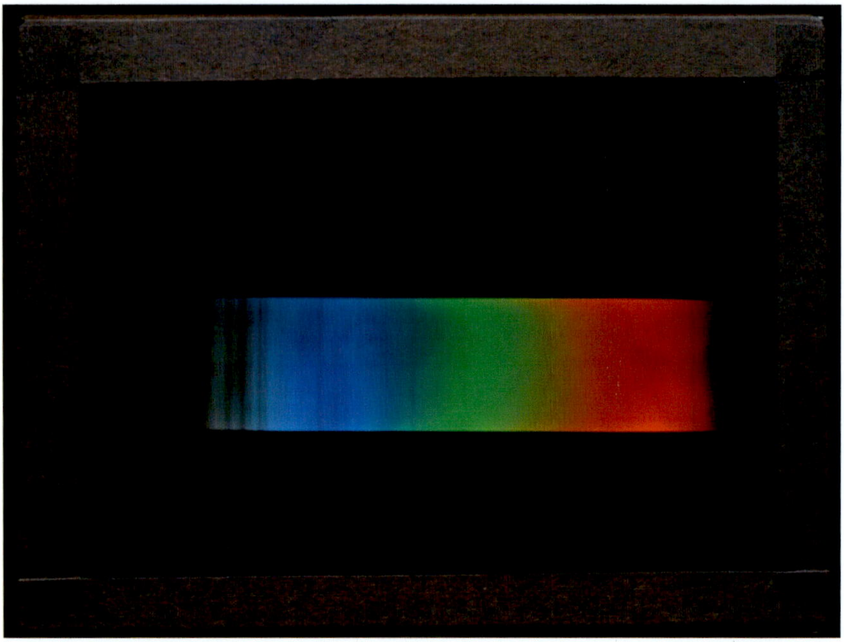

Fig. 8. Cutout from a Lippmann plate of the solar spectrum mounted on paper, original size 63 x 85 mm, with a superimposed glass wedge of thickness 7–18 mm, 1900, published in Faber 2003: 142. Note that this Neuhauss spectrum only shows the prime colours blue, green and red, but no yellow.

stuffed parrot, Neuhauss often photographed the solar spectrum, created by a professional grating spectroscope, which generated normalised spectra that allowed a direct and precise determination of wavelengths, not just of vague colour ranges (Neuhauss, 1894–1900). He was also in close contact with Lippmann, with whom he corresponded and whom he also once visited in Paris in his apartment, 10 rue de l'Éperon.

Together with an expert in histology and microscopic preparation (the Berlin anatomist Dr. Flatau), Neuhauss produced roughly one hundred microhistological cuts through small samples of red to reddish yellow sections of developed Lippmann plates. The wavelength λ of light in this spectral region varies between 7,600 and 6,600 Å, so the distance between adjacent zones of constructive interference ($1/2\ \lambda$) should be within the range 0.00038 to 0.00033 mm. Taking into account the different refraction index $n = 1.5$ of air-dried silver bromide and unavoidable effects of expansion in the photo layer, which had to be embedded in paraffin to allow histological

preserved; see https://holowiki.org/data/lippmann/Neuhauss_Journals.pdf; cf. also Neuhauss (1898), Lehmann (1906), Bjelkhagen (1999b, 276).

microcutting, Neuhauss obtained reasonable agreement between theoretical estimates and microscopically observed distances between adjacent layers (Neuhauss 1898, 60–72).

Since Neuhauss's plate (enlarged four thousand times) is not very clear, I have selected plates obtained roughly ten years later by Hans Lehmann (1875–1917), who followed up on Neuhauss's work with improved techniques and ultra-fine gelatin plates (Lehmann 1906, 60–73). Whereas Neuhauss's albumen-based plates had an expansion factor of only 30 per cent, Lehmann's gelatin emulsions had an expansion factor of four to five times. This, in turn, made it easier for him to also measure layer distances for colours with wavelengths shorter than in the red. Preparations of his microscopic samples of various colour ranges, not only of pure spectrum colours but also of mixed colour regions, were done by the anatomist and prosector Dr. Alexander A. Böhm (?–1912) in Munich, and the microphotographs were taken and enlarged by Lehmann's Zeiss colleague, Dr. Köhler, in Jena, using the best available Zeiss microscope with an aperture of 1.4 and a 2,000–fold optical enlargement. Together with the post-exposure expansion of the gelatin, this yielded a 7,000– to 10,000–fold increase in the distances between adjacent Zenker layers and thus to very clear photographs (Fig. 9). In the end, Lehmann achieved good agreement between theoretical estimates and microscopic observations for homogeneous red, blue, and ultraviolet (cf. Fig. 9b for homogeneous red).

Theory also explained why Lippmann photography yielded particularly convincing images for pure spectrum colours, since only one pattern of regular line distances was needed to record them, whereas mixtures of spectrum colours led to superpositions of two or more of these patterns, thus markedly weakening the colour intensity. Not surprisingly, white (i.e., the complete superposition of *all* spectrum colours) was, for a long time, the most problematic. It often only manifested itself as dirty grey in Lippmann plates.

Incidentally, enlargements of Lehmann's excellent photographs also revealed the photographic grain size, estimated by Lehmann (1906, 67–70) to be in the order of magnitude of 0.3 to 0.4 μm. The post-exposure expansion of Lehmann's type of granulation by factor three to four meant that no structures in the photographic layer smaller than 0.1 μm = 1000 Å could be seen, but all visible light is in wavelengths larger than that.

The neuroanatomist and neurohistologist Santiago Ramón y Cajal (1852–1934) in Madrid obtained estimates for the grain size in Lippmann plates even lower, by a factor of roughly ten – in other words, 0.02 to 0.05 μm – but he added the reservation that subtleties in photochemistry and microtomic preparation of samples and observations so close to the limit of

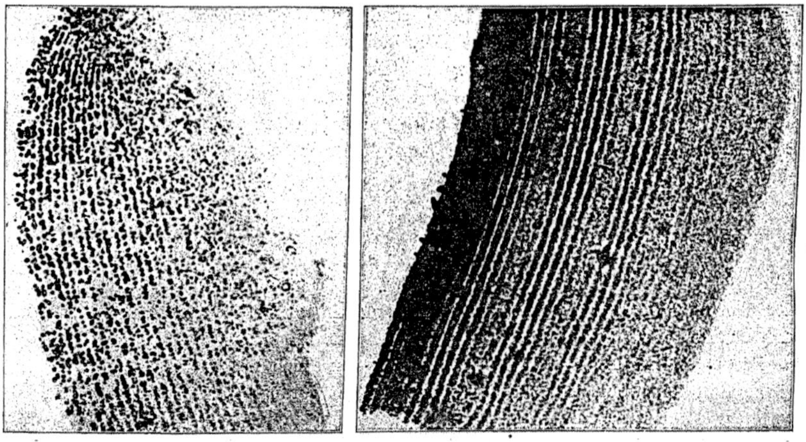

Fig. 9a. **Left**: Homogeneous red with λ = 6250 Å in 7100 times enlargement, showing a very even distancing, compressed at the top by the histological cut. **Right**: Spectral mixture of two wavelengths (λ = 5630 and 4820 Å) with a periodically larger interval between line groups.

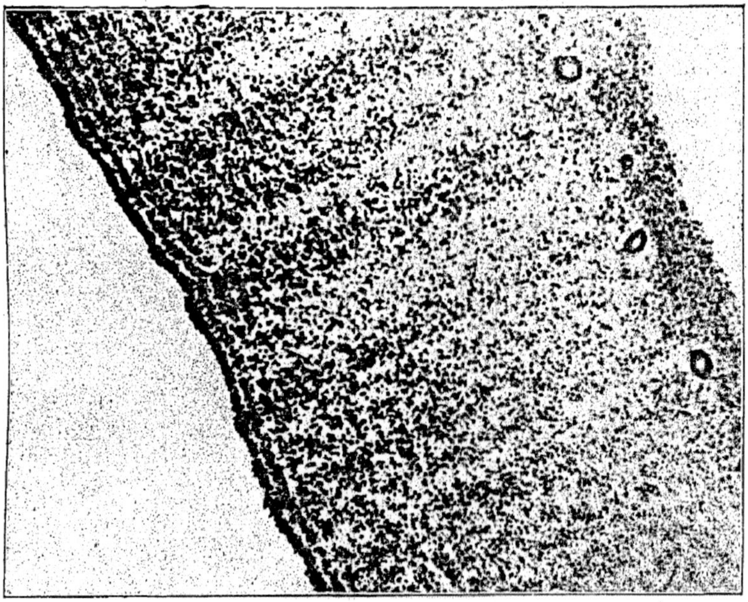

Fig. 9b. Homogeneous colour mixture of a yellow and white flower leaf. From Lehmann 1908: 282, 285

optical resolution might easily lead to quantitative errors (Cajal 1907, 217 and 215). Ramón y Cajal was an expert in fine structures of the nervous system, for which he often conducted subtle neurohistological examinations of microscopic samples cut by an ultra-fine microtome – an expertise which

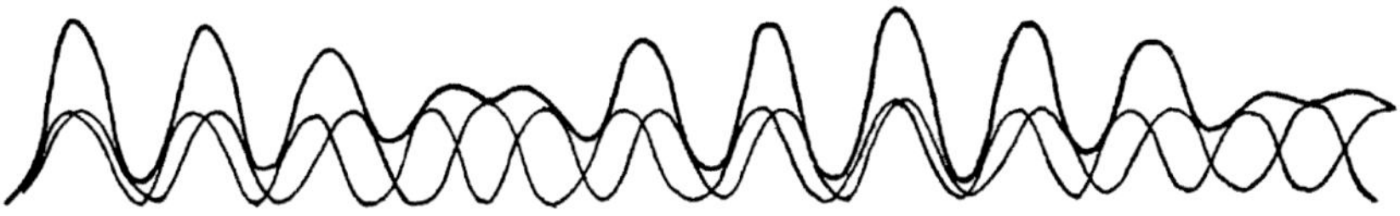

Fig. 10. Superpositioning of two different wavelengths leads to a periodic succession of groups of strong interference (here groups of 3 & 5), interspersed with areas of low interference. From Ives 1908: 336.

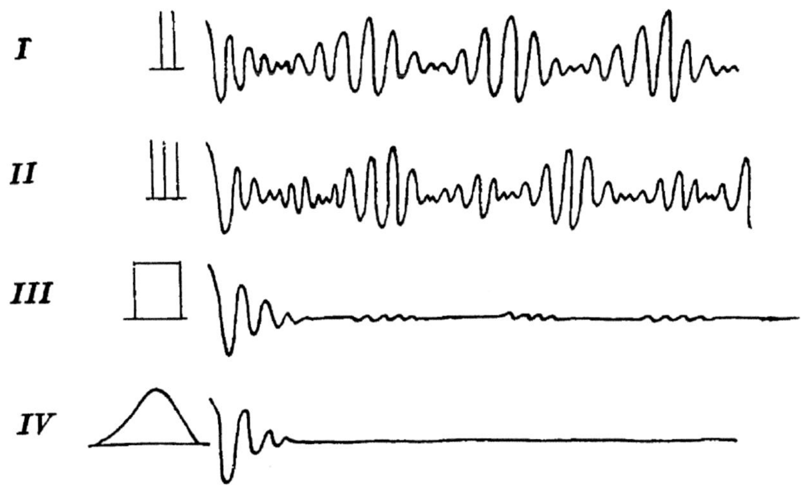

Fig. 11. Interference patterns for two and three interfering spectrum lines, for a whole spectrum band, and for continuous light coming from natural objects. From Ives (1908, 340).

came in handy in preparing microtomic cuts of small sections of Lippmann plates (Ramón y Cajal 1907, Triarhou and Cerro 2008).[7]

Cajal's cuts were so sharp and precise that he could even estimate the number of Zenker layers. Normal plates usually had no more than five to eight, whereas some of his intensely coloured plates had thireteen or more (Cajal 1907, 221). His gelatin-based samples were solid enough to permit fine cutting completely through the layer, so he could also study the spatial distribution of these Zenker layers, which provided the explanation for why some plates showed anomalous effects such as disturbing reflections, inappropriate colours, or a lack of colour intensity or brilliance. Only roughly a third of the gelatin layers exhibits such metallic Zenker layers, which are most distinct near the free

7 I also recommend the many publications by Javier DeFelipe on Cajal's art of experimentation and drawing.

surface, whereas the deeper ones become increasingly diffuse and indistinct. Careful removal of the uppermost layers by alcohol already destroyed the colour effect, even though the deeper layers were unaffected. According to Cajal, only the first and second laminae are responsible for the colours in Lippmann's heliochromes, whereas all additional laminae only have faint, intensifying action. In areas of pure white, the Zenker layers were not traceable at all. Thus, according to Cajal, the colour white in well-done Lippmann plates is not due to interferometric effects. Intentional removal of the uppermost surface layer and subsequent microtomic cutting and microscopic examination confirmed many of his claims. They mark the contemporary climax of research in the structure of Lippmann plates (Ramón y Cajal 1907, 227, 233ff.).[8]

Optimisation of the "Lippmann Process"

Lippmann's technique of obtaining "photochromic" images was soon optimised, by himself and roughly a dozen other experienced practitioners of photography (Berget, Krone, Neuhauss) and a few scientists (besides Lippmann, Otto Wiener, H. W. Vogel, H. Lehmann, E. Valenta), mostly specialists in photochemistry or physical optics (Wiener 1890, 1895, 1899, Berget 1891, Krone 1892, Vogel 1893, Zenker 1891, 1893, Valenta 1894). Most of the publications about Lippmann photography that appeared between 1891 and 1912 dealt with improved recipes for making the ultra-homogeneous light-sensitive layer, for sensitising or developing the plates (Lippmann 1892, 1894, Krone 1894, Neuhauss 1898, Cajal 1907, Ives 1908, Lehmann 1909c, Valenta 1912, 60ff.), or other efforts to make minor practical improvements – many of them ultimately only employed by their inventors. Lippmann photography requires mercury as the reflecting surface (as Lippmann himself described and used). Alternative reflectors suggested by other photographers result in reduced quality. Exposure time also plays a crucial role. A sharply defined spectrum furthermore requires a fixed incidence angle of the solar ray. Hence, a coelostat is essential for exposures requiring longer times, because the position of the Sun in the sky changes rapidly.[9] Since heliostats use a mirror, usually made out of glass, the light intensity is reduced and other disturbing factors come into play.

8 His papers originally written in Spanish were quickly translated into German, French, and English, not surprisingly so, since he had just obtained his Nobel Prize for Physiology or Medicine in 1906.

9 Lippmann himself had invented such a coelostat in which the image of the sky remains absolutely unchanged during long exposition sessions: see Lippmann (1893).

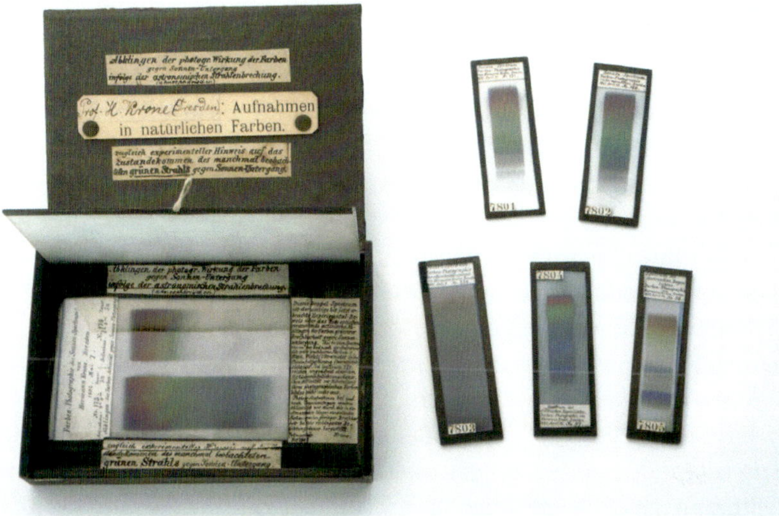

Fig. 12. A few arc spectra recorded by Hermann Krone using the Lippmann process; cf. Thomas ed. Beauty of Another Order, Yale Univ. Press 1997: 74, Hentschel 2002 pl IV, and Lehmann 2008.

In cases where mercury is inaccessible, Hermann Krone (1827–1916) in Dresden first described how to obtain satisfactory results using a glass plate covered with black velvet instead of applying mercury on the reflective plane (Krone 1892, 1894, 64–66). Because nowadays work with mercury is also prohibited for health reasons, modern replicators of Lippmann's process also have to resort to these alternatives (Bjelkhagen 1999, 277f., 2003). In 1905, Hans Lehmann tried to revert to metallic mirrors or silvered glass (or celluloid foil of only 0.01 mm thickness), pressed onto the Lippmann plate in a special "pressure cassette" in order to prevent air bubbles from forming between the surfaces and generating Newtonian rings (Lehmann 1909a, 785).

If one compares Krone's spectrum plates in Fig. 12 with Lippmann's (Fig. 7) or Neuhauss's (Fig. 8), the Krone spectra are definitely fainter, but the intermediary spectrum colours are more visible (Neuhauss 1905, 40, Lehmann 1909a). None of these improvements led to a substantial reduction in the effort required to produce an interferential Lippmann photograph in true colours. They rather increased the complexity of the technique and the improvements were consequently practised by only a few obsessed converts (Donath 1906, Valenta 1912, Bjelkhagen 1999, 2003).

This also meant that relatively few products were developed for photographers, since the few practitioners of Lippmann's technique, optimised it at their own workbenches and in their own chemical laboratories. One exception is the company of Carl Zeiss, which specialised in high-end optical instrumentation. Lehmann, a spectroscopist, physicist and instrument maker who was working for Zeiss at the time, developed a plate cassette that was adaptable to any photographic camera. The mercury for the reflective surface could be poured in shortly before the exposure. An apparatus for easier viewing of a finished Lippmann plate had a flat ten-degree prism installed in direct contact with the reflective surface to avoid unwanted shifts in colour and reflections of stray light. A correction filter for the photographic objective to absorb ultraviolet light, which overexposes the plate, and an adapter for conventional projection apparatus allowed one to project finished Lippmann plates onto a wall for presentation to larger audiences in substantial magnification (Lehmann 1907, 1908, 1909b, 1911).

A few dry plate manufacturers, such as Kranseder & Cie in Munich or Richard Jahr (1881–1938) in Dresden, provided photographic plates with an ultra-fine and homogenous photosensitive emulsion for their local customers (Lehmann 1909c). A few other instrument makers provided utensils for their local clients (Neuhauss 1898, 33, Lehmann 1908, 1909c, Bjelkhagen 1999, 277), but that was insignificant in the enormous market of more conventional photographic techniques. Lippmann's "procédé interférentiel" thus never attained important economic scale. It had other advantages, though:

- Lippmann photographs had no visible grain and could thus be extremely sharp.
- Provided the lighting conditions were optimal and the timing was perfect and the (manual) production and development of the plates were sufficiently routine, the colours were very intense, vivid, and natural (especially the pure spectrum colours and metallic reflections).
- Finished Lippmann plates do not fade upon further exposure to light (quite unlike Becquerel's plates and most modern colour prints, which tend to pale or manifest disturbing colour distortions after fifteen to twenty years), since the plates consist of colourless, pigment-free matter – in this respect, similar to mother-of-pearl or soap film (Fournier 1991, 260).

Preconditions for successful Lippmann photographs

In 1892, Hermann Krone, one of the most successful appliers of Lippmann's process, summarised preconditions for its optimal use (Krone 1892b, 1894; cf. also Valenta 1894, 63ff., 1912, 67ff.):

> 1) Very fine granulation and perfect homogeneity of the light-sensitive layer. The result is very low sensitivity, which calls for very long exposure times.

2) Limitation of the thickness of the image layer. It has to be thick enough to record enough interferometric layers and not be too thin. In particular, the layer must be absolutely evenly thick throughout.[10]
3) The trueness in colour in a photographed spectrum is not always absolute, but relative to the neighbouring areas. In other words, the least refrangible parts of the spectrum are redder than the less refrangible parts, which are comparatively more blue (Zenker 1891, 296, 1900, 107, Krone 1894, 48, Valenta 1894, 7, Ives 1908, 331ff.).
4) Colour faithfulness also depends crucially on the temperature during the drying of the photographic layer, on the exposition time, and on the developer chemicals used.
5) Colours in Lippmann photographs are also dependent on their water content. Further drying shortens the distances between the Zenker layers and thus leads to a reduction in wavelength, i.e., to a colour shift towards the blue, whereas exposure to humid air does the contrary. These changes can also happen during storage long after the exposure.
6) An intense recording of the solar spectrum should only be done when the Sun is not too near the horizon; otherwise, the actinism of the solar rays is reduced and the photographic result is less satisfactory, being either greyish or greenish.
7) Infrared parts of the solar spectrum not visible to the human eye often show as dark purple whereas ultraviolet parts beyond the H-group manifest a strange yellowish-pink lavender (Wiener 1899 and Valenta 1912, 78).
8) In the electric arc spectrum, a region beyond the H-group suddenly becomes a much more intense than in the blue-violet part of the visible spectrum.[11]

Limitations of Interferential Photography

This long list already shows how difficult it was for ordinary photographers to apply the Lippmann process. The low light sensitivity of his sensitised albumen or gelatin coatings implied extremely long exposure times of

10 Krone considered unevenness to be the main source of grey or colour-shift areas on the photographs. Ives (1908, 326–30) later showed that, contrary to Lippmann's original specifications, thin sensitive layers of no more than 30 half-λ gave the best results, and Ramón y Cajal (1907) provided microphotographic evidence.
11 Such overexposures were later prevented by means of UV filters regularly used in Lippmann photography.

several hours, compelling Lippmann to confine his subject matter to the most luminous spectra and to other mostly immobile, strongly illuminated objects. The introduction of gelatin emulsions reduced these exposure times to the order of minutes, but a further diminution to just a few seconds was impossible (Krone 1894, 67–76, Neuhauss 1898, 2–28, Bjelkhagen 1999, 275). Well aware of this disadvantage, Lippmann ends no less than his Nobel speech thus: "The length of exposure (one minute in sunlight) is still too long for the portrait. It was fifteen minutes when I first began my work. Progress may continue. Life is short and progress is slow." (Lippmann 1908)

Hence, most motifs of Lippmann plates are inanimate, static objects such as landscapes or still lifes. Even the famous parrot of 1899 – reproduced so often in publications about the Lippmann process – was stuffed! Not only the laypeople but even the few experts often complained about the many fruitless or unsatisfactory trials they had to do before they finally obtained colour-true Lippmann plates free from striae, grey or green clouds, and other disturbing effects. True colours crucially depended on exact timing, perfect emulsion quality and optimal thickness of the photosensitive layer and on the sensitisers. Areas of mixed (non-spectrum) colour or white were especially troublesome (Ives 1908, 341–43, 326, Cajal 1907, 222). In addition, Lippmann images were tricky to use, since one has to view them in parallel diffuse light, shielded from any oblique light, and under exactly the right angle to see all the colours properly. When looked at from the side, all the colours shift, a phenomenon known as *iridescence*: orange then appears to be yellow, yellow in turn is green, and green is blue (Ramón y Cajal 1907, 225). Reflections off the gelatin surface of the emulsion can also disturb and even cause colour distortion, because of a phase shift at the gelatin/mercury interface (Bjelkhagen 2003, 14), and if the emulsion was transparent enough to look through the plate rather than by reflection, complementary colours appeared (Neuhauss 1905, 273). Actually, this latter effect was yet another demonstration that Lippmann photographs are interference images, not based on any bodily colour pigment inside the plate.

Furthermore, they could not easily be copied or printed – each Lippmann photograph is thus unique and prized accordingly.[12] Finally, Lippmann images are sensitive to changes in temperature and air humidity, so they are also quite demanding for archivists and museum curators.[13] Altogether,

12 In 1905, Lippmann proposed a complex procedure for copying Lippmann plates, but this was never widely practiced; cf., e.g.,Neuhauss (1905, 273).

13 See, e.g., Fournier and Burnett (1996) and Bjelkhagen (1999). On the other hand, Lippmann photographs contain no dyes or pigments.

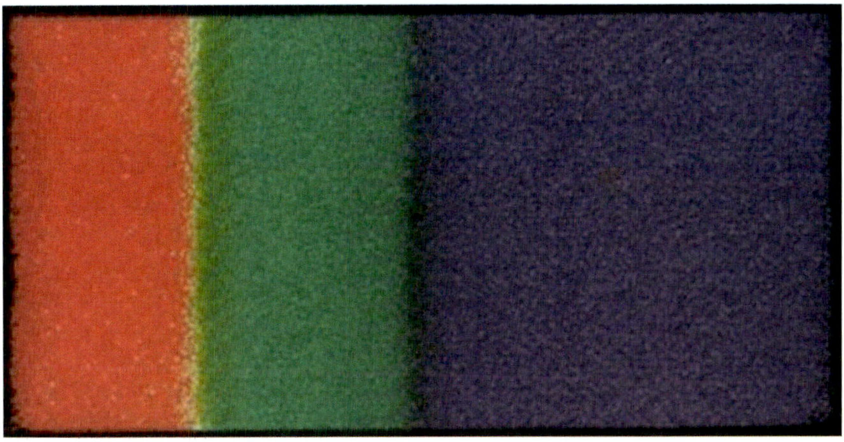

Fig. 13. An autochrome colour spectrum shows how coarse by comparison the colour gradations between the three print colours of violet, green and red are in comparison with Lippmann's and how monochromatic these three broad areas are; like in Neuhauss's spectrum the yellow-orange area is far too slim, and whereas the Neuhauss spectrum shows no violet, the Lumière spectrum has no indigo or blue; Autochrome plates may have been easier to print but were much less useful for spectroscopic purposes of true colour representation of a spectrum. (From a colour plate in Valenta 1912: pl. V)

the practical difficulties of Lippmann's interferential technique were so inhibiting that it was never widely used.[14] After 1907, the mass production of the Lumière brothers' autochrome plates led to a rapid decline in interest in Lippmann's interferometric photography, even though the colour faithfulness of autochrome plates is actually far poorer than Lippmann plates (see Fig. 13).

Summary

Lippmann's interferential photography was a link between scientists exploring photographic processes and advanced practitioners using photographic processes to maximum effect. For the normal photographer, the Lippmann process was not an option, not only because of its many limitations in the complex production of an image but also because of its high rate of defective results. This was aggravated by the iridescence of these images, which only yielded the proper colours when viewed under exactly the right angle and perfect lighting conditions. All of these practical limitations left Lippmann's technique in the hands of a few experts, who were not inordinately bothered by them. For these select few, the Lippmann

14 See especially Bellone and Fellot (1981) and Bjelkhagen (2003).

process was at least the "theoretically most interesting and very elegant solution" (Lehmann 1907, 842) to the long-standing problem of how to obtain colour photographs. For the expert, it was important as "brilliant evidence of the modern theory of light, even if it never was to become practically important" (Lehmann 1908, 280).[15] For Ramón y Cajal (1907, 213), Lippmann plates were an excellent apparatus to record light waves, analogous to a phonograph recording sound waves. The ups and downs on the shellac disc were thus analogous to the patterns of Zenker layers in the Lippmann plate.

For these few specialists, the many practical problems of Lippmann photography were only a challenge in instrumentation and experimentation. The interferometrical Lippmann technique was quite in line with their efforts to turn the science-oriented part of "serious" photography into a new discipline of scientific photography alongside the existing subdisciplines of photochemistry, physical optics, and optical technology. One indicator of this trend was the founding of new journals, such as the *Zeitschrift für wissenschaftliche Photographie, Photophysik und Photochemie* in 1902 and *La Photographie des Couleurs et la Revue des Sciences Photographiques et de leur Applications*, edited by Charles Mendel.[16] In German-speaking lands, Hermann Wilhelm Vogel at the Technische Hochschule Berlin-Charlottenburg, Hermann Krone in Dresden, Eduard Valenta at the Vienna Technical University, and Hans Lehmann at Zeiss in Jena were the main protagonists in this contemporary endeavour to make photography "scientific." Photography was more than a mere practical art, so far totally dominated by empirical know-how and accidental discoveries that had driven its earlier development. These investigators wanted to explain, to understand theoretically, what happens during the photographic recording and – if possible – to foresee further advances on the basis of insights provided by scientific theory. Seen from this point of view, "Becquerel's and Lippmann's methods deserve a strong scientific interest" even though their practical value is quite low (Donath 1906, 54). Unaware of how much they perpetuate the trend of scientification of photography around 1900, the historians of photography Bertrand Lavédrine and Jean-Paul Gandolfo (2015, 56, my emphasis) state: "Lippmann's invention was notable for the originality of his scientific thinking that underpinned his technique […]

15 In a similar tone, see Krone (1894) and Valenta (1894).

16 For more on the contemporary efforts at institutionalisation, see Wilder 2009 and Hentschel 2015. A detailed list of German, Austrian, and international journals of photography with links to scanned issues is provided here: https://de.wikisource.org/wiki/Zeitschriften_(Photographie)

[and] *remains to this day the only photographic process that was developed based entirely on theoretical considerations.*"

Between 1901 and 1908, Lippmann was nominated twenty-three times by one Swiss colleague and by eleven French compatriots for the Nobel prize. His nominators often applauded that his procedure "allows the simultaneous recording of all the properties of a ray of light,"[17] but for several years, the Nobel committee did not consider his "elegant demonstration of the wave theory of light" prizeworthy enough to be ranked as "epoch-making progress" (Mitchell 2010, 331–32). In 1907, fourteen members of the French Académie des Science signed their coordinated nomination of Lippmann, this time "for the entirety of his contributions to the physical sciences" – however, in the end, the American Albert A. Michelson got the prize for his highly impressive and multifarious experimental work in interferometry and metrology. In 1908, Max Planck, who had introduced the quantisation of light in 1900, was Lippmann's strongest competitor, with strong backing by physico-chemist Svante Arrhenius (1859–1927), but the chairman of the Nobel committee, experimental physicist Knut Angström (1857–1910), in the end overturned the decision in favour of Lippmann, arguing that "the art of photography concerns one of the most important advances in human culture" and that the French physicist's work inspired "a large literature important scientifically as well as practically" (Mitchell 2010, n.83).[18] Lippmann's endeavours were thus crowned with the 1908 Nobel Prize in Physics "for his method of reproducing colours photographically based on the phenomenon of interference." In today's short summary of Lippmann's work on the website of the Nobel Foundation, it is conceded that "the method never came into widespread usage, but lived on as a physics experiment" (Lippmann 1908/09). Lippmann's Nobel prize and Ramón y Cajal's work in Madrid, also awarded a Nobel prize in Medicine in 1906, both fit into this scheme perfectly. Nevertheless, few universities founded chairs in the new field – contemporary efforts to institutionalise their field failed. Scientific photography in general and Lippmann interferometric photography in particular remained a "research technology" in Terry Shinn's sense (Shinn and Joerges 2001, Hentschel 2012), providing useful, often even crucial means for experimenters and observers in a broad range of disciplines. The field of spectroscopy is similar in this regard, because it also never became

17 For the full list of nominators, see: https://www.nobelprize.org/nomination/archive/show_people.php?id=5533

18 Final voting was 46:14 in favour of Lippmann versus Planck, who eventually got the Nobel Prize in Physics for the year 1918.

a discipline but remained a widely used, generic research tool (Hentschel 2005, 2015). Despite its usefulness for a few experts as a research technology for inquiries into processes of photochemistry, Lippmann photography's complexity and its rare use outside science certainly is another reason for its historiographic neglect or disqualification as a dead end – something that this anthology seeks to remedy.

Works Cited

Becquerel, Edmond. 1839. "Mémoire sur le Rayonnement Chimique qui Accompagne la Lumière Electrique", *Comptes Rendus Hébdomadaires de l'Académie des Sciences, Paris* (in the following abbreviated CRAS), 11, 702–703.

Becquerel, Edmond. 1842. "Mémoire sur la Constitution du Spectre", (a) *Bibliothèque Universelle de Genève* 40, 341–67; (b) English trans. in Taylors *Scientific Memoirs* 3 (1843): 537–557.

Becquerel, Edmond. 1847. "L'image Photographique Colorée du Spectre Solaire", (a) *CRAS* 26, 181–183; (b) *Annales de Chimie* 22 (1848), 451–459.

Becquerel, Edmond. 1848. "De l'Image Photochromatique du Spectre Solaire", (a) CRAS 28 200–209; (b) *Annales de Chimie* 25 (1849): 447–474.

Becquerel, Edmond. 1868. *La Lumière, ses Causes et ses Effets*, Paris: Didot.

Bellone, Roger & Luc Fellot,1981. *Histoire Mondiale de la Photographie en Couleurs*, Paris: Hachette.

Berget, Alphonse.1891. *Photographie des Couleurs par la Méthode Interférentielle de M. Lippmann*, Paris: Gauthier-Villars.

Bjelkhagen, Hans I. 1999. "Lippmann Photography: Reviving an Early Colour Process." *History of Photography* 23, 274–80.

Bjelkhagen, Hans I. 2003. "Lippmann Photography: Its History and Recent Developments." *PhotoHistorian*: 11–19.

Buchwald, Jed Z. 1989. *The Rise of the Wave Theory of Light: Optical Theory and Experiment in the Early Nineteenth Century*, Chicago: University of Chicago Press.

Cornu, Marie Alfred.1891 "Sur une éxperience récente [by O. Wiener 1890]", *CRAS* 112, 189–189 (see also *ibid.*, 325–331, 365–370. 383 for the ensuing debate with Poincaré & Potier),

Donath, Bruno.1906. *Die Grundlagen der Farbenphotographie*, Braunschweig: Vieweg.

Eder, Josef Maria.1945. *History of Photography*, New York: Dover.

Faber, Monika. 2003. "Josef Maria Eder und die Wissenschaftliche Fotografie 1855–1918" in *Das Auge und der Apparat. Eine Geschichte der Fotografie an der Sammlung der Albertina*, edited by Monika Faber & Klaus Albrecht Schroeder, Vienna: Albertina & Paris: Bibliothèque Nationale de France, 142–169.

Fournier, J.-M. 1991. "La Photographie en Couleurs de Type Lippmann; Cent Ans de Physique et de Technologie", *Journal of Optics* 22, 259–266.

Fournier, J.M. & P.L. Burnett.1996. "Color Rendition and Archival Properties of Lippmann Photographs", *Journal of Imaging Science and Technology* 38, 507–512.

Harrison, W. Jerome. 1887. *A History of Photography. Written as a Practical Guide and an Introduction to its Latest Developments*, New York: Scovil Manufacturing Co.

Hentschel, Klaus. 1999. "Photographic Mapping of the Solar Spectrum, 1864–1900", *Journal for the History of Astronomy* 30, 93–119 (I), 201–224 (II).

Hentschel, Klaus. 2002a, *Mapping the Spectrum. Techniques of Visual Representation in Research and Teaching.* Oxford: Oxford University Press.

Hentschel, Klaus. 2002b. "Why Not One More Imponderable? John William Draper and his "Tithonic" Rays", *Foundations of Chemistry* 4, 5–59.

Hentschel, Klaus. 2005. "Wissenschaftliche Photographie als Visuelle Kultur: Die Erforschung und Dokumentation von Spektren", *Berichte zur Wissenschaftsgeschichte* 28, 193–214.

Hentschel, Klaus, ed. 2012. *Zur Geschichte von Forschungstechnologien: Generizität, Interstitialität, Transfer.* Diepholz & Berlin: GNT-Verlag.

Hentschel, Klaus. 2015. "(Scientific) Photography as a Research-enabling Technology – Not a Discipline", in: *Frame and Focus. Photography as a Schooling Issue* edited by Maren Gröning, Vienna: Bonartes, 98–122.

Hesse, Wolfgang, ed. 1998. *Hermann Krone, Historisches Lehrmuseum für Photographie: Experiment, Kunst, Massenmedium,* Dresden: Kupferstich-Kabinett der Staatlichen Kunstsammlungen Dresden sowie Technische Universität Dresden.

Ives, Herbert E. 1908. "An Experimental Study of the Lippmann Color Photograph", *Astrophysical Journal* 27, 325–352.

Krone, Hermann. 1892a. "Weiteres über Farbenphotographie des Sonnenspectrums", *Deutsche Photographen-Zeitung* 9, 29–30.

Krone, Herman. 1892b. "Über Farbenphotogramme von Spectren", *Annalen der Physik und Chemie* (Ser. 3) 46, 426–430.

Krone, Herman. 1894. *Die Darstellung der Natürlichen Farben Durch Photographie*, Weimar: Deutsche Photographen.

Lavédrine, Bertrand & Jean-Paul Gandolfo. 2015. *The Lumiere Autochrome: History, Technology, and Preservation.* Los Angeles: Getty Publishing.

Lehmann, Hans. 1906. *Beiträge zur Theorie und Praxis der Direkten Farbenphotographie mittels stehender Lichtwellen nach Lippmanns Methode*, Freiburg: Troemer.

Lehmann, Hans.1907. "Über die direkten Verfahren der Farbenphotographie nach Lippmann und Lumière", *Physikalische Zeitschrift* 8, 842–849.

Lehmann, Hans. 1908. "Über die Struktur entwickelter Lippmann-Platten", *Archiv für Optik* 1, 280–285.

Lehmann, Hans. 1909a. "Interferenzfarbenphotographie mit festem Metallspiegel", *Physikalische Zeitschrift* 10, 784–787.

Lehmann, Hans. 1909b. "Die Praxis der Interferenzfarbenphotographie", *Photographische Rundschau* 23, 125–134.

Lehmann, Hans. 1909c. „Quelques Remarques sur le Procédé Interférentiel", *La Photographie des Couleurs et la Revue des Sciences Photographiques et de leur Applications,* June, 132–134.

Lehmann, Hans. 1911. "Perfectionnements à la Methode Interférentielle", *La Photographie des Couleurs et la Revue des Sciences Photographiques et de leur Applications*, Feb. 25–32 & Mar. 1911, pp. 45–52.

Lehmann, Hans-Ulrich. 2008. *Hermann Krone – die Photographien: Werkverzeichnis.* Dresden: Staatliche Kunstsammlungen Dresden.

Lippmann, Gabriel. 1891. "La Photographie des Couleurs", *Comptes Rendus des scéances hebdomadaires de l'Académie des Sciences* 112, 274–275 (with additional observations by Edmond Becquerel, appended *ibid.*, p. 275; in German trans. in Valenta 1894, 21–23.

Lippmann, Gabriel. 1892a.–"Sur la Photographie des Couleurs [deuxième note]" in *Comptes Rendus des Scéances Hebdomadaires de l'Académie des Sciences* 114, 961–962.

Lippmann, Gabriel. 1892b. "Photographies Colorées du Spectre sur Albumine et sur Gélatine Bichromatées", *Comptes Rendus des Scéances Hébdomadaires de l'Académie des Sciences* 115, 575.

Lippmann, Gabriel. 1893. "*Sur un Coelostat, ou Appareil à Mirroir, Donnant un image du Ciel Immobile par Rapport à la Terre", Comptes rendus des Scéances Hébdomadaires de l'Académie des Sciences* 120, 1015–1019.

Lippmann, Gabriel. 1894. "Sur la Theorie de la Photographie des Couleurs Simples et Composées par la Methode Interférentielle", (a) *Comptes Rendus des Séances Hébdomadaires de l'Académie des Sciences* 118, 92–102; (b) *Journal de Physique Théorique et Appliquée,* 3 (1), 978–107; (c) in Engl. trans: *On Colour Photography by the Interferential Method, Proceedings of the Royal Society of London* 60 (359/367) 10–13.

Lippmann, Gabriel. 1905. "Photographies en Couleurs du Spectre Négatives par Transmission", *Comptes Rendus Hebdomadaires des Séances de l'Académie des Sciences* 140, 1508–1509.

Lippmann, Gabriel. 1909. "Colour photography (= Nobel lecture, Dec. 14, 1908)", in: *Les Prix Nobel en 1908,* Stockholm: Norstedt, reprinted in: *Nobel Lectures, Physics 1901–1921,* Amsterdam: Elsevier, 1967, online at https://www.nobelprize.org/prizes/physics/1908/lippmann/lecture/

Livingston, Dorothy Michelson. 1973. *The Master of Light: A Biography of Albert A. Michelson,* New York: Scribner's.

Lumière, August & Louis. 1893. "Mittheilung über die Photographie in den natürlichen Farben", *Photographisches Archiv* 34, 164–166. (French orig. in *Bull. Soc. franc. de Photogr*, 249)

Meslin, Georges. 1892. "Sur la Photographie des Couleurs", *Annales de Chemie et de Physique* 27, 369–391.

Michelson, Albert A. 1903. *Light Waves and Their Use*, Chicago: University of Chicago Press.

Mitchell, Daniel Jon. 2010. "Reflecting Nature: Chemistry and Comprehensibility in Gabriel Lippmann's 'Physical' Method of Photographing Colours", *Notes and Records of the Royal Society of London* 64, 319–337.

Neuhauss, Richard. *Journal 1891–1900* (*with notes on gelatine emulsions, colour photography and Lippmann photography*), preserved at the Preus museum in Horten, Norway, and published in transcribed form at https://holowiki.org/data/lippmann/Neuhauss_Journals.pdf

Neuhauss, Richard. 1898. *Die Farbenphotographie nach Lippmann's Verfahren: Neue Untersuchungen und Ergebnisse.* Halle: Knapp.

Neuhauss, Richard. 1905. "Eine Erweiterung von Lippmanns Farbenverfahren", *Photographische Rundschau* 19, 40, 226, 273–274.

Pier, J.P. & J.A. Massard. 1997. *Gabriel Lippmann. Commémoration par la section des Sciences Naturelles Physiques et Mathématiques de l'Institut Grand-Ducal de Luxembourg du 150e Anniversaire du Savant né au Luxembourg, Laureate du Prix Nobel en 1908.* Luxembourg: Section des Sciences Naturelles, Physiques et Mathematiques de L'institut Grand-Ducal de Luxembourg.

Ramón y Cajal, S. 1907. "Die Struktur der Lippmannschen Photochromien", *Zeitschrift für wissenschaftliche Photographie, Photophysik und Photochemie* 5, 213–245; in Engl. transl.: The structure of Lippmann Heliochromes, *British Journal of Photography*, Aug. 27, 59–79 given in the *Annual Reports of the Board of Regents of the Smithsonian Institution* 1908 (orig. Spanish 1906).

Lord Rayleigh, 1887. "On the maintenance of Vibrations by Forces of Double Frequency, and on the Propagation of Waves Through a Medium Endowed with a Periodic Structure", *Philosophical Magazine* (Ser. 5) 24, 145–159.

Schütt, F. 1896. "Innerer Bau und Optisches Verhalten der Lippmann'schen Photographien in natürlichen Farben", *Annalen der Physik* (Ser. 3) 57, 533–554.

de Seauve, Victor, et al. 2020. "Spectroscopies and Electron Microscopies Unravel the Origin of the First Colour Photographs", *Angewandte Chemie* 132, 9198–9204.

Shinn, Terry & Bernward Joerges. 2001. *Instrumentation Between Science, State and Industry*, Dordrecht: Springer.

Staley, Richard. 2003. "The Interferometer and the Spectroscope: Michelson's Standards and the Spectroscopic Community", *Nuncius* 18, 779–801.

Triarhou, Lazarus C. & Manuel del Cerro 2008. "The Structure of Lippmann Heliochromes: Cajal and the 1908 Nobel Prize in Physics", *Journal of Chemical Neuroanatomy* 35, 1–11.

Valenta, Eduard. 1894. *Die Photographie in Natürlichen Farben: mit Besonderer Berücksichtigung des Lippmann'schen Verfahrens,* Halle: Knapp.

Vogel, Hermann Wilhelm. 1893. "Ueber Photographie natürlicher Farben", *Jahrbuch der Photographie und Reproductionstechnik* 7, 285–287.

Wagener, Renée. 2016. "Gabriel Lippmann, ein moderner Wissenschaftler", *Ons Stad* 113, 38–40.

Wiener, Otto. 1890. "Stehende Lichtwellen und die Schwingungsrichtung polarisirten Lichtes", *Annalen der Physik* (Ser. 3) 40, 203-243 & pl. II.

Wiener, Otto. 1895. Farbenphotographie durch Körperfarben und mechanische Farbenanpassung in der Natur, *Annalen der Physik* (Ser. 3) 55, 225–281.

Wiener, Otto. 1899. "Ursache und Beseitigung eines Fehlers bei der Lippmann'schen Farbenphotographie, Zugleich ein Beitrag zu ihrer Theorie", *Annalen der Physik* (Ser. 3) 69, 488–530.

Wilder, Kelley. 2009a. "Photography and the Art of Science", *Visual Studies* 24, 163–168.

Wilder, Kelley. 2009b. *Photography and Science.* London: Reaktion Books.

Zenker, Wilhelm. 1868. *Lehrbuch der Photochromie.* Braunschweig: Vieweg, 1900.

Zenker, Wilhelm. 1891. "Die Entstehung der Farben in der Photochromie", *Jahrbuch für Photographie und Reproductionstechnik* 5, 293–303.

Zenker, Wilhelm. 1893. "Ueber die Entstehung der Farben im Lippmann'schen Spectrum", *Jahrbuch für Photographie und Reproductionstechnik* 7, 114–121.

About the Author

Prof. Dr. **Klaus Hentschel** double-majored in physics and philosophy at the University of Hamburg, submitted his thesis in 1989 on philosophical (mis)-interpretations of Einstein's theory of relativity, and took his habilitation degree in 1995 on the interplay between scientific instrument design, experimental practice, and theory formation in spectroscopy and astrophysics. He is professor of the history of science and technology at the University of Stuttgart since 2006.

3. Colour Under the Microscope: Santiago Ramón y Cajal Does 'Histology' on Lippmann Heliochromes

Lazaros C. Triarhou and Manuel del Cerro

Abstract
This paper highlights Gabriel Lippmann's discovery of colour photography and the microscopic analyses performed upon the photographs by Santiago Ramón y Cajal (1852–1934), professor of anatomy and histology in Madrid, acknowledged "father of modern neuroscience", and avid photographer, in a historic and a modern perspective. Cajal placed sections of Lippmann photographs under the microscope to study the structure of their materiality. He focused on the laminae of Zenker, which produced mixed colours, especially white. His technical studies culminated in a 1908 article in the *Annual Report of the Board of Regents of the Smithsonian Institution* and in his subsequent Spanish classic monograph, *Photography in Colours*, expounding upon the theoretical physicochemical principles and practical applications of the "art of Daguerre." The authors thus explore Lippmann's reception in Spain.

Keywords: Santiago Ramón y Cajal, interferential colour photography, Gabriel Lippmann, histology, microscopy

Had the versatile Santiago Ramón y Cajal (1852–1934) not been absorbed by the celebrated research career that led to the elucidation of the microscopic structure of the nervous system and secured him the 1906 Nobel Prize in Physiology or Medicine, he might have established the first successful film manufacturing industry of Spain. In his *Recollections* (Ramón Cajal 1988), Cajal confesses his long-standing devotion to the art of Daguerre. He recounts the circumstances under which he was concomitantly initiated as a teenager

Hannouch, H. (ed.), *Gabriel Lippmann's Colour Photography: Science, Media, Museums.*
Taylor & Francis Group, 2022
DOI 10.5117/9789463728553_CH03

to both anatomy and photography in 1868. At the age of twenty-seven, he was an early visionary of instant photography and had successfully cultivated the art to the point of producing ultra-rapid gelatin-bromide plates of his own formulation, aided by his wife and laboratory assistant Silveria Fañanás (1854–1930).

Cajal was convinced that a clear understanding of the theory is half of the pleasure of photography, the other half being the practical remedy of failures of each photographic operation (Ramón Cajal 1988). As a means of recreation and relief from his neurobiological work, Cajal engaged in studies on theoretical and practical issues of photography (Ramón Cajal 1906b, 1907a, 1907c); he published his results in the official journal of the Royal Photographic Society of Madrid (Ramón Cajal 1903, 1904, 1906a, 1907b). These endeavours culminated in a voluminous technical monograph on colour photography (Ramón Cajal 1912), one of the early pioneering works on the fundamental physical and chemical principles behind this particular application of science. Moreover, his daughter Fe conserved an unpublished handwritten Cajal manuscript on the history of photography (Ramón Cajal 1870). Throughout the twentieth century, Spanish authors have paid tribute to Cajal as a photographer, testifying to the fascination with his investigation of the medium (Tello Muñoz 1906, de Madrid 1952, Tomero Alarcón 1966, López Mondéjar 1982, Sougez 1982, Romero 1984, Larrea 1989, López Piñero et al. 2000, Romero Santamaría 2002, Márquez 2004, López-Cantos 2016).

From all the chapters in his monograph on colour photography (Ramón Cajal 1912), Cajal takes especial pride in section II, which covers the principles and practical rules of Lippmann's method. The Luxembourger Gabriel Lippmann (1848–1921), professor of mathematical and experimental physics at the Sorbonne, had developed a method for the photographic reproduction of colours on theoretical grounds, based on the phenomenon of wave interference (Lippmann 1891, 1894). For that development, he was awarded the 1908 Nobel Prize in Physics; to place it in a historical context, in 1908, the Nobel Prize in Physiology or Medicine was awarded to Élie Metchnikoff (1845–1916) and Paul Ehrlich (1854–1915) for their work in immunity, and the Nobel Prize in Chemistry to Ernest Rutherford (1871–1937) for his investigations into the disintegration of the elements.

For the preceding few years, Cajal had been studying cross sections of Lippmann heliochromes under the microscope, with special emphasis on the structure of the laminae (or interference fringes) described by Wilhelm Zenker (1829–99) in Berlin, which produce mixed colours, and white in particular. Those experiments led to a series of technical papers by Cajal. A comprehensive study appeared in the *Review of the Royal Academy of*

Sciences of Madrid (Ramón Cajal 1906c) and duly in a German translation (Ramón Cajal 1907d). Also, the *Annual Report of the Board of Regents of the Smithsonian Institution* in Washington, DC hosted the English version of Cajal's article on Lippmann heliochromes (Ramón Cajal 1908), reprinted from the colour photography supplement of the *British Journal of Photography* (Ramón Cajal 1907e), a study well known and cited in articles by scientists working in the field of interference colour photography. The circulation of knowledge generated by Cajal in a variety of languages and contexts not only reflects its broad reception but also the sustained interest in his particular 'histological' approach to colour photography.

In the present chapter, we reintroduce that side of Cajal's productivity, by tracing his investigation of Lippmann's heliochromes through his transposition of histological practices from the realm of neurobiology to colour photography. We also pay homage to his restless search for explanations through subjecting scientific problems to the rigour of the experimental test.

The Evolution of Colour Photography

The physico-chemical foundations of photography, the science and art of freezing space in time, have been evolving ever since the Swedish pharmacist Carl Wilhelm Scheele (1742–86), in addition to discovering the atomic elements oxygen, nitrogen, barium, manganese, molybdenum, tungsten, and chlorine, observed in 1777 that silver chloride can be decomposed to metallic silver by exposure to light (Harrison 1888). The French chemist Joseph-Nicéphore Niépce (1765–1833) – or perhaps his nephew Claude-Félix Abel Niépce de Saint-Victor (1805–70) – likely succeeded in obtaining the first permanent picture in 1827, before his subsequent collaboration with Louis-Jacques Mandé Daguerre, whose renowned photographic process came into the public arena in 1837 (Towler 1864, Roebuck 1919, Alschuler 2002).

The idea of the direct method of colour photography was promoted between 1839 and 1848 by Edmond Becquerel (1820–91), who recorded spectral colours by means of silver chloride and silver bromide salts, albeit in a non-permanent manner, thence known as *les fantômes de spectres* (Becquerel 1868, Pagneux et al. 2000). In 1850, the Reverend Levi Hill (1856), working in New York State, may have produced, through an accidental combination of chemicals that he was not able to replicate for the rest of his life, a daguerreotype in natural colour that he termed a "Hillotype." A report of the United States Senate dated March 3, 1853 (32nd Congress, 2nd Session, Rep. Com. No. 427) supports the originality of Hill's specimens, which "afford sufficient proofs that the inventor has solved the problem of photographic colouration." Hill introduced the term "heliochrome" (from the Greek, helios ἥλιος=sun and chrome χρῶμα"=colour), meaning the reproduction of the natural colours of an object.

One of the first registrations of a colour photograph on a special lacquered paper is credited to Louis-Alphonse Poitevin (1819–82) in 1865 (Pagneux et al. 2000). On November 23, 1868, Louis Ducos du Hauron (1837–1920), author of one of the most important technical books on the process of trichromatic photography (Ducos du Hauron 1869), patented his indirect method in France under number 83.061. The following year, Charles Cros (1842–88) also suggested the idea of using three different layers of emulsion, each being sensitive to one of the three fundamental colours, red, yellow, and blue, which formed the basis for "subtractive" processes (Alschuler 2002). On the other hand, "additive" processes are based on the separation of colours through screens, ideas originally outlined in du Hauron's patents (Alschuler 2002).

Autochromes (< Gk. αὐτός=self + χρῶμα=colour) are a wash-off dye transfer process invented by Louis and Auguste Lumière of Lyon, France in 1903 and marketed in 1907. They are positive colour transparencies on glass, produced through a reversal process from a dyed starch-based negative glass plate coated with shellac and a silver bromide photosensitive emulsion. Autochromes can be viewed through a "diascope," a special flat-case device consisting of a folding cover, raised to view the image and lowered to protect it by means of sunlight reflected off the white base through the ground glass diffuser that holds the slide and reflected on a mirror. Additional details on the historical evolution of colour photography can be found in Ives (1894), Bolas (1900), Ives (1913), Roebuck (1919), Pagneux et al. (2000), Alschuler (2002), and Biedermann (2005).

Gabriel Lippmann and the Theory of Wave Interference

In one of those circumstances where theoretical reasoning based on well-known scientific facts may lead to the working out of a successful process (Senior 1900), Lippmann reported on February 2, 1891 his success in recording the colour spectrum in a permanent manner. It was based in part on Maxwell's electromagnetic theory and (according to German historians and photographers) on Zenker's explanation of interference fringes – which, besides Lippmann heliochromy, also anticipated reflection holography (Pagneux et al. 2000, Burder and Green 2004). For neuroscience, 1891 was the hallmark year when Wilhelm Waldeyer (1836–1921) in Berlin

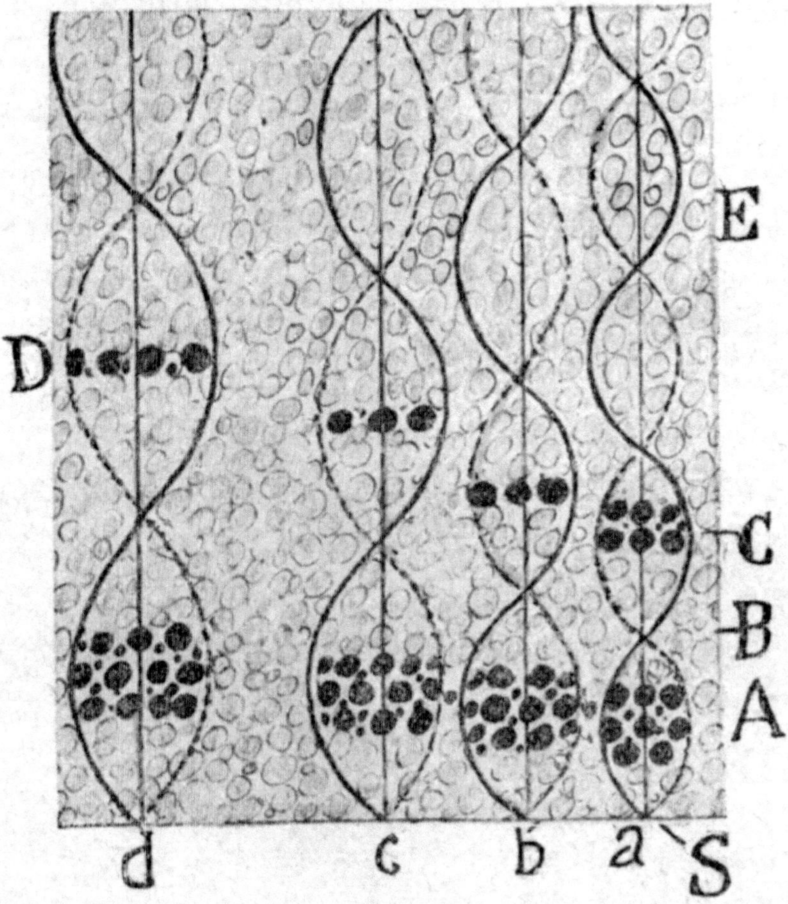

Fig. 14a. Schematic drawing to show that, given the dimension of the silver grains, the development must produce a fusion of the first laminae of Zenker. Abbreviations: A, specular lamina; B, first nodal space; C, second, thinner lamina of Zenker, produced by a short wave; D, deep laminae of Zenker created by a thick wave, the red, for example; E, region of the gelatin lacking laminae; a, violet wave; b, blue wave; c, amarilio green; d, red. From Ramón Cajal (1912).

published his article that applied the Homeric word "neuron" to popularise the neuron doctrine (Waldeyer 1891).

The Lippmann process, which yields colours remarkable for their brilliancy and clearness, is the only method of producing photographs of objects that show colours without the use of dyes, in contrast to three-colour processes (Senior 1900, Roebuck 1919). It is based on the principle that when two waves of the same length go in opposite directions, they interfere and set up standing waves, in other words, places where they neutralise or augment each other (Fig. 14a). This is achieved by means of a layer of mercury in a special

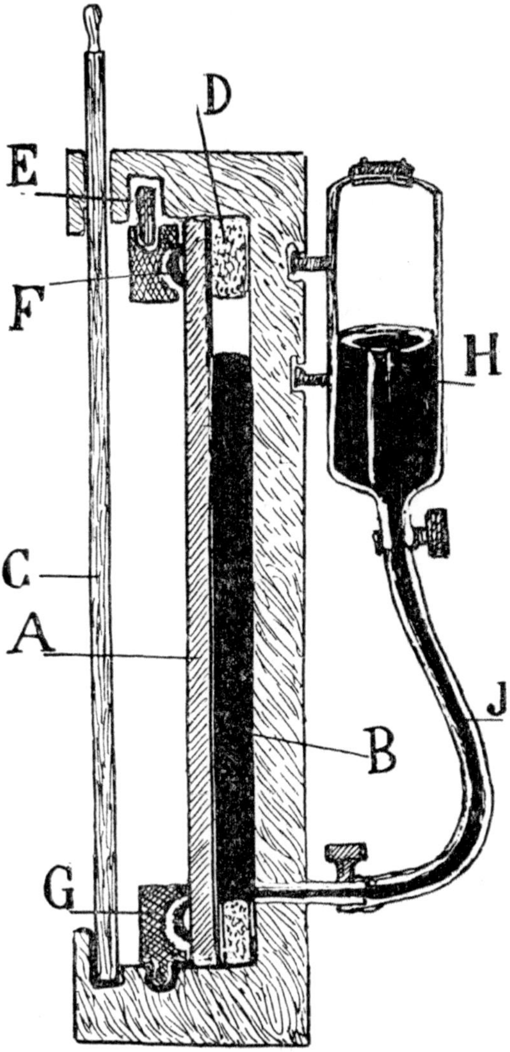

Fig. 14b. Schematic drawing of the mercury chassis. Abbreviations: A, sensitive plate; B, mercury layer; C, covers of the chassis; D, square gasket of chamois, located between the plate and the frame of the chassis and intended to seal the intercalated space in the hermetically closed cuvette; G, metallic clamp for pressure; F, springs for this clamp; H, mercury deposit; J, rubber tube. All four figures handdrawn by Ramón Cajal (1912). Note Cajal's use of the arrows, lettering, and general patterns that are typical of his renowned neurohistological drawings.

holder placed behind the photographic emulsion of an orthochromatic or panchromatic glass plate (i.e., sensitive to light rays through the entire visible spectrum). The emulsion side of the glass plate faces away from the lens and is in optical contact with the mercury mirror (Fig. 14b).

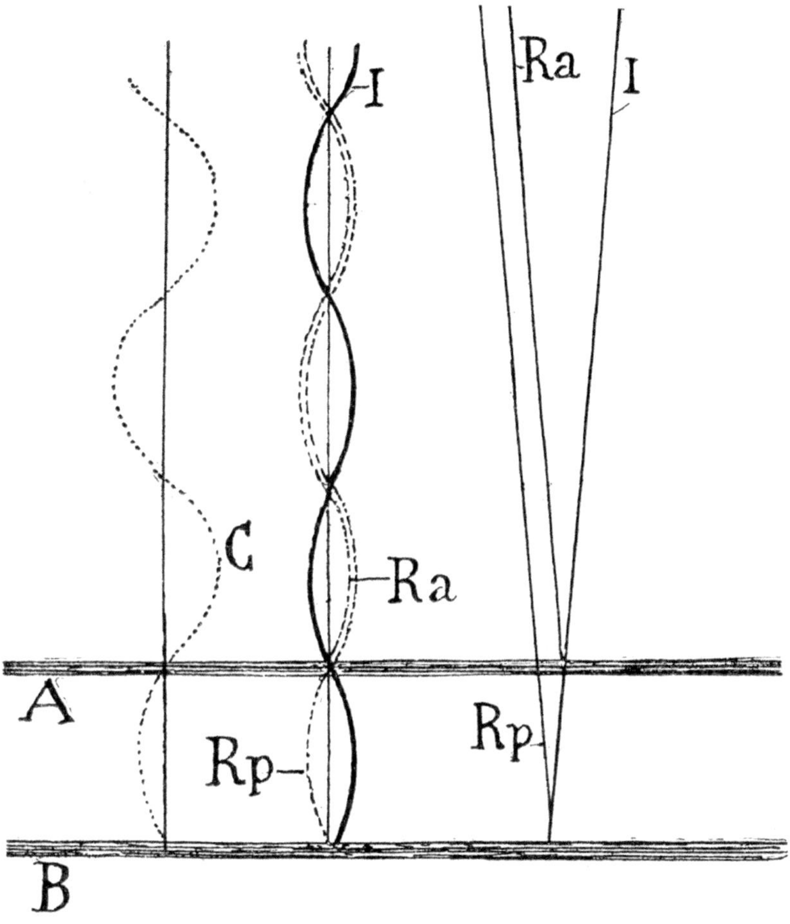

Fig. 14c. Pathway of luminous rays when hitting against two refringent laminae of negligible thickness. Abbreviations: A, anterior lamina; B, posterior lamina. To the right is simply reproduced the direction of the radiation, showing the coincidence of the two reflected rays, the anterior and the posterior (I). In the centre a representation of the undulatory movement of these rays, the incident I, and the two reflected: the posterior (Rp) and the anterior (Ra). Finally, in the left is shown the isolated ray reflected (C), the sum of the amplitudes of the two reflected undulations, the anterior and the posterior. From Ramón Cajal (1912).

Light rays reflected off the mercury layer interfere with incident light rays, forming regular standing waves and latent images at different depths, according to colour. After development, the film contains a series of layers of silver deposit where the light acted and of intermediate places where the silver is not reduced. In viewing the picture, colour is reproduced through light falling on the multiple layers and the resulting interference fringes throughout the thickness of the emulsion that are reflected to the eye

(Fig. 14c). Light reflected from two layers adds together to produce more light only if the two waves are in phase, and this will occur when the distance between layers is half a wavelength of the light used. As a result, colours that are not of the right wavelength to agree with the distance between the layers will cancel out by interference, while light that agrees with this wavelength will be increased by each successive addition of more light from the deeper layers (Niewenglowski 1894, Berthier 1895, Berget 1901, Ramón Cajal 1912, Roebuck 1919, Fournier 1991, Burder and Green 2004). The space between layers is determined by the light that fell on the film; therefore, the film will look the same colour when viewed afterwards from the emulsion side by ordinary reflection, while the image appears negative by transmitted light (Senior 1900).

Due to its technical complexity, especially the extreme slowness of the emulsions used in Lippmann heliochromes, it was a challenge to find manufacturers capable of meeting its needs. For example, the lenses must have a large aperture and the shortest possible focus; suitable designs are the Zeiss Anastigmat *f*=6.3, Goerz Doppelanastigmat *f*=7.7, and Voigtländer Collinear *f*=6.3 (Senior 1900). Lippmann heliochromes can also display harmonics of the original source colour shifting, such as infrared into blue (Alschuler 2002), in a way that one cannot avoid noticing a similarity with the "Aliquot" system in the sound domain, patented by the Leipzig piano maker Julius Blüthner (1824–1910), whereby a fourth string, placed above the regular triplet, is tuned an octave higher (first harmonic) to enhance, through resonance, the volume and "singing" quality of tones in the upper octaves (Dolge 1911).

In essence, Lippmann's method is a symmetrical two-step process of interference-mediated encoding and reconstruction (Biedermann 2005) to generate colour photographs on a black-and-white medium (Alschuler 2002). Colours appear to the eye by the abstraction of certain colours from the components of white due to interference or the superposition of direct and reflected waves (Senior 1900). As Lippmann remarked in his Nobel Lecture on December 14, 1908, such a phenomenon of selective light reflection is also observable in the case of soap bubbles and mother-of-pearl (Nobel Foundation 1967). Certain insects have evolved periodic lamellae through natural selection, operating on the basis of similar physical principles of optical scattering and diffraction (Biedermann 2005). In a different context, in the history of visual physiology, when the two prevailing views were the three-colour "physicist" theory of Young and Helmholtz and the four-colour "physiologist" theory of Hering, an interference theory of colour vision was presented by William Forbes (1885–1968) of Cornell University, based on the similarities that Cajal had described in the appearance and connections of cones in the fovea with the rods found in the retina of humans (Forbes 1928).

In his effort to record and present images, Lippmann used a revolutionary approach to fundamental physics, rather than following an evolutionary progress in engineering, as did the Hungarian physicist Dennis Gabor (1900–76) decades later, who was awarded the 1971 Nobel Prize in Physics for developing the holographic method, and thus extending photography in three- dimensional space (Biedermann 2005). The ideas of Lippmann and Gabor contributed to the evolution of three-dimensional holograms by the Soviet physicist Yuri Nikolaevich Denisyuk (1927–2006) in 1962 (Denisyuk 1991). Kubota (1991) further describes a new technique for recording Lippmann colour holograms – using three recording wavelengths corresponding to the three primary colours – by means of a holographic stereogram.

Cajal's Microscopy Experiments on Lippmann Heliochromes

Cajal used Lippmann heliochromy to photographically document histological sections of normal and pathological nervous tissues (Ramón Cajal 1908, 242, n. *b*), and for leisure photography such as his image of a still life (*bodegón*) featuring an arrangement of bottles, apples, flowers, and chemicals from circa 1912. He explains the interferential method in the first two chapters of section II in his book on colour photography (Ramón Cajal 1912, 203–53), including the theoretical bases of the method, the structure of the image, simple and composite colours, the necessary instrumentation, the making of the transparent emulsion, and light exposure factors (Fig. 14d and Fig. 15).

The question that Cajal addressed experimentally (Ramón Cajal 1908) was whether one could actually document the existence of Zenker laminae by means of light microscopy. The laminae, which are responsible for colour quality and brilliance, were central to the experiments of interferential photographers who sought to understood on the most minute of levels how colour emerges and, when it does, how can it be improved upon. In fact, Cajal's was one of only a handful of studies to address this problem; others were those published by Richard Neuhauss (1898), Edgar Senior (1900), Hans Lehmann (1907a, 1907b), Herbert Ives (1908), Eduard Valenta (1912), and Wilhelm Scheffer (1917). These authors used thin strips cut off the edge of the film emulsion and examined them under the microscope to analyse the alternating opaque and transparent layers running parallel to its surface.

To obtain sections of Lippmann heliochromes for microscopic study ("some hundreds of very careful observations"), Cajal used a method of

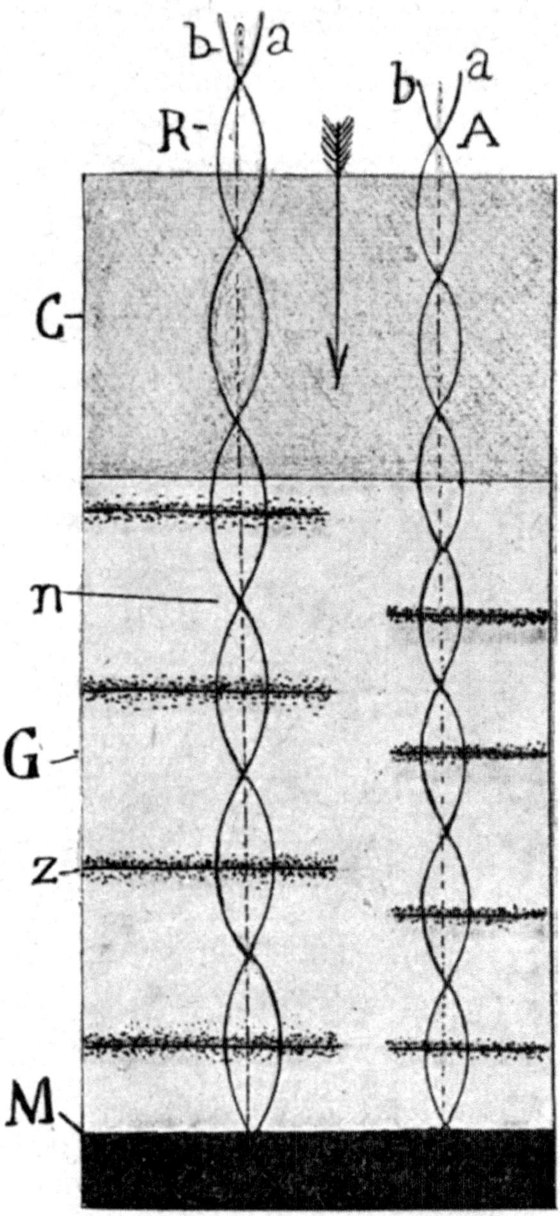

Fig. 14d. Pathway of luminous rays when hitting against two refringent laminae of negligible thickness. Abbreviations: A, anterior lamina; B, posterior lamina. To the right is simply reproduced the direction of the radiation, showing the coincidence of the two reflected rays, the anterior and the posterior (I). In the centre a representation of the undulatory movement of these rays, the incident I, and the two reflected: the posterior (Rp) and the anterior (Ra). Finally, in the left is shown the isolated ray reflected (C), the sum of the amplitudes of the two reflected undulations, the anterior and the posterior. From Ramón Cajal (1912).

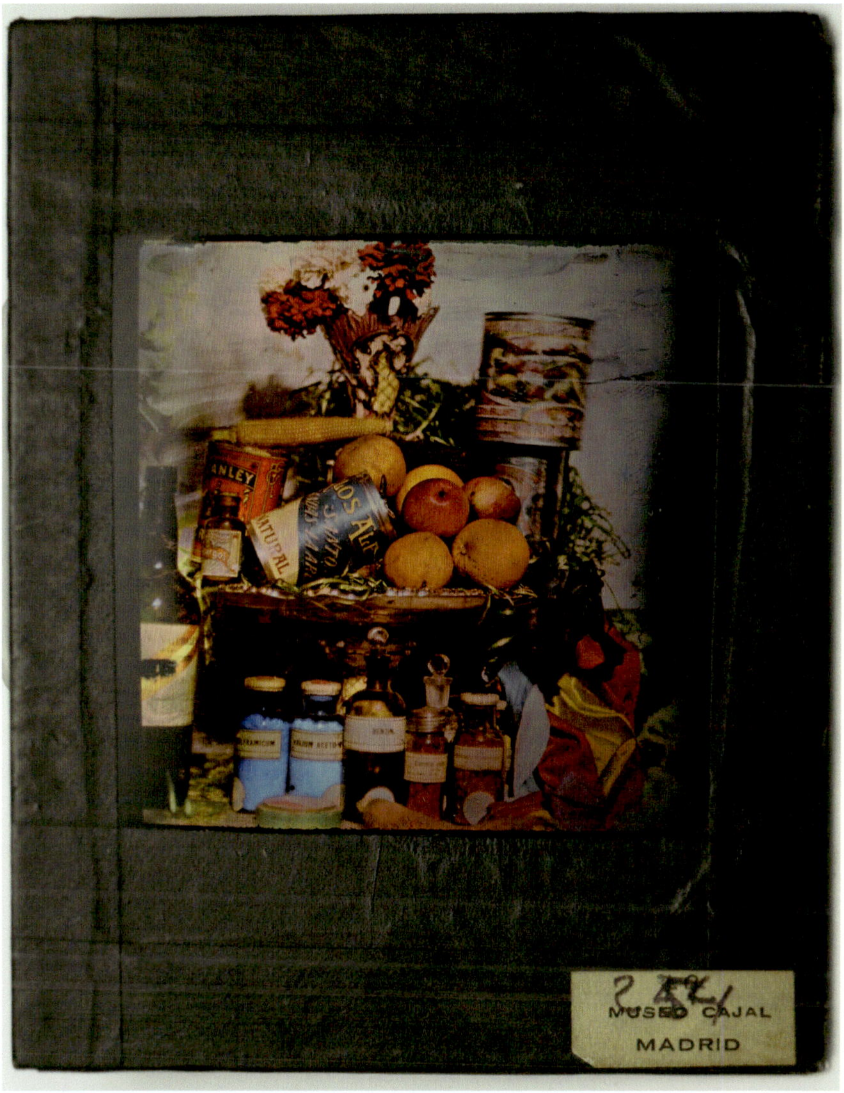

Fig. 15. Interference colour photograph of still nature (*bodegón*) made by Cajal circa 1912; original size 90×90 mm. Inventory number 1798. Courtesy of the Cajal Legacy. Instituto Cajal (CSIC), Madrid.

soaking the photographic plates in water, stripping the film off by means of a freshly broken piece of glass (Ramón Cajal 1908, 240) and immersing the stripped film sequentially in an aqueous alcohol solution, absolute ethanol, and celloidin. Cross sections of the film were cut at a right angle and left to swell either in water or in glycerine. Occasionally, Cajal stained film specimens with hydrophobic aniline dyes, a practice that was already

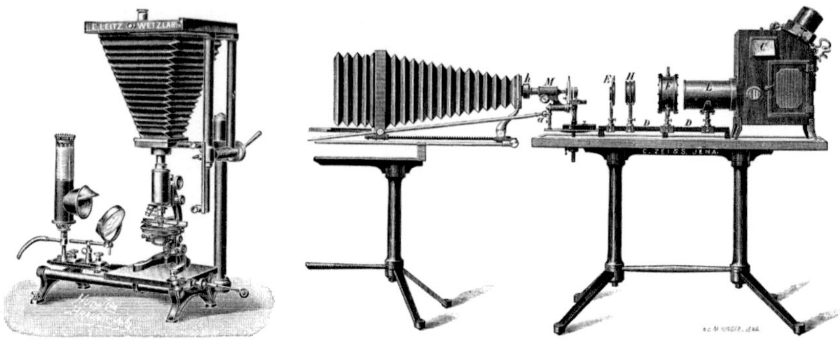

Fig. 16. Vertical Leitz (left) and horizontal Zeiss (right) photomicrography apparatus, similar to those Cajal used to record his microscopic specimens. Abbreviations: M, microscope; h, microscope-camera light-tight adapter ring; A, on-off switch arrangement; E, microscope condenser; F, optical filter; H, cuvette filter, filled with a cupric salt solution to filter out infrared; L, light source; D, base of optical bench. Sources: left, from Hager (1912); right, from an original 1892 lithograph.

known to him as a neurobiologist using histological analysis. An alternative method was to hack the damp film with a scalpel and to mount it under a glass coverslip.

One serious difficulty was the following. Since Zenker laminae are half a wavelength (1/2 λ) apart, the Zeiss Apochromat objective with a numerical aperture (N.A.) of 1.40 that was available then, and which Cajal used (he did not dare use a N.A. 1.60 lens that would require monobromonaphthaline immersion) was right at the limits of microscopic resolution (the theoretically possible resolving power of an optical microscope – based on Ernst Abbe's calculations – is given by $\delta=\lambda/2\alpha$, where δ is the distance between two image points, λ the wavelength of light, and α the N.A. of the objective).

The type of apparatus Cajal used for microphotography included extension bellows and glass negative plates; the bellows-mounted negative holder could be placed either vertically or horizontally (Fig. 16 left), attached to a rail for stability purposes (Hager 1912). Such arrangements were similar to the ones that Cajal was using for his neurohistological work, orienting, for example, the axis of a Zeiss or a Leitz 13×18 cm plate camera at the same angle as the lower body of the microscope (Fig. 16 right). The length of the bellows could vary from 25 to 100 cm, and exposure times from one to sixty minutes, depending on the focal length and N.A. of the objective, specimen properties, source of light, and plate sensitivity. The circulation of instruments and experimental methods between neurohistology and colour photography testifies to how these two sciences informed each other, expanding what can be known about the nervous system as well as the Zenker laminae at the minutest level of resolution.

Regarding the transparent albumin or gelatin emulsion, Lippmann had contended that it had no grains. However, Neuhauss proved the existence of grains after development that he estimated to range from 100–300 nm in size, whereas Cajal considered the spherical grains to be closer to 20–50 nm in diameter, which was necessary to register violet ($1/2\,\lambda = 171$ nm).

Cajal's results from his analyses of Lippmann heliochromes were published in 1908. On examining plates in pure spectral colour, different dark and transparent (exposed and unexposed) layers can be seen. They include a "limiting zone" between the free surface and the first dark lamina, more of Zenker laminae (consisting of metallic precipitate and ranging in number from four to thirteen or more), and the clear, unexposed intervals or space (Fig. 17). The first lamina is the most distinct (Fig. 17B and 17H). The first interval is formed by the surface of the gelatin and is the most colourless. Beyond the second lamina and interval, contrasts become filled with precipitate and less distinct. In all, colour brilliancy depends on the quality, rather than quantity, of intervals and laminae, especially the two closest to the free surface. The formation of pure white results from image intensification. Whites consist of three regions, the "mirror zone," the laminated zone, and a diffuse-reflection "rear zone"; behind the mirror zone, there is a fine interval and a series of dark and thin, closely compacted stripes (Fig. 17E). Conditions to obtain whites are a metallic reflecting precipitate in the limiting zone and complete opacity of the first lamina. Compound colours mixed with white possess a better mirror zone, as well as laminae with intervals corresponding to the principal colour (Fig. 17F). On the other hand, dark tones mixed with black depend on the fineness and transparency of the laminae (Fig. 17P); colour tones may become falsified in darker parts of the plate. Single (Fig. 17H) or double (Fig. 17I) intensification makes grains larger and give pictures a milky fog appearance relative to non-intensified photochromes (Fig. 17G). Overexposed pictures show a lustreless white, and pure – but hard, silhouetted – colours; in those cases, both the first lamina and the contrast between laminae and intervals are substantially reduced (Fig. 17L and 17M).

It is possible to expose plates without a mercury mirror. In such cases, much longer exposures are necessary; the results are better for shorter wavelengths, from violet to green. Laminae are composed of very fine grains, such as the section through the blue shown in Fig. 17Q. Modern experimenters, such as Alschuler (2002), have been also omitting the mirror, at the same time using commercial holographic emulsions. Therefore,

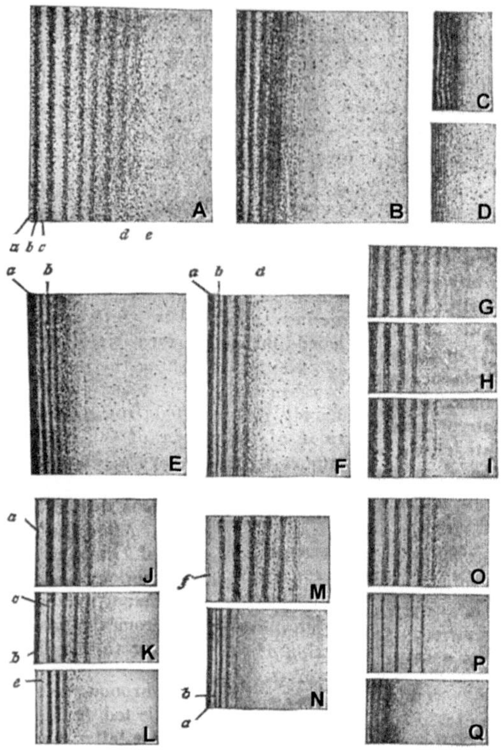

Fig. 17. Cajal's results from experimenting with Lippmann heliochromic plates at the microscope. (A) Section through pure or almost pure red. Swollen in water and examination with a Zeiss apochromat, N.A. 1.40, 2 mm focus. Central white light, a the limiting zone, b first Zenker lamina, c second interval, d deeper lying laminae, with indefinite edges, e unlaminated zone. (B) Section through the blue, in the reproduction the deeper lying laminae are badly drawn, conditions of examination similar to (A). (C) Section through the red in dry gelatin – that is gelatine not swollen in water; examination in Canada balsam, central monochromatic light. (D) Section through greenish yellow, same conditions as in (C); the limiting zone and the grains in individual laminae cannot be seen. (E) Section through pure brilliant white; swollen gelatine, a opaque mirror-zone, b the fine stripes lying under the mirror-zone. (F) Section through yellowish white, a mirror-zone, c fine stripes, d laminae corresponding to the yellow. (G I) The action of intensification on colour. (G) Unintensified colour, the stripes being too dark in the reproduction. (H) The same colour intensified once in a sublimate bath. (I) After two intensifications. It will be observed how the scarcely visible grain of (G) becomes thick and dark in (I). (J) Red; the thickness of the first laminae was reduced by friction, so that blue and green stripes appear. (K) Section through bright green, which by overexposure and overdevelopment has become white; b mirror-zone, c fine stripes 26 belonging to the white; the other laminae belong to the green. (L) Section through overexposed blue; the paleness of the laminae e and the absence of the mirror-zone will be noticed. (M) Section through overexposed orange; the first lamina f is wanting, and the second is also rather pale. (N) Section through bright blue mixed with white; a mirror-zone, b fine secondary laminae. (O) Section through bright lemon yellow; the first laminae represent the phase of conversion into the mirror-zone. (P) Section through underexposed and overdeveloped green, which corresponds to the shadow side of an orange; the thinness and transparency of the laminae, which are somewhat too dark in the reproduction, will be noted. (Q) Section through the blue in a plate exposed without a mercury mirror. Individual frames adapted and rearranged from Ramón Cajal (1908).

studying the laminae means being able to reclaim the entire photographic process and to gain insight into the experimental conditions under which the colours emerged. In several instances, Cajal tested his hypotheses through chemical experiments, for example, placing pictures in a benzol tank to spectroscopically examine light reflected from the whites, rubbing whites with alcohol to cause the mirror zone to disappear, and using reducers, such as potassium cyanide, to obviate plate faults caused by humidity.

Cajal summarised his findings as follows (Ramón Cajal 1908):

> (a) As had been already recognised by Neuhauss, spectral colours are produced by a series of metallic laminae, separated with colourless intervals from each other. Such films occupy one third or half of the thickness of the gelatin. Near the free surface, they are sharply defined and distinctly separate; the deeper they are, the more diffuse and indistinct they become.
> (b) Generally, there is a clear zone between the first laminae and the surface, corresponding to the first opposite phase or "null-point." Frequently, through intensification, this contracts considerably or disappears completely.
> (c) The colours of natural objects give pictures, the structure of which generally agrees with that of spectral colours.
> (d) The production of white is due to the formation of a dense metallic lamina, the "mirror zone," with great reflective power, composed of an opaque dark, closely compacted precipitate. There are some fine, closely contiguous stripes, probably corresponding to short waves of the visible spectrum.
> (e) Colours mixed with white show with their own laminae a thin surface film, filled with a metallic precipitate, the mirror zone.
> (f) In certain instances, colours mixed with white show two kinds of laminae: large stripes separated from each another, belonging to the long waves of the predominant colour, and finer pale films corresponding to shorter wavelengths.
> (g) Interference phenomena that produce colours in Lippmann heliochromes can be actually ascribed to the action of rays reflected from the first and second laminae. The remaining laminae only have a faint, intensifying action. Pure spectral colours are an exception; if the metallic precipitate is rather transparent, the deeper lying zones (or laminae) may also act.
> (h) Good colour rendition is mainly caused by the correct limiting and perfect transparency of the upper laminae and by the normal value of the intervals. Factors such as long exposure, overdevelopment (Fig. 7K), and incorrect intensification, which upset the ratio of the two first laminae regarding intensity and thickness, or fixation and dampness (especially using too dry plates in humid weather), which affect interval size, alter true

> colours and may result in a falsification of tonalities. The two greatest difficulties of the Lippmann process are, first, that the distance of individual laminae produced in the plate by coloured light during exposure must be maintained strictly, in spite of the contracting action of fixation and the expanding action of intensification, and second, that too little transparency and too great thickness of the first lamina must be avoided, although a certain opacity is essential for the correct reproduction of bright tones.

Cajal suggested that workers in interferential photography ought to achieve a perfect balance between those last two opposing requirements through careful and clean work.

Conclusion

The present chapter underlines Cajal's original contributions between 1906 and 1908 on a diversified topic, bearing on the physical principles of light wave interference and its relation to the capture of colour, the subject that earned Gabriel Lippmann the 1908 Nobel Prize in Physics. Lippmann's method is considered to be of immense scientific interest, but difficult for routine use, because it requires patience and special skills. Its difficulties are associated with the necessity to use a thin emulsion, which will not be opaque to light falling upon it; the need to use a reflector consisting of a mercury layer in immediate contact with the film; long exposure times owing to the slow, fine-grained emulsion; the need to avoid long film development to prevent the film from becoming too dense; and the sensitivity of the film to treatment, because, for example, the treatment of the film may lead to a loss of silver from undeveloped layers and bring the layers of reduced silver closer together, causing colour shifts (Roebuck 1919).

More than a mere hobby, photography was, for Cajal, a lifelong passion, of equal importance to him as neuroanatomy (Ramón Cajal 1988). His deep understanding of the chemical basis of darkroom techniques proved an expedient tool for inventing or modifying staining and impregnation methods for nervous tissues, especially during the era of the wide use of metallic salts in histology. Stressing the similarities between histological and photographic permutations, Márquez (2004) even suggested that the working knowledge and profound understanding of the principles of photochemical processes were the necessary preconditions that allowed Cajal to achieve his neurobiological successes. The fact that one of Cajal's prime pursuits, interference heliochromy, eventually led to the Nobel Prize in Physics for its inventor, Lippmann, attests that Cajal grappled with technical problems at the core of scientific evolution in his epoch. Cajal's technical publications on the principles of colour photography between 1904 and 1912 document yet another domain, where this rounded investigator excelled by contributing original finds to general science. Centenary celebrations honouring Cajal began in 1952, marking his birth (Marañón 1952, Trueta 1952), followed by centenaries for the 1888 discovery of the cerebellar (Llinás and Sotelo 1992) and retinal (Piccolino 1988) circuits, the 1894 Croonian lecture (Jones 1994), publication of *Histology* (Marijuán 2001, Andres-Barquin 2001), the chemotactic hypothesis (Sotelo 2002), and the Nobel Prize in Physiology or Medicine shared with Golgi (CSIC 2006, DeFelipe and Bentivoglio 2006, Glickstein 2006, Grant et al. 2006, López-Muñoz et al. 2006, Grant 2007). Our original paper (Triarhou and del Cerro 2008) marked yet another Cajal centenary; although not a neuroanatomical article sensu stricto, it might appear to contain little that is of direct biological content. Nonetheless, it reflects the breadth in the workings of the mind of Cajal, one of the greatest neuroanatomists, and the fundamental tools of physics, chemistry, and microscopy that enabled him to decipher the structure of the nervous system. As a matter of fact, Cajal's fundamental understanding of the theoretical bases of photography was a precondition for his success in devising and modifying neurohistological methods. This facet of Ramón y Cajal may have remained little known or virtually forgotten among neuroscientists, perhaps overshadowed by his epoch-making biological discoveries.

Acknowledgments

The authors gratefully acknowledge the constructive comments of Dr. William R. Alschuler, which led to an improved manuscript. We thank Dr. Hanin Hannouch for her kind invitation to participate in the Lippmann Centenary Volume and the *Journal of Chemical Neuroanatomy* for hosting the original version of our manuscript (Triarhou and del Cerro 2008), on the occasion of the centenaries of Cajal's 1908 paper and Lippmann's Nobel Prize.

Works Cited

Alschuler, William. 2002. *Lippmann Photography and the Glory of Frozen Light: Eternal Photographic Color Real and False*. Proceedings of the 31st Applied Imagery Pattern Recognition Workshop, IEEE Computer Society. doi:10.1109/AIPR.2002.1182248.

Andres-Barquin, Pedro José. 2001. "Ramón y Cajal: A Century After the Publication of his Masterpiece." *Endeavour,* 25, 13–17.

Becquerel, Edmond. 1868. *La Lumière, ses Causes et ses Effets.* Paris: Firmin Didot, 2 vols.

Berget, Alphonse. 1901. *La Photographie des Couleurs par la Méthode Interférentielle de M. Lippmann.* Paris: Gauthier-Villars Imprimeur-Librairie.

Berthier, Alphonse. 1895. *Manuel de Photochromie Interférentielle. Procédés de Reproduction Directe des Couleurs.* Paris: Gauthier-Villars et Fils Imprimeurs-Libraire.

Biedermann, Klaus. 2005. "Lippmann's and Gabor's Revolutionary Approach to Imaging." https://www.nobelprize.org/prizes/themes/lippmanns-and-gabors-revolutionary-approach-to-imaging/

Bjelkhagen, Hans. 2004. "References on Lippmann Photography and Earlier Techniques." https://holowiki.org/forum/viewtopic.php?t=5638.

Bolas, Thomas. 1900. Historical Development of Heliochromy and General Survey of Processes. In *A Handbook of Photography in Colours,* edited by Thomas Bolas, Alexander A.K. Tallent, and Edgar Senior. London: Marion and Company, 1–87.

Burder, David G., and Darran Green. 2004. *"Natural" Colour Photography – Heliochromes, Daguerreotypes, and Lippmann Photography Revisited.* The Royal Photographic Society, Holography Group, 28th February 2004, 1–3. www.casedimage.com/forum/BurderRPSHeliochromes.pdf.

C.S.I.C., 2006. Centenario de la Concesión del Premio Nobel a don Santiago Ramón y Cajal. Consejo Superior de Investigaciones Científicas, Madrid. www.csic.es/monograficos/cajal.

DeFelipe, Javier, and Marina Bentivoglio. 2006. *Society for Neuroscience Marks 1906 Nobel Prize Awarded to Neuroscientists Camillo Golgi and Santiago Ramón y Cajal.* Centennial Celebration at the Society for Neuroscience Headquarters, Washington, DC, December 7, 2006. www.sfn.org/index.cfm?pagename=news_120106.

de Madrid, Rafael, 1952. "Cajal, fotógrafo." *Arte Fotográfico,* 1: 131–132.

Denisyuk, Yuri Nikolaevich. 1991. "Imaging Properties of Light Intensity Waves: The Development of the Initial Lippmann Ideas." *Journal of Optics,* 22, 275–280.

Dolge, Alfred. 1911. *Pianos and their Makers: A Comprehensive History of the Development of the Piano.* Covina California: Covina Publishing Company.

Ducos du Hauron, Louis. 1869. *Les couleurs en Photographie: Solution du Problème.* Paris: A. Marion Éditeur.

Forbes, William Trowbridge Merrifield. 1928. "An Interference Theory of Color Vision." *The American Journal of Psychology,* 40, 1–25.

Fournier, Jean-Marc. 1991. "Le Photographie en Couleur de Type Lippmann: Cent Ans de Physique et de Technologie." *Journal of Optics,* 22, 259–266.

Glickstein, Mitch. 2006. "Golgi and Cajal: The Neuron Doctrine and the 100th Anniversary of the 1906 Nobel Prize." *Current Biology,* 16, R147–R151.

Grant, Gunnar. 2007. “How the 1906 Nobel Prize in Physiology or Medicine was Shared between Golgi and Cajal.” *Brain Research Reviews*, 55, 490–498

Grant, Gunnar, Sten Grillner, Thomas Hökfelt, Edward G. Jones, and Larry W. Swanson. 2006. “A New Century of Neuroscience: Celebrating the 1906 Nobel Prize.” *A Cajal Club Symposium Co-Hosted by Karolinska Institutet*, The Nobel Forum, Stockholm, June 11–13.

Hager, Hermann. 1912. *Das Mikroskop und seine Anwendung: Handbuch der Praktischen Mikroskopie und Anleitung zu mikroskopischen Untersuchungen*. Berlin: Verlag von Julius Springer.

Harrison, William Jerome. 1888. *A History of Photography, Written as a Practical Guide and an Introduction to its Latest Developments*. London: Percy Lund and Company.

Hill, Levi L. 1856. *A Treatise on Heliochromy; or, the Production of Pictures, by Means of Light, in Natural Colors*. New York: Robinson and Caswell.

Ives, Frederic Eugene. 1894. “Photography in the Colors of Nature.” In Board of Regents of the Smithsonian Institution, *Annual Report for the Year 1893*. Washington, DC: Government Printing Office, 151–162.

Ives, Herbert E. 1908. “An Experimental Study of the Lippmann Color Photograph.” *Astrophysical Journal*, 27, 325–352.

Ives, Herbert E., 1913. “The Present Condition of Color Photography.” *Journal of Physical Chemistry*, 17, 41–46.

Jones, Edward G. 1994. “Santiago Ramón y Cajal and the Croonian Lecture, March 1894.” *Trends in Neurosciences*, 17, 190–192.

Kubota, Toshihiro, 1991. “Lippmann Color Holography.” *Journal of Optics*, 22, 267–273.

Larrea, Rafael, 1989. “Cajal, Fotógrafo.” *Madrid Médico*, 10, 67–68.

Lehmann, Hans, 1907a. “Über eine neue kornlöse Platte für Lippmann-Photographie.” *Zeitschrift für die Wissenschaftliche Photographie*, 5, 279–283.

Lehmann, Hans, 1907b. “Über die direkten Verfahren der Farbenphotographie nach Lippmann und Lumière.” *Physikalische Zeitschrift*, 8: 842–849.

Lippmann, Gabriel. 1891. La Photographie des Couleurs. *Comptes Rendus de l’Académie des Sciences*, 112, 274–275.

Lippmann, Gabriel. 1894. “Sur la Théorie de la Photographie des Couleurs Simples et Composées par la Méthode Interférentielle”. *Journal de Physique*, 3, 97–107.

Llinás, Rodolfo, and Constantino Sotelo. 1992. *The Cerebellum Revisited*. New York-Berlin-Heidelberg: Springer-Verlag.

López Mondéjar, Publio. 1982. “Ramón y Cajal, Fotógrafo.” *Nueva Lente*, 155, 9–14.

López-Cantos, Francisco. 2016. “Photography in the Boundaries of the Visible: From Santiago Ramón y Cajal to Cecil Frank Powell.” *Cogent Arts and Humanities*, 3, 1182715. https://doi.org/10.1080/23311983.2016.1182715.

López-Muñoz, Francisco, Jesús Boya, and Cecilio Alamo. 2006. "Neuron theory, the Cornerstone of Neuroscience, on the Centenary of the Nobel Prize Award to Santiago Ramón y Cajal." *Brain Research Bulletin*, 70, 391–405.

López Piñero, José Maria, María Luz Terrada Ferrandis, M.L., Rodríguez Quiroga. 2000. *Bibliografía Cajaliana: Ediciones de los Escritos de Santiago Ramón y Cajal y Estudios sobre su Vida y Obra*. Valenica: Albatros – Artes Gráficas Soler.

Marañón, Gregorio, 1952. "Reflexiónes Ante el Homenaje a Cajal en su Primer Centenario." In *Revista IBYS: Número Homenaje a Cajal*. Madrid: El Instituto de Biología y Sueroterapía, 31–36.

Marijuán, Pedro C. 2001. *Cajal and Consciousness: Scientific Approaches to Consciousness on the Centennial of Ramón y Cajal's Textura*. New York: Annals of the New York Academy of Sciences, Volume 929.

Márquez, Miguel Bobo. 2004. "Santiago Ramón y Cajal: Algo más que un Fotógrafo." *Ámbitos: Revista* Andaluza *de Comunicación*, 11/12, 139–153.

Mueller, Martin, 2006. Lippmann Photography: A Holography Database. www.holographyforum.org/HoloWiki/index.php/Lippmann_Papers

Neuhauss, Richard. 1898. *Die Farbenphotographie nach Lippmann's Verfahren. Neue Untersuchungen und Ergebnisse*. Halle: Verlag von Wilhelm Knapp.

Niewenglowski, Gaston-Henri. 1894. Die neuesten Fortschritte der Chromophotographie und der Theorie der Lippmann'schen Versuche. *Eder's Jahrbuch für Photographie und Reproduktionstechnik*, 8, 73–82.

Nobel Foundation. 1967. Nobel Lectures, Physics, 1901–1921. Amsterdam: Elsevier Publishing Company.

Pagneux, Marc, Sylvie Aubenas, Daniel Girardin, and Christophe Brandt. 2000. Fotographie: Der lange Weg zur Farbe. DU—*Die Zeitschrift der Kultur*, 708, 12–53.

Piccolino, Marco. 1988. "Cajal and the Retina: A 100–Year Retrospective." *Trends in Neurosciences*, 11, 521–525.

Ramón Cajal, Santiago. 1870. "Historia de la Fotografía." In García Durán Muñoz, and Francisco Alonso Burón, editors. 1978. *Ramón y Cajal, Tomo II: Escritos Inéditos*, Zaragoza: Institución Fernando el Católico, 79–89.

Ramón Cajal, Santiago. 1903. "Anatomía de las Placas Fotográficas." *La Fotografía*, 17, 133–145.

Ramón Cajal, Santiago. 1904. "La Fotografía Cromática de Puntos Coloreados." *La Fotografía*, 37, 5–8.

Ramón Cajal, Santiago. 1906a. "Una Modificación al Proceder Fotocrómico de Lumière a la Fécula." *La Fotografía*, 61, 101–114.

Ramón Cajal, Santiago. 1906b. "Reglas Prácticas Sobre la Fotocromía Interferencial de Lippmann." *Ciencia Popular*, 6, 1–8; 7, 97–100.

Ramón Cajal, Santiago. 1906c. "Estructura de las Imágenes Fotocrómicas de G. Lippmann." *Revista de la Real Academia de Ciencias. Exactas Físicas y Naturales*, 4, 1–47.

Ramón Cajal, Santiago. 1907a. "Notes Microphotographiques." *Anales de la Sociedad Española de* Física y Química, 5, 204–212.

Ramón Cajal, Santiago. 1907b. "Las Placas Autócromas Lumière y el Problema de las Copias Múltiples." *La Fotografía,* 73, 1–15.

Ramón Cajal, Santiago. 1907c. "Chromomicrophotographie par la Méthode Interférentielle." *La Photographie des Couleurs,* 2, 97–101.

Ramón Cajal, Santiago. 1907d. "Die Struktur der Lippmannschen Photochromien." *Zeitschrift für Wissenschaftliche Photographie,* 5, 213–245.

Ramón Cajal, Santiago. 1907e. "The Structure of Lippmann Heliochromes." *British Journal of Photography* [Supplement] Aug. 2, pp. 59–61; Sep. 6, pp. 69–71; Oct. 4, pp. 77–79. www.holographyforum.org/lippmann.

Ramón Cajal, Santiago. 1908. "The structure of Lippmann Heliochromes." In *Board of Regents of the Smithsonian Institution, Annual Report for the Year 1907.* Washington, DC: Government Printing Office, 239–259.

Ramón Cajal, Santiago. 1912. *La Fotografía de los Colores: Fundamentos Científicos y Reglas Prácticas.* Madrid: Nicolás Moya.

Ramón Cajal, Santiago. 1988. *Recollections of My Life.* Birmingham, Alabama: Gryphon Editions.

Roebuck, John Ransom. 1919. *The Science and Practice of Photography: An Elementary Textbook on the Scientific Theory and a Laboratory Manual.* New York and London: D. Appleton and Company.

Romero, Alfredo. 1984. *Fotografía Aragonesa/1: Ramón y Cajal.* Zaragoza: Diputación Provincial de Zaragoza.

Romero Santamaría, Alfredo, 2002. "Cajal Como Fotógrafo." In *Visiones: Santiago Ramón y Cajal 150 Aniversario,* edited by Ismael Grasa Adé and Alfredo Romero Santamaría. Zaragoza: Diputación de Zaragoza, 115–131.

Scheffer, Wilhelm. 1917. *Die Grundlagen der Photographie.* Berlin: Union Deutsche Verlagsgesellschaft.

Senior, Edgar, 1900. "Lippmann's Process of Interference Heliochromy." In *Handbook of Photography in Colours,* edited by Thomas Bolas, Alexander A.K. Tallent and Edgar Senior. London: Marion and Company. 315–343.

Sotelo, Constantino. 2002. "The Chemotactic Hypothesis of Cajal: A Century Behind." *Progress in Brain Research,* 136, 11–20.

Sougez, Marie-Loup. 1982. "Aportaciones Técnicas de Cajal en Fotografía." *Nueva Lente,* 155, 20–21.

Tello Muñoz, Jorge Francisco. 1906. "Cajal, Fotógrafo." *Clínica y Laboratorio,* 2, 316–319.

Tomero Alarcón, José. 1966. "Cajal, Fotógrafo." *Arte Fotográfico,* 15, 305–308.

Towler, John. 1864. "Heliochromy, or the Art of Taking Photographs in Natural Colors." In *The Silver Sunbeam: A Practical and Theoretical Text-Book on Sun*

Drawing and Photographic Printing, Comprehending All the Wet and Dry Processes at Present Known. New York: Joseph H. Ladd.

Triarhou, Lazaros C., and Manuel del Cerro. 2008. "The Structure of Lippmann Heliochromes: Cajal and the 1908 Nobel Prize in Physics." *Journal of Chemical Neuroanatomy*, 35, 1–11.

Trueta, Joseph. 1952. "Ramón y Cajal: The First Centenary of his Birth." *The Lancet*, 260, 281–283.

Valenta, Eduard. 1912. *Die Photographie in natürlichen Farben mit besonderer Berücksichtigung des Lippmannschen Verfahrens sowie jener Methoden, welche bei einmaliger Belichtung ein Bild in Farben liefern, zweite Auflage*. Halle: Verlag von Wilhelm Knapp.

Waldeyer, Wilhelm. 1891. "Über einige neuere Forschungen im Gebiete der Anatomie des Centralnervensystems." *Deutsche Medizinische Wochenschrift*, 17, 1213–1218, 1244–1246, 1287–1289, 1331–1332, 1352–1356.

About the Authors

Lazaros C. Triarhou is professor of neuroscience at the Department of Psychology, Aristotelian University Faculty of Philosophy in Thessaloniki, Greece. He obtained his MD from the Aristotelian University and continued his training in the United States, where he completed a MSc in neuroscience at the Center for Brain Research at the University of Rochester, New York, and a PhD in medical neurobiology at Indiana–Purdue University in Indianapolis.

Manuel del Cerro is professor emeritus of ophthalmology and neurobiology at the University of Rochester, New York. A native of Argentina, he obtained his BA and BSc degrees from the Buenos Aires National College and his MD from the University of Buenos Aires Medical School. He is one of the most active microscope collectors in the United States, and he is the founder of the Microscope Historical Society.

4. Lost in Description: The Misunderstanding of Frederic Ives (1856–1937) and the Experiments of his son Herbert (1882–1953) on Lippmann's Interference Photography

Susan Gamble

Abstract

Published in 1908, Herbert Ives' *An Experimental Study of the Lippmann Colour Photograph* translated Gabriel Lippmann's work from French to English while modernising the terminology for the twentieth-century optical physicist. In this article, I explain how Herbert Ives' paper, established Lippmann's process within the discipline of physics, but concealed the practical assistance offered to Lippmann by the industrialist Lumière brothers. I also explore the sceptical criticisms of his father, American artisan-inventor Frederic Ives whose own colour process competed with Lippmann's for display and publicity. At the end of the nineteenth century, these competing processes exemplified divisions between artisan-photographers and professional experimenters. Whereas Frederic Ives relied solely on patents and demonstrations to promote his inventions both in the USA and in Europe, Lippmann deployed wave theory to an audience at the *Académie des Sciences* in Paris.

Keywords: Frederic Ives, Herbert Ives, Gabriel Lippmann, history of science, translation, Lumière brothers

Gabriel Lippmann had gained credibility within the scientific community when he revealed his theory of the solar spectrum to the Académie in 1891. The spectrum provided a photographic image of an instrumentally isolated object, the sun, via a spectroscope with the ability to record the characteristic Fraunhofer black lines, familiar to astronomers. It demonstrated the full potential of instantaneous *natural colour* photography: a photographic process without the artifice of pigment or the intervention of colour separations. Like a daguerreotype, the Lippmann photograph had to be held in the available light at the right angle for the image of the spectrum to be visible. For a larger audience, light reflecting from the surface of the plate could reconstruct the image to project it onto a white board or screen. This projection might appear to the audience to be an aesthetically satisfying simulacrum of nature's "rainbow." When Lippmann presented this photograph, his first claim for this invention within his paper was the provision of a photographic image that would last indefinitely; it would not fade or decay under light (Lippmann 1891, 274): "Je me suis proposé d'obtenir sur une plaque photographique l'image du spectre avec ses couleurs, de telle façon que cette image demeurât désormais *fixée* et pût rester exposée indéfiniment au grand jour sans s'altérer."[1]

This was possible because the photograph contained no dyes or light-sensitive remnants of organic chemistry; interference was a physical, not a chemical phenomenon. The interference process has to physically affect or "shape" the gelatin emulsion to generate an image, rather than provide a tonal graduation of silver salts. Nor could the image be manipulated or retouched; it presented a significant advantage to the scientific community in its quest for objective documentation, according to Lippmann himself. However, to gain wider public attention and to prove the viability of the process, pictorial imagery was required. The lay public did not perceive the solar spectrum as a representational image. In May 1892, less than a year after the publication of his original theory and using photographic plates supplied by the Lumières, Lippmann presented pictorial colour photographs to the Académie des Sciences, depicting a bowl of oranges, a stained-glass window, a parrot, and a group of flags. These were also presented publicly alongside work by other French artisan inventors in the Champ de Mars exhibition in Paris

Lippmann did not have the facilities to produce reliable batches of the emulsion-coated plates needed for pictorial experiments. His laboratory was considered "rudimentary," according to his biographer: it was a former glassworks situated behind a wine merchant (Lebon 1911, 4–5). Historians of science Robert Fox and Anna Guagnini also support this view, citing an eminent French educational administrator on the squalor of "dilapidated attics and hangars of the court Saint-Jacques where Lippmann had produced his first steps in colour photography" (Fox and Guagnini 1999, 57). By contrast, the factory owned by the Lumière brothers was a controlled environment. By

1 I set out to obtain on a photographic plate the spectrum with its colours, in such a way that this image would henceforth remain fixed and be exposed indefinitely to broad daylight without altering. (Translation by the author)

1894, they employed two hundred skilled, mostly female handlers, producing "Blue Label" photographic plates (Rittuad-Hutinet 1995, 14). This relationship between savant and commerce was one motivated by technical requirements on Lippmann's behalf and by the Lumière brothers' desire to align themselves with an increasingly professional scientific elite, all of which served both parties' desires to develop French imaging technology. This relationship lasted three years, for which Lippmann nominated the Lumières for a prize to the Paris Académie des sciences.[2] The prize brought them recognition from an elite professional group that the Lumière brothers did not previously have direct access to and further benefited their mutual interests.

The resources provided by the Lumières in producing the emulsion, which is possibly one of the most difficult and mysterious aspects of the Lippmann process to the studio photographer, created a false impression of the ease with which Lippmann was able to produce his own pictorial photographs. If the Lumières produced a pictorial image on a Lippmann plate themselves, they did not publish it before Lippmann. According to historical witness Eduard Valenta, then professor of photochemistry at the Versuchsanstalt für Photographie und Reproductionsverfahren in Vienna, still in 1892, the Lumières announced a chemical formula for making a silver bromide Lippmann emulsion at the Société des Sciences Industrielles in Lyon, which would have made a pictorial image possible, but they only demonstrated a Lippmann spectrum (Valenta 1894, 47). The historian and photochemist Josef Eder commented that this publication by the Lumières was not widely transmitted within the photographic community at the time, and it contained no detailed description on preparing the plate with this formula (Eder 1945, 670). In this way, the emulsion remained an enigma to artisan photographers (Valenta 1894, 47).[3]

It was the Lumières that were credited, by Eder, with producing that first "satisfactory" pictorial images (Eder 1945, 670). This presentation of Lippmann photographs was given in Geneva, on the evening of August 22, 1893, at the International Photography Congress. These were projected with the Lumières invention: the Megascope and the metallic screen. These images demonstrated Lippmann's invention as a technology capable of

2 Auguste and Louis Lumière, Letters, 17. Lippmann to Lumieres: "At tomorrow's session of the Acadèmie I shall present your note together with its pretty photographs. I am sure you will be most successful. I should be grateful if you would let me have the pictures by Tuesday. They will be returned to you after the sitting [...] Yours ever, G. Lippmann."

3 Valenta wrote the formula in the journal *Photographischen Correspondenz*, 1892. no. 432, and then himself went on to produce and demonstrate what he described as "a brilliant" projected spectrum to a large audience in Vienna.

successfully appealing to an urbane, cultivated, and bourgeois group. The local Geneva newspaper reporter wrote:

> [...] a series of landscapes, views of villas and boathouses, as well as some portraits. The results are truly amazing. The countryside matches what the eye perceives of nature with groves of trees the most beautiful green, houses with white facades against a blue sky scattered with clouds. (Chronique locale 1893)

Among these images was the first human portrait; the Lumières had finally resolved a depiction of a young girl. These images were compared to the established medium of painting, as digital photographs or inkjet prints are often today. The Lippmann photographs were described as having "[...] the effects of vivid watercolours" and of "luminous watercolours," soft gentle colours associated with roses, or a young girl's skin, were the ideal in nature and in artifice (Vidal 1893, 194–97). The Lumières produced a display set of Lippmann photographs, which they toured in London and the United States; these accompanied their presentations and probably those of Lippmann after 1892. The images were intended to elevate the medium so that it might be appreciated in the realm of French society, where it was necessary to maintain a highly aesthetic state of appeal. Their scientific elegance did not suffice for the widespread acclaim Lippmann sought. Married to Laurence Cherbuliez (1857–1934), the daughter of the French novelist Victor Cherbuliez, Lippmann mixed in a "milieu de haute culture et de haut savoir." According to his colleague Daniel Berthelot, Lippmann's society included men of literature as well as scientists "les plus fameux: Ernest Renan, Hippolyte Taine, Paul Dubois, Gaston Paris [...]" (Berthelot 1921, 2). Lippmann posed for the "legendary" photographer Nadar (Gaspard-Felix Tournachon, 1820–1910) and invited the Nadars to his home to dine at 10, rue de l'Eperon (Berthelot 1921, 2).[4]

The American Frederic Ives Relocates to London

In 1892, Frederic Ives, the American inventor of colour photography and subsequent critic of Lippmann's process, arrived in London. Born into a New England, self-reliant puritan farming community, Ives worked with a religious dedication and a commitment to continue in his chosen vocation in the

4 Nadar, L. to Madame Lippmann, MS Bibliothèque Nationale Richelieu. cote (reference) number 24276. Item number 415. 1899.

face of any adversity (Roberts-Miller 1999, 9).[5] He came to exhibit at The Royal Institution, invited by Cameron Swan, son of Sir Joseph Wilson Swan (1828–1914), inventor of a light bulb and a photographic dry plate. Ives also sought to promote and market his colour photographic process, which he called *composite heliochromy* (Ives 1928, 39). In 1893, an interviewer asked: "Why do you call your process of colour photography by such a long name?" Ives replied:

> Because, popularly speaking, it is not colour photography at all, and I do not wish to mislead. Colour photography, in the sense of reproducing directly the colours of nature in a camera, I believe to be an impossibility; and certainly no such picture has ever been produced. (*Photographic News* 1893, 555)

Ives states clearly his own belief in the impossibility of such a colour photographic process, such as Lippmann's, ever existing. Intent on manufacturing a tabletop instrument and its subsequent stereo update, the Kromskop, in Europe, Ives took up residence in London in 1894. He brought with him his wife Mary, his son Herbert, whom Ives sent to the elite school Harrow, plus a young assistant, Bill Jennings. It was Jennings whose defining remark on the Kromskop, "That name killed it,"[6] summed up their investment of time and struggle with this particular product. For promoters of invention like Ives, as well as potential buyers, London was "[...] a social summit [...] accessible to worldwide ambition" (Crook 1999, 195). London provided, through its many societies with royal patronage, arenas for display and elite social events that did not exist for Ives in the United States. Such high-society occasions were requisite for the attention of the press and public and Britain had the most permissive commercial regulations in Europe; sterling was the international currency. Ives sold both his instruments and sets of his accompanying photographs through the London private gallery system.[7]

This robust spirit was evident in Ives's public demonstrations. They took the form of projected lantern slides with aphorisms intended to motivate

5 The exact quote reads: "the insistence on one's religious duty to pursue one's worldly calling."

6 W. N. Jennings (Coast-to-Coast Photo-Service) to Friend Fred, November 25, 1930, Collection of F. & H. Ives, Manuscript Division of the Library of Congress. Jennings recalls the events and clients of London 1894: "[...] Herr K----- The morning all Germany blew in the shop–Fur lined coat to his feet [...] Looks long and earnestly in to Kromskop (that name killed it)–"

7 Mary Olmstead Ives, to (an unidentified) Dear Friend, January 22, 1897, copy of a letter, from the Collection of F. & H. Ives, Manuscript Division of the Library of Congress. "Mr Ives invention has been on exhibition at the St Georges Gallery the past month, and has received a great deal of attention. The first order came from N [W?] H. Kendall the actor [...]"

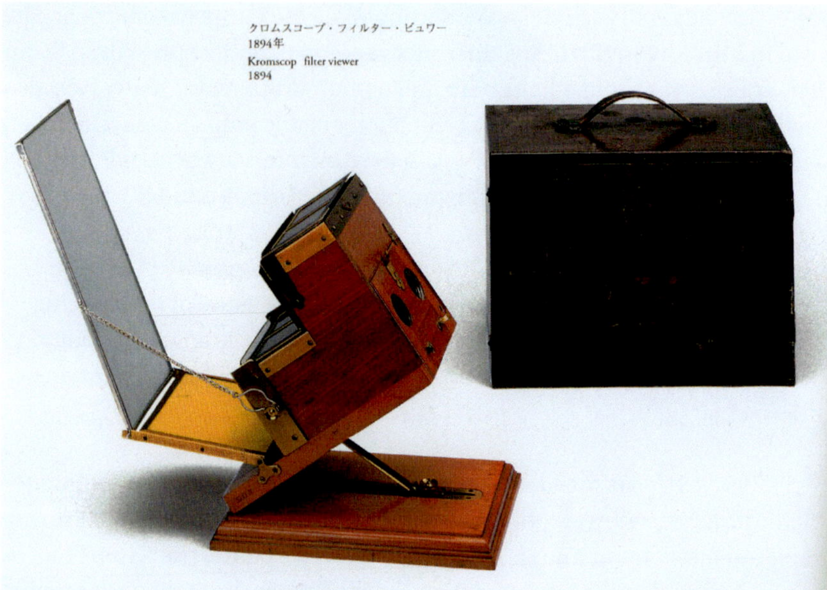

Fig. 18. Kromskop, stereo viewer, 1894, by Frederic Ives, in the collection of the Tokyo Metropolitan Museum of Photography. Sold with images of vases and masks.

a "mind-and-body unison" as the means to commercial success: "Mental Sunshine Creates Physical Health: I Am Optimistic in Thought and Constructive in Action: Day-by-Day; I am Renewing and Vitalizing Every Organ of My Body."[8] In this direct manner, Ives did not separate his personality or physical identity from his mechanical inventions in promoting his ideas to others. In his own photography and instruments, Ives relied on superimposing three images optically over each other with prisms, rendered with elegant mechanics. Any inspection of the Kromskop would reveal Ives's technique as a visually transparent one. However, it was this apparently mundane and pragmatic approach to creating an image that was disappointing to the public and press in the late nineteenth century, as is evident in *The Photographic News* report on Ives winning a gold medal for his composite heliochromy in 1893 (*Photographic News* 1893, 13):

> Mr. Ives' process is not the kind of photography for which the world has been looking, and may never find, nor is it "photography in natural colours" [...] In a technical and scientific sense "natural colours" are those which

8 Frederic E. Ives, Undated Lantern Slides (1992.c047.024: Photography Collection of the Smithsonian Museum of American History).

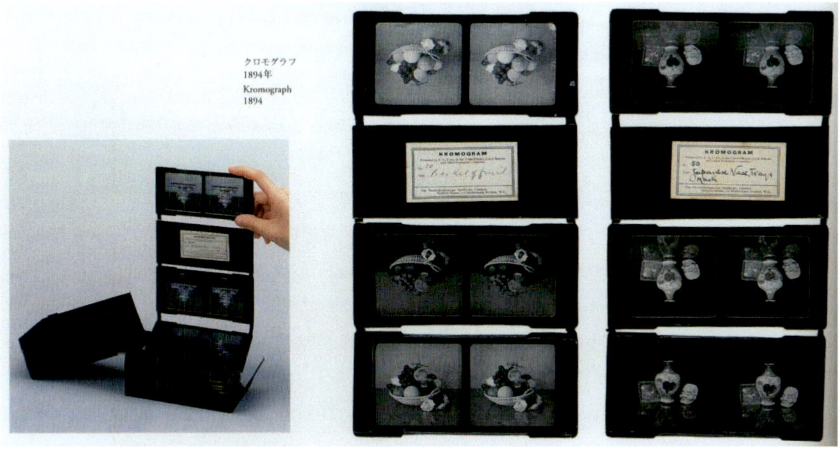

Fig. 19. *Kromographs* by F. Ives, three linked sets of stereo pairs, taken through three different colour filters. When slotted into the *Kromskop*, a system of prisms combined the three images. In the collection of the Tokyo Metropolitan Museum of Photography.

> are produced in any substance by the direct effect of light itself, acting according to the laws of nature.

Ives had succeeded without theory, rather with demonstration, basing his inventions on a practical and craft understanding of geometrical optics. For example, Ives was to later admit unapologetically his ignorance of Maxwell's colour theories from which – to professionals in the field – his own process was seen to have been derived (Ives 1928, 57):

> Unknown to me, Prof. James Clerk-Maxwell had proposed a method of colour reproduction by making three negatives to represent the "primary" colours, and projecting positives from these three through three magic lanterns, each by light of the colour which it recorded.

Unfortunately for Ives, he was to find himself framed as the typical "Yankee" inventor by the French-speaking photographic community. Accompanying the presentation of the Lumières' projected display in Geneva, the *Revue de Photographie* had republished and translated into French a letter to the editor that cast Ives as the antagonist, while claiming the letter to have been first reproduced in the *Photographic News*, June 3, 1892 (although no issue with this date appears to exist) (Ives 1893, 191–94).

The *Revue de Photographie* took the licence to reprint the corpus of favourable quotations by Ives, extracted from the English press on Lippmann photography. To achieve this, the *Revue de Photographie* gave Ives's criticism a defined theory that he never possessed, achieved by employing a bricolage of keenly quoted phrases edited from comments and interviews with Ives, found in more than just one issue of the *Photographic News*. This list produced a miscomprehension of Lippmann's theory. Both these photography journals are typical of the press in this period, which Walter Benjamin described as having a peculiar lack of authority, because at any moment, "[...] increasing numbers of readers became writers" (Benjamin 1968, 234). Unwittingly, Ives had damned Lippmann's work before having seen it himself: "A scientific friend of mine who has been to Paris to see it says the results are pure humbug. The colours got by it are due to interference of light by thin films on plates backed with mercury" (Mr. Ives and Photochromy 1893, 554). Ives refers here to the visual phenomena seen on the surfaces of blown soap bubbles, or in Newton's rings between thin sheets of glass, both popular nineteenth-century educational experiments. These were typical exhibits in the London display arena Ives operated in, as well as images Lippmann himself alluded to in describing his process: "The colours reflected by the film are due to interference: they are of the same kind as those reflected by soap bubbles or by Newton's rings" (Lippmann 1896, 12). To Ives, it seemed that Lippmann had concocted a theory to claim authorship over familiar and naturally occurring events. Wave theory was humbug to Ives. It was outside the pragmatic understanding of normative photography. The rules of this photographic practice and its subsequent expectations were those that had already gained acceptance by Ives and his associates. As historian of photography Ariane Isler-de Jongh has pointed out, individual inventors, such as Ives, pioneering colour photographic techniques were hampered by a lack of communication and foreign language skills (Isler-de Jongh 1994, 115). Whereas educated professionals, such as Lippmann who was fluent in French, English, and German, could rely on the conduits of communication established for their disciplines.

This literary exercise in the *Revue de Photographie*, was perhaps, for the French-speaking press, a strategy designed to divest themselves of responsibility for any technical misunderstanding of Lippmann's process. They also achieved this by adopting a connoisseurship of aesthetic appreciation for the Lumières' imagery. For the non-scientific audience, Lippmann could not rely on his rhetoric outside his peer group or the terminology of physics or the eloquence of the French language in explanations to the English. English science was not presented so poetically; to the French, language was perceived as the main instrument to its cultural dissemination (Keiger 2001, 14). To the French – as the friend of Lippmann and realist author on religion and science Ernest Renan saw it – English Protestant science was inelegant. Renan wrote that it was, "so lacking in loftiness, in philosophy" (Renan 1891, 16) and under a state church, it was, "[a] kind of petty process to knock a little bit of understanding into folk" (Renan 1891, 17). For the English audience, Lippmann attempted to provide a visual demonstration of the event of interference. In 1897, Thomas Bolas, the English photographic inventor and fellow of the Chemical Society, reported with some scepticism on such a model demonstrated to a group of photographers at the Royal Photographic Society:

> [...] a pretty illustration of the formation of so-called standing waves by reflection is obtained if a rope as thick as ordinary clothes-line and some thirty feet long [...] is fastened by one end to a nail and the other end is held in the hand and set in motion so as to produce waves [...] (Bolas 1900, 9)

This model failed to provide these practitioners with a convincing demonstration of what may be occurring within the emulsion. Models were part of the descriptive process and these also needed to match their audience's expectations. This was coupled with the disparaging view of the French towards the English photography community. French art critic Hippolyte Taine, an acquaintance of Lippmann, had previously described English photographers as "[...] unable to speak their own language correctly, and at best figures of fun" (Taine 1957, 208). However, Frederic Ives was to point out that this successful display by the Lumières was too late; the reception of Lippmann's invention was already tainted by the initial inadequacy of Lippmann's pictorial images when seen outside of the Académie des Sciences (Ives 1893, 795–96):

> [...] when it was announced this year [1893] that the brothers Lumière had succeeded in so far obtaining really satisfactory colour photographs of natural landscapes, people who have seen the photographs for which such extravagant and inaccurate claims were made a year before were naturally and very justly sceptical.

To experience the invention's attributes, the Lippmann photograph required witnessing first-hand under controlled conditions, which audiences outside of university or private laboratories had little access to. Neither Lippmann's luminescent full-colour recording by "interference," nor Frederic Ives's stereo colour invention were reproducible in print. These inventions relied heavily on written reports – descriptions of the subjective viewing experience – in popular and professional journals. Ives defended his position. He saw himself as intentionally picked on merely, "[...] because I am so inoffensive and not a member of a close corporation like the French savants" (Mr. Ives and Photochromy 1893, 554). Judgements from an outsider were deemed to be of less importance than those from within the same community. This form of peer protection worked for both Lippmann and Ives, who represented two extreme social and disciplinary groups that had approached the same problem.

Herbert Ives Modernises Lippmann's Process

In 1908, the year Lippmann was to receive his Nobel Prize, Herbert Ives published a timely review of Lippmann's theory, "An Experimental Study of the Lippmann Color Photograph," in the *Astrophysical Journal*. Herbert Ives had undertaken this research for his PhD at Johns Hopkins University, an institution considered at the time to have pioneered the concept of a research university (Thomson 1936, 169). That Herbert Ives was a PhD student was no doubt instrumental to the paper having a single author, at a time when sole authorship was becoming rarer and constituted just over 10 per cent of articles (De Solla Price 1963, 87).[9]

The supervisor of Herbert Ives was Robert Wood, a spectroscopist and authority on physical optics, who also invented a colour photographic technique. Wood shared some of Frederic Ives's deeply held scepticism towards French scientific methods. Wood had denounced the French physicist and fellow academician René Blondlot's observations of Blondlot's so-called N-Rays, in an article in *Nature* published in 1904 (Wood 1904, 530).[10] Like the Lippmann photograph, this was an example of the disputes occurring in the late nineteenth century between the demonstration and theory of invisible radiations. Apparently, only French experimenters trained by Blondlot could repeat his published experiment of 1903, whereas the Americans, British, and Germans had failed (Johnston 2001, 59). For both Wood and Herbert Ives, attempting to recreate the excellence of the interference photographs attributed to Lippmann or proving the Lippmann technique to be false were two different research outcomes that would have appealed to them. This research pursuit was facilitated by the theoretical assistance from Wood and

9 Citing unpublished research by Badash, L. at Yale University: "Data from *Chemical Abstracts* show that in 1900 more than 10 percent of papers have a single author almost the rest were pairs."

10 See also Malcolm Ashmore, 1993, "Theatre of the Blind," *Social Studies of Science*, 23: 61–106; F. Sean Johnston, *A History of Light and Colour Measurement: Science in the Shadows* (Bristol and Philadelphia: Institute of Physics Publishing, 2001), 59–60.

practical help in equipment building from Frederic Ives. On completing his studies that year, Herbert Ives gained a qualification, which was by 1910, to become the standard credential for an American research scientist seeking corporate employment (Kohler 1996, 103).

The paper by Herbert Ives translated Lippmann's original publication into the plainer analytical language of the modern American physicist. The significance of this paper, for the French, is indicated by Herbert Ives winning a medal awarded by the Paris Académie des Sciences. Herbert Ives began his career in the modern spirit of international exchange and cooperation, an aim of the new Nobel Prize. Herbert's findings were to support Lippmann's process as a viable one. Herbert Ives set out to approach the theoretical and practical conditions Lippmann demanded, with the intention of discovering the difficulties posed for others in replicating this process. He first analysed the published results obtained by those few who had worked with Lippmann's process at the time of its original publication:

> Good results have been obtained by the process as worked by these and other experimenters, (Wiener, Neuhaus, Valenta, Lehmann) but its difficulties have been found so great as to prevent its wide use. Some discrepancies with the theory have been found, and compromises with the best conditions as indicated by theory have been found necessary in practice. (Ives 1908, 325–26)

Herbert, like earlier experimenters, sought the archaeology, the historical record of this physical process by inspecting the structure of the "laminae" formed within the emulsion under a microscope. He confirmed the complexity of many laminae required to replicate the colour white and the difficulty in recording it successfully. However, where Lippmann had described the events in terms of those found in *nature*, Herbert Ives modelled the mixing of colour on concepts established within the discipline of physics. Ives now perceived the events within the emulsion as being similar to "[…] the interference fringes that would occur in a Michelson interferometer" (Ives 1908, 325–26). In his paper, he reproduces illustrations from Albert A. Michelson's two papers to describe these events.[11] Thus, he introduced another historical standpoint, one of interferometry, into the review. With this change of description, Herbert Ives retains Lippmann's concept within the discipline of physics as opposed to photography.

11 See A. A. Michelson, "On the Application of Interference Methods to Spectroscopic Measurement," Phil. Mag., 5, no. 31 (1891). A. A. Michelson, "On the Application of Interference Methods to Spectroscopic Measurement," Phil. Mag. 5, no. 34 (1893).

Herbert Ives also confirmed the patient labours and care required to produce these photographs (Ives 1908, 345):

> A larger percentage of failures is to be expected in any process of colour photography than with black and white, since the eye is more sensitive to errors of treatment where colour occurs. The sensitiveness of the Lippmann process to slight deviations from correct conditions is, however, much greater than the three-colour methods, and good results come only from repeated patient trials. When obtained, they are extremely dependent on correct viewing conditions to appear to any advantage.

The less efficient the image is in terms of brightness, the stronger the reconstruction light will need to be to see any results, and then this will only be visible at the correct viewing angle:

> It is of extreme importance that they be observed by parallel light and shielded from all side light. The best conditions are given by a small opening in a wall facing a brilliant white sky. If the observer stands with their back to the opening and holds the picture at arm's length, reflecting the sky, the image appears at its best.

Herbert Ives confirms the reliance of the viewer on specific viewing conditions; according to his paper, he has to rely on sunlight – whereas the Lumière brothers had developed the Megascope. Although this paper analyses the formation of the image, Herbert Ives also discusses the difficulties he experienced in reproducing the material substrate: the glass and emulsion onto which the photograph was recorded. He describes his attempts to duplicate the material quality of the Lippmann emulsion from supplies he sourced from local American stores:

> [...] the gelatin used was Elmer & Amend's "Gold Label", Nelson's "No.1", or a department store gelatin recommended as the best for puddings, etc., which was found very hard and free from grease. The emulsion was flowed on pieces of crystal plate glass cut three by three inches. (Ives 1908, 326–327).

Herbert Ives, and his supervisor Wood, might have assumed that the facilities at Johns Hopkins University would have been comparable, or possibly superior, to those of Lippmann's laboratory at the Sorbonne University, Paris. Wood and Herbert Ives would have been unaware of the intervention of the Lumière brothers and their factory resources in producing emulsion-coated plates for Lippmann. Evidence in the form of Lippmann photographs taken by Herbert Ives of the university courtyard from a window, now in the collection of George Eastman House, indicate that Herbert Ives failed to achieve any photographs comparable to those now credited to Lippmann, although he did achieve excellent results isolating and recording single colours using various light sources of differing wavelengths. The poorer pictorial results by Herbert Ives are evidence of the superior ability of the Lumière-produced Lippmann emulsion to record all the colours evenly in one exposure and the peculiar success of the Lippmann-Lumière collaboration. Now, in 1908, as Lippmann photography was again brought to the attention of the scientific community through Lippmann receiving the Nobel Prize in Physics, Herbert Ives's paper exposed a new audience of American astrophysicists to this medium. Yet, simultaneously, his publication inadvertently concealed the early Lippmann-Lumière collaboration, through the implication of Ives's complete command and scrutiny of the process. This overlooked aspect of Lippmann's success – the collaboration with the Lumière brothers – was both hidden and carried forward again, with the "modernising" review by Herbert Ives.

A New Use and Discipline for Lippmann Photography

In his pursuit of Lippmann photography, Ives applied the process to the emerging sub-discipline of photometry, a vocational field pioneered by Johns Hopkins University to improve electrical tuition (Johnston 2001, 237–54).[12] Ives published a journal paper for the newly formed Optical Society of America, a group that initially started at Eastman Kodak in 1916, and one that would include photometry in its journal (Johnston 2001, 87–88). This is one example of a discipline that had grown, from what historian and scientist Derek De Solla Price described in 1963 as the formation of "invisible colleges" (De Solla Price 1963, 83). These were newly formed collectives of practitioners who were restricted to a group able to sustain a close correspondence. The larger the original scientific discipline, the more likely subgroups were to appear and these, De Solla Price states, can award their own "status pay-offs" (De Solla Price 1963, 85). Early on in his career, Herbert Ives found a supportive arena and status within such a group.

In his paper, Herbert Ives suggested that recording isolated colours or wavelengths with the Lippmann process was more applicable to

12 For a discussion of photometry as a peripheral science, see also the chapter "An Undisciplined Science," 237–54.

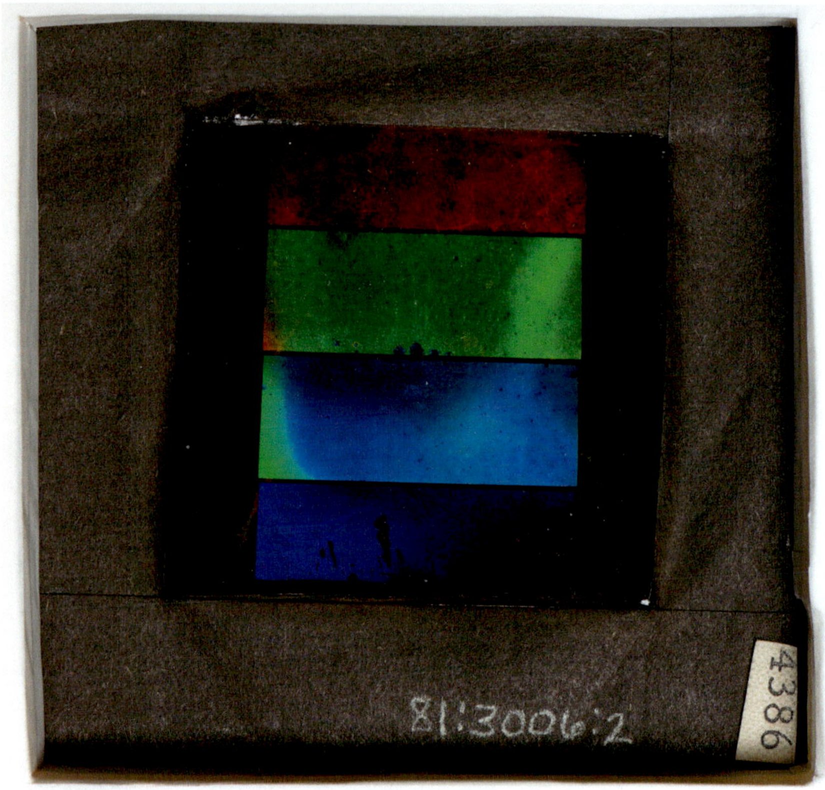

Fig. 20. As a PhD student, in 1906 Herbert Ives succeeded in recording isolated colour or wavelength with the Lippmann process. Courtesy of the George Eastman House Museum, Rochester, New York.

the interests of modern photometric research science than producing representational photographs. Ives reaffirms that Lippmann's process belonged more to the realm of physics and less to photography. He also speculated on recording monochromatic light sources to use as filters, which could then be applied tools within spectral analysis, employing a Lippmann photograph in stellar spectroscopy, for example, to isolate a specific waveband rather than disperse out the full spectrum (in the manner of a Lippmann solar spectrum). From what might have been considered a failure – the inability to produce full-colour pictorial images – Ives, in producing isolated recordings of differing wavelengths, retrieved new possibilities from his review (Ives 1917). In principle, if Ives had introduced a "subject," a speck of dust or a two-dimensional slide, into a monochromatic recording, it might have appeared (under a microscope) three-dimensional – in the 1940s, Dennis Gabor had only used a monochromatic source for his early experiments with what he named as the *hologram* (Gabor 1949, 54).

Conclusion

When the Lippmann photograph and Frederic Ives's Kromskop appeared together on display at the Royal Society in London in 1896, the attention of the press and public had already shifted to another new photographic process: the X-ray. On presenting his process to the Royal Society, Lippmann was asked by Lord Rayleigh if this would "[...] combine with the new Röntgen X-light" (*Nature* 1896, 12–13). What, to a photographer, might be perceived as an inherent defect, its changeable colour when illuminated at differing angles could be used to bring invisible X-rays into the visible spectrum. The brilliant colour of well-differentiated wavelength was another attribute to be employed by science, as demonstrated by Herbert Ives's paper, but considered a problem to the photographer. As scientific disciplines, such as microscopy, increasingly required imaging in parts of the spectrum not visible to the human eye, this problem of changing the wavelength continued into the mid-twentieth century. Dennis Gabor's 1949 proposal to bring an image produced in the electron microscope, from a part of the electromagnetic spectrum, into the visible was one solution. The three-dimensional hologram became its popular incarnation. However, neither Lippmann's nor Frederic Ives's inventions were to survive as solutions to pictorial colour photography at the end of the nineteenth century. Photography became increasingly absorbed into academic subdisciplines and the sites of publication increasingly professionalised. The endeavours of independent inventors, such as Frederic Ives, which had previously relied on photographic societies and journals for the dissemination of their products, were harder to sustain; these sites lacked the credibility given to academic science publications. A Lippmann emulsion became a name given to generic photographic emulsions highly sensitive to the entire visible spectrum, probably in limited use for high-resolution scientific imaging: spectroscopy and astronomy. In 1933, this was still in use at the Institute du Radium, Paris, and historian of science Peter Galison cited Marietta Blau, an 'emulsion physicist', as remarking that: "[...] Lippmann plates, for example, often failed to register a single hit by an alpha" (Galison 1997, 151).[13] But the name Lippmann was eventually dropped from these plates

13 See also Peter L. Galison, "Marietta Blau: Between Nazis and Nuclei," *Physics Today* 50, no. 11 (1997).

in favour of manufacturers' trade names or numbers. Both these inventions were absorbed into larger corporate products: Frederic Ives as an independent inventor happily sold patents to Kodak Eastman, and Kodak was to produce a Lippmann emulsion-coated plate. Although the Nobel Prize did not have in 1908 the prestige and fame it later gained, its award signified the reception and credibility attained within the scientific community. As historian of science Derek J. De Solla Price pointed out in 1963: "In fields tending to honour their pioneers by eponymic fame––name laws, name constants, name species [...] their chance of citation increases" (De Solla Price 1963, 81).

Acknowledgement

This research was originally funded by the Arts and Humanities Research Board, UK.

Works Cited

"Mr Ives and Photochromy." *The Photographic News*, (1893), September 1, 555.

Benjamin, Walter. 1968. " The Work of Art in the Age of Mechanical Reproduction." *In Illuminations*, edited by Hannah Arendt, 217–252. New York: Harcourt, Brace & World, Inc.

Berthelot, Daniel. 1921. *Funeral of Gabriel Lippmann*. Paris: Institute de France Academie de Sciences.

Bolas, Thomas. 1900. "Historical Development of Heliochromy and General Survey of Processes." In *A Handbook of Photography in Colours*, edited by Thomas Bolas, Alexander A.K. Tallent, and Edgar Senior, 1–87. London: Marion & Co.

Chronique locale. "Photographie En Couleurs." *Journal de Genève*, 24 August 1893.

Crook, J. Mordaunt 1937–. 2002.1999. *The rise of the Nouveaux Riches: Style and Status in Victorian and Edwardian architecture*. London: John Murray.

Eder, Josef Maria. 1945. *History of Photography*, New York: Dover.

Johnston, Sean F. 2001. *A History of Light and Colour Measurement: Science in the Shadows*. Bristol and Philadelphia: Institute of Physics Publishing.

F.E. Ives, "Photographie Des Couleurs," Revue de Photographie, Organe Mensuel de la Société Genovoise de Photographie [reprinted letter from Photographic-News] (2 June 1892? not found by author)(1893): 191–194.

Fox, Robert, and Anna Guagnini. 1999. *Laboratories, Workshops, and Sites: Concepts and Practices of Research in Industrial Europe, 1800–1914*. Berkeley: Office for History of Science and Technology, University of California.

Frederic E. Ives, *Undated Lantern Slides* (1992.c047.024: Photography Collection of the Smithsonian Museum of American History).

Gabor, D. 1949. "Microscopy by Reconstructed Wave-Fronts". *Proceedings of the Royal Society of London. Series A, Mathematical and Physical Sciences.* 197 (1051): 454–487.

Galison, Peter Louis. 2005. *Image and Logic: a Material Culture of Microphysics.* Chicago: University of Chicago Press.

H. Ives. 1917. "Lippmann Color Photographs as Sources of Monochromatic Light in Photometry and Optical Pyrometry." *Journal of the Optical Society of America* 1, March-May.

Isler-de Jongh, Ariane: The Origins of Colour Photography. Scientific, Technical and Artistic Interactions, in: *History of Photography*, Vol.18 1994, No.2,111–119.

Ives, Frederic Eugene. 1934. *The Autobiography of an Amateur Inventor.* Philadelphia: Priv. Print: Innes & Sons.

Ives, Herbert E. 1893. "The Lumière-Lippmann Colour Photography." *The Photographic News London*, 15 December 1893, 795–796.

Ives, Herbert E. 1908. "An Experimental Study of the Lippmann colour Photograph." *The Astrophysical Journal* 27. 325–352.

J.J. Thomson, Sir. 1936. *Recollections and Reflections.* London: G. Bell and Sons, Ltd.

Keiger, John F. V. 2001. *France and the world since 1870.* London: Arnold.

Lebon, Ernest: *Savants du jour: Gabriel Lippmann. Biographie, bibliographie, analytique des écrits*, Paris: Gauthier-Villars. 1911.

Léon Vidal. 1893. "In Response to F. Ives", *Revue de Photographie, organe mensuel de la Société Genovoise de Photographie.* 194–197.

Lipmann, Gabriel. 1896. "On Colour Photography by the Interferential Method", *Proceedings of the Royal Society,* 60. 12.

Lippmann, Gabriel. 1891a. "La Photographie des Couleurs", *Comptes Rendus des Scéances Hebdomadaires de l'Académie des Sciences* 112 (5), 274–275.

Lumière, Auguste, Louis Lumière, Jacques Rittaud-Hutinet, and Yvelise Dentzer. 1995. *Letters.* London: Faber and Faber.

Mary Olmstead Ives, to (an unidentified) Dear Friend, January 22 1897, copy of a letter, from the Collection of F. & H. Ives, Manuscript Division of the Library of Congress.

Nadar, L. to Madame Lippmann, MS Bibliothèque Nationale Richelieu. cote (reference) number 24276. Item number 415. 1899.

Philadelphia Photographic Society Correspondent, "A Gold Medal for Mr. Ives", *The Photographic News*, January 6, 1893, 13.

Price, Derek J. de Solla. 1963. *Little Science, Big Science.* New York: Columbia University Press.

Renan, Joseph Ernest. 1891. *The Future of Science. Ideas of 1848. Translated from the French.* London: Chapman and Hall Ltd.

Robert Kohler. 1996. "The PhD Machine: Building on the Collegiate Base." In *The Scientific Enterprise in America*, edited by Ronald L. Numbers, *Readings from Isis*. Chicago and London: University of Chicago Press.

Roberts-Miller, Patricia. 1999. *Voices in the Wilderness: Public Discourse and the Paradox of Puritan Rhetoric*. Alabama: The University of Alabama Press.

Staff reporter. "Professor Lippmann's Presentation at the Royal Society, April 23." *Nature* 154, no. 1384 (May 1896): 12–13.

Staff Reporter. "Mr Ives and Photochromy". The Photographic News, September 1 (taken from The

Taine, Hippolyte. 1957. *Notes on England*. London: Thames and Hudson.

Valenta, Eduard. 1894. *Die Photographie in Natürlichen Farben unter besonderer Berücksichtigung des Lippman'schen Verfahrens*. Halle: W. Knapp.

W.N. Jennings (Coast-to-Coast Photo-Service) to Friend Fred, 25 November 1930, Collection of F. & H. Ives, Manuscript Division of the Library of Congress.

Wood, R. 1904. "The N-Rays." *Nature* 70: 530–531.

About the Author

Susan Gamble is a visual artist who works in photo-technology, including holography. With her partner, Michael Wenyon, they are UNESCO laureates for their work in technology. Gamble has a BA in fine arts from Goldsmiths College, London and a PhD in the history of science from Cambridge University. She was a fellow at the Center for Advanced Visual Studies at MIT. Wenyon & Gamble have work in the photography collection of the V&A Museum, the National Portrait Gallery in Washington, DC, and the MIT Museum; they have exhibited at the Whitney Museum in New York, the Tokyo Museum of Photography, the Calouste Gulbenkian Museum in Lisbon, and many other venues. Gamble has previously written on this topic.

Part II

Media History, Aesthetics, and Culture

5. Lippmann Photography and the Problems of Media Historiography

Jens Schröter

Abstract
Lippmann photography is the first imaging technology to exploit wave-optical effects. It forces a revision of media-historical models, such as the approach of Jonathan Crary, which is based solely on the emergence of physiological optics around 1820. First, this leads to a differentiated model of the history of optical media. Second, Lippmann photography, through its practical obsolescence and its reappearance in discourses on security technologies at the end of the 1990s, raises fundamental questions about the relationship between media technologies and practices. The same applies to the second development in the field of optical media, which is centrally based on wave optics, namely holography. This paper aims to trace and discuss theoretically the history of knowledge of wave-optical images based on wave optics media and its implications.

Keywords: Jonathan Crary, holography, media historiography, reproducibility, wave optics

> [...] waves of light which would rush through the film at an enormous speed and get away into space without leaving any impression, were stopped by some special kind of film and went surging up and down in confinement – making strata [...] "supairposeetion [*sic*] of strata" [...] no Englishman could move his hands with that smoothness, making you see. "Violet subchloride of silver." [...] When the colour photographs came, Miriam was too happy for thought.
> (Dorothy Richardson 1979, 106–7)[1]

1 Spelling according to the original. In all probability, this passage contains the only mention of the Lippmann method in literature. My gratitude goes to Nicola Glaubitz for the reference to this text.

Hannouch, H. (ed.), *Gabriel Lippmann's Colour Photography: Science, Media, Museums.*
Taylor & Francis Group, 2022
DOI 10.5117/9789463728553_CH05

Introduction

Lippmann photography is an interesting case insofar as its existence poses interesting problems writing media history. I will show this by discussing in detail an important proposal by Jonathan Crary. He constructs the history of vision and visual media along a rupture between geometrical and physiological optics. This is one, and a particularly helpful, construction of the history of visual media (see Schröter 2014). But it has its problems – and one of those is Lippmann photography, which Crary does not even mention. The first part of this article will introduce Crary's model, and the second one will discuss its relation to Lippmann photography, making clear that his "successionist" approach does not work. The third part addresses the point that media do not simply survive or not, but sometimes come back after being forgotten – as is the case with Lippmann photography. The fourth part contains a conclusion.

Problems of Linear Histories I: Crary's Model

Crary's widely read *Techniques of the Observer. On Vision and Modernity in the Nineteenth Century* was published in 1990, effectively promoting the discussions on a history of vision.[2] One of its reviewers called it a "sort of bible of critical research in perception" (Apel 2002, 38, my translation). Another termed it one of the "most powerful studies in the realm of a cultural history of perception from a discourse analytical perspective" (Stiegler 2004, 287, my translation). Not only did it almost immediately play a prominent role in the theories of arts and media; recently, several branches of literary studies also discovered Crary's concepts, using them to approach such different subject matters as Brockes's baroque poetry or the psychological novels by Henry James.[3]

Wherein then lies the importance of *Techniques of the Observer*? On the one hand, Crary has given a new impulse to theories of art since he

2 See Crary (1990). Earlier, Crary had developed some theses in individual essays that were then combined in the book; see Crary (1985, 1988a, 1988b).

3 Some examples would be Hick (1994), one of the earliest German texts in which Crary is mentioned; Hick (1999) also essentially relies on Crary, as does Stiegler (2001). Steigerwald (2000) analyses Brockes's "techniques of the observer," and Brosch (2000) discusses the "*Veränderung der Wahrnehmung im 19. Jahrhundert*" (change of perception in the nineteenth century, my translation, part of the title) with regard to Henry James's writings by more or less projecting Crary's theses onto James while Kosenina (2001) discusses "microscope and (portable) 'Zograscope' as tools and metaphors of poetry" (my translation) by rather loosely connecting back to Crary.

proposed writing a critique of the long tradition of the "grand narratives." On the other hand, he generally offers a new approach by attempting to transcend a history of technically generated images or optical media. He tries to write a history of vision with regard to a history of the observer by connecting elements of art history with those of the history of technology against the background of a rather generalising characterisation of modernity as a capitalist formation. His arguments, which build mainly on Foucault and Deleuze, are as inspiring as they are problematic. Therefore, I will first introduce – but also criticise – his concept, especially with regard to the programmatic introduction to *Techniques of the Observer*, pointing out the perplexing problems of his approach. These problems motivated me to think about an alternative model of media history.

"This is a book about vision and its historical construction" is the opening statement of *Techniques of the Observer*. The first page leaves no doubt that the motivation of this historical study is a most profound media upheaval of the twentieth century. Referring to "computer-generated imagery," Crary lists "[c]omputer-aided design, synthetic holography, flight simulators, computer animation, robotic image recognition, ray tracing, texture mapping, motion control, virtual environment helmets, magnetic resonance imaging, and multispectral sensors"[4] (Crary 1990, 1). He argues that this diffusion of manifold new forms of imagery and practices with new media make the culturally established meanings of the observer and representation obsolete. Crary is using a strong rhetoric of ruptures.

Yet, *Techniques of the Observer* does not really deal with the new pictorial forms that it only mentions in the beginning. It focuses on a rupture in the first third of the nineteenth century, an upheaval that produces "crucial preconditions" (Crary 1990, 3) for today's transformations. This rupture around 1820 can neither be reduced to a historical event in technology nor to certain art historical changes; it separates "Renaissance, or *classical* models of vision and of the observer" from a "modern and heterogeneous regime of vision"[5] (Crary 1990, 3) – a regime in which the body of the observer plays

4 Indeed, it doesn't seem to be a coincidence that *Techniques of the Observer* opens with such a heterogeneous list. It is heterogeneous, because methods of computer graphics ("ray tracing") appear next to technical apparatuses ("flight simulators") and new fields of practice ("computer aided design"). The whole study more or less follows Foucault's *The Order of Things*, which, on its first page, also begins with a heterogeneous enumeration (Foucault 1966, 14), where Foucault refers to Jorge Louis Borges's (fictitious) "Chinese Encyclopaedia." Crary's reference to *The Order of Things* will still have to be discussed regarding its implications.

5 On the term "regime of vision," see Jay (1988).

a central role. It is as important here that the *heterogeneity* of the modern regime of vision is mentioned, as it is that the author talks of classical *models* of vision in the plural. This is pointed out specifically here, because later on, Crary will not mention the *synchronic* plurality and heterogeneity of regimes of vision again. Instead, when continuing, he problematically homogenises both the classical (see Atherton 1997) and the modern regimes of vision.

He assumes a "crucial systemic shift" (Crary 1990, 5) to an *embodied observer* through whom, in the end, both modern painting and the optical media like photography and film should emerge (even though Crary completely leaves out the questions of gender and race).[6] The rupture then is one from a "pervasive suppression of subjectivity in vision in seventeenth- and eighteenth-century thought" to "models of subjective vision" (Crary 1990, 9).

Crary explicitly does not see this disruptive transformation of the status of the observer as one exclusive to the realm of (visual) perception. He considers visual perception a nebulous term, particularly when it is reduced to "changing forms of artworks over time"[7] (Crary 1990, 5). He argues, in contrast, that the observer is a position where vision historically materialises. The observer is seen as the effect of an "irreducibly heterogeneous system of discursive, social, technological, and institutional relations. There is no observing subject prior to this continually shifting field"[8] (Crary 1990, 6). Crary reveals his methodological sources in a footnote to this last sentence: Michel Foucault, whose genealogical or archaeological method can be understood as an analysis of the historical conditions for the formation of subjectivity instead of as a type of historiography that presupposes the subject.[9] This pointed reference to (a specific reading of) Foucault's approach is not without consequences in Crary's methodological design. He particularly refers to Foucault's book *Les mots et les choses* (1966) in which Foucault also argues using strong rhetoric of ruptures. Closely aligning himself to this, Crary diagnoses a "general break or discontinuity at the beginning of the nineteenth century" (Crary 1990, 7).

Since the modern observer supposedly is embodied, Crary characterises this break consequently as the "passage from geometrical optics of

6 On this criticism, see Williams (1995) and Hentschel (2001).

7 This remark is directed against the concept of a "history of vision," as it has been used in traditionally formalistic histories of art (see specifically Wölfflin 1950; see also Crary 1990, 36, critically on "Woelfflinism") where it was seen only as a history of *artistic* images.

8 It should be asked whether the description of the field as "continually shifting" is consistent with assuming *one* regime of modern vision.

9 See Foucault (1984) on the term "genealogy." See also Paech (2003) on the term "dispositive."

the seventeenth and eighteenth centuries to physiological optics, which dominated both scientific and philosophical discussion of vision in the nineteenth century" (Crary 1990, 16). This is the main reason why Crary's book is so significant for the design of this essay: he attempts to deduce the history of optical media – and also of the first three-dimensional image, stereoscopy – from the history of various optics, something that is not done by other authors.[10] Such a deduction seems useful; for example, recognising "physiological optics" allows for plausible conclusions:

> efficiency and rationalization in many areas of human activity depended on information about the capacities of the human eye. [...] Retinal afterimages, peripheral vision, binocular vision, and thresholds of attention all were studied in terms of determining quantifiable norms and parameters.[11] (Crary 1990, 16)

According to Crary, some of the optical tools, which he subsequently discusses more closely, evolved within the frame of these determinations.[12] Examples are stereoscopy as a by-product of analysing binocular vision and cinema as a consequence of the study of the "persistence of vision." Consequently, photography is not the optical technology that embodies most succinctly the regime of modern vision after 1820 – as one could believe – but *stereoscopy* (see Crary 1990, 116). The camera obscura, contrarily, is assigned to the status of the observer in the seventeenth or eighteenth century (Crary 1990, 8).[13] It is hardly surprising then that Crary can in no way accept the

10 See Hick (1999). Somewhat imprecisely, she adds up historical materials that, in the end, lead into the invention of the cinema. (For its criticism, see Geimer 2000). Kittler (2010, 19–46) begins his study with a preface and a discussion of theoretical considerations. In the German text, the word "Optik" (optics) appears only once, but in the English translation of the book it is translated into "visual," removing Kittler's argument even more from the history of optics. Wave optics is mentioned only en passant and with reference to photography, which is in no way essentially connected to it (Kittler 2010, 124). His study is interested in other matters such as the history of optical media as a history of escalation through military influence (Kittler 2010, 41–43; see on this, Stiegler 2003, 70). See also Mann (2000) for a historical overview of other optical instruments, which also does not relate these to a history of optical knowledge.

11 The knowledge emerging in this way for Crary is just as fundamental regarding the shift to modern art (Crary 1990, 9). This directly makes sense for movements like pointillism insofar as, for example, Seurat had based his theories directly on the findings of physiological optics; cf. Schug (1997); see Crary (1999, 152–76).

12 Cf. also Rieger (2001, 2002, 2003); Waldenfels (1999) talks of an "abstractive and normalizing corporeality" (169, my translation) of the modern observer in Crary's book.

13 On excluding the body in the camera obscura paradigm, see Crary (1990, 41). Incidentally, Crary does not equate this paradigm with linear perspective (34, 41), even though he does not

customary narrative[14] according to which photography, officially introduced in 1839, developed as a direct outcome of the earlier camera obscura.

Problems of Linear Histories II: The Emergence of Lippmann Photography

Crary substantiates the alleged transition from geometrical to physiological optics with a passage from Goethe's *Theory of Colours*, published in 1810 (see Crary 1990, 67–71):

> Let a room be made as dark as possible; let there be a circular opening in the window-shutter about three inches in diameter, which may be closed or not at pleasure. The sun being suffered to shine through this on a white surface, let the spectator from some little distance fix his eye on the bright circle thus admitted. The hole being then closed, let him look towards the darkest part of the room; a circular image will now be seen to float before him. (Goethe 2000, 16)

Goethe meticulously measures the following after-images with a clock:

> I looked on the bright circle five seconds, and then, having closed the aperture, saw the coloured visionary circle floating before me. After thirteen seconds it was altogether red; twentynine [*sic*!] seconds next elapsed till the whole was blue [...]. (Goethe 2000, 17)

And so on. Crary argues that this is striking proof for his claim. The opening in a dark room through which the light is falling – indicating the camera obscura dispositive – is suddenly shut by Goethe in order to focus his attention to the emerging after-images of changing colours flowing into each other. The paradigm of the camera obscura ends here, and now the attention

seem to rigorously maintain to this differentiation. See, for example, page 86, where he talks of the "point-to-point epistemological setup" of the camera obscura (which corresponds exactly to linear perspective), or when his criticism of traditional narratives in the history of art and media quietly relocates their reference to perspective (3–4) into a reference to the camera obscura (26). The connection between linear perspective and the camera obscura will be discussed at a later point – here, it should only be mentioned that both belong to the series of geometrical optics. I take the notion of "series" from Foucault (1969).

14 See, for example, Frizot (1998, in particular 17–18).

focuses on the exactly measured physical elements of vision – physiological optics. Quod erat demonstrandum.[15]

But it seems that Crary's invocation of Goethe simplifies too much. A single glance at the table of contents of the *Theory of Colours* (Goethe 2000) shows that the "physiological colours" are only one section. In other chapters, Goethe also turns to chemical and physical processes and the colour effects created by them. As a general overview suggests, for Goethe, several forms of knowledge coincide and overlap[16] so that one cannot really talk of a clear transition to a discourse that centres on the body. There is one passage in particular not quoted by Crary, which can be found in the section "In place of the promised supplementary part" (Goethe 1989, my translation) in the second volume of the *Theory of Colours*. It presents a historical survey of the different theoretical positions on colour. The passage is not by Goethe himself, but by the physicist Thomas Johann Seebeck (1770–1831) with whom Goethe maintained an intense friendship.[17] Goethe introduces Seebeck's contribution of about twenty pages with the following words:

> Even though for the reasons mentioned above we are refraining from reflecting upon all that has happened in the last twenty years in our field, we should not ignore the most important point that Herschel once again reminded us of, and by that we mean the effect of coloured lights on glow stones, metallic oxides and plants – a chapter that we have only sketchily presented in our design but that has to become increasingly more important in chemistry. We cannot fulfil our duty in a better way than by inserting an extensive essay by Dr. *Seebeck* from Jena [...]. (Goethe 1989, 927, my translation)

Even though this addition is obviously not by Goethe himself, he nevertheless agrees with it completely. The passage I am referring to can be found in the subsection "On the chemical activities of light and of coloured illumination" (Goethe 1989, my translation):

> It is one of the most important discoveries of modern times that with the exterior changes of the body through sunlight, which we have known about for a long time, quite often an interior change is also connected to

15 Additional witnesses named by Crary are Maine de Biran and Schopenhauer; see Crary (1990, 72–79).
16 By the way, this is also true for colour in Immanuel Kant whom Crary (1990, 74) opposes to Goethe; see Schröter (2000).
17 On Goethe and Seebeck, see Nielsen (1989, 1990).

> it, a change in the chemical components. [...] One of the most sensitive substances reacting to sunlight is horn silver [i.e., silver chloride (AgCl)]; as is well known it is white when in a precipitate state and turns grey quite quickly and finally black, losing the biggest part of its acidity, if not all of it.[18] (Seebeck in Goethe 1989, 936, my translation)

Obviously, Seebeck experimented with silver halides or, to be more exact, with silver chloride, in other words, one of those substances that, only a few years later, would be used for the development of (black-and-white) photography. It has to do with light effects on chemicals and therefore it cannot be a question of physiological optics. In his experiments with horn silver, Seebeck made an astonishing observation:

> When I let the spectral colours of a fault-free prism fall [...] onto white, still damp horn silver spread onto paper, keeping it in the same place for fifteen to twenty minutes by way of an appropriate apparatus I found the horn silver in the following changed state. It had become reddish-brown in the violet realm (at times more violet, at times more blue) and this colouring even extended beyond the limits of the violet indicated earlier, but it was not stronger than in the violet. In the blue realm of the spectrum the horn silver had become a pure blue and this colour extended into the green, diminishing and becoming lighter. In the yellow the horn silver for the most part seemed almost unchanged, in some cases it seemed somewhat more yellow than before; in the red, however, and in several cases even beyond the red, it had taken on mostly a red of roses or hydrangeas. In some of the prisms (those in which the strongest warming had taken place outside of the red) this reddening fell completely outside of the red realm of the spectrum.[19] (Seebeck in Goethe 1989, 937, my translation)

One has to closely examine what is happening here: Seebeck obtains *colours* even though he has covered the paper with horn silver (silver chloride) only. No colourants and therefore no three-colour processes, as required by colour film as it was used in the twentieth century, are involved (and even today's digital cameras rely on three-colour separation). Goethe's *Theory of Colours* is presented by Crary as evidence for the transition to physiological optics, but

18 It was Johann Heinrich Schulze who made the discovery of the light sensitivity of silver salts in 1727.

19 The last sentence points to infrared rays. Later, Seebeck reports on several variations of this experiment. See also Eder (1978, 153–55).

it contains elements that clearly don't fit into that narrative. This observation was still being made in various other areas in the nineteenth century. In the celebrated talk on August 19, 1839, before the Chamber of Deputies, François Arago referred to Seebeck's observations, to similar findings by Nicéphore Niépce (cf. Niépce 1839, 43–44), as well as to experiments by Sir John F. W. Herschel (cf. Arago 1839, 36). In a paper received by the Royal Society in 1840, Herschel[20] reports on an experiment of July 9, 1839:

> A highly concentrated spectrum was therefore formed, by receiving the rays after emergence from the prism [...] on a large crown lens, and receiving the focus on paper prepared as in Art. 31. The result was equally striking and unexpected. A very intense photographic impression of the spectrum was rapidly formed, which, when withdrawn and viewed in moderate daylight, was found to be coloured with sombre, but unequivocal tints, imitating those of the spectrum itself.[21]

Herschel had also made a table in which he recorded the results (Herschel 1840, 18). Nevertheless, he has to concede, "I have not succeeded in fixing these tints" (Herschel 1840, 19). Also Henry Fox Talbot occasionally observed during his photographic experiments that especially the blue sky would be reproduced in blue.[22] The best examples of these "chemical colour photographs," as Connes (1987, 154) translates the German word *Photochromie* (sometimes also called *heliochromie*), were produced by Becquerel and Niépce de Saint-Victor.[23] Instead of paper soaked with silver chloride, they used chlorinated silver plates (i.e., a similar method as Daguerre's).[24] At least one of Becquerel's chemical colour photographs representing a parrot was shown in Paris at the Exposition Universelle in 1855: it was presented in a dark tent, lit only with candles. "The

20 Herschel, by the way, was the inventor of the term *photography* (March 14, 1839; see also Eder 1978, 259–60) as well as the discoverer of the fixative sodium thiosulfate ($Na_2S_2O_3$; see Eder 1978, 334–36), which is still in use. Herschel was also the first user of the terms "negative" and "positive" (see Herschel 1840, p. 3), in use throughout the photographic process of chemical photography in the twentieth century.

21 Herschel (1840, 18–19). In the "Art. 31" (10) mentioned in the quote, the preparation of a silver chloride solution is described. Again, only silver chloride is used, but no colourants.

22 See Talbot (1851, 627): "Il y a cependant une exception, c'est la couleur du ciel, qui s'est reproduite plusieurs fois dans mes expériences d'un azur très-naturel." (There is however an exception, it is the color of the sky, which was reproduced several times in my experiments and is of a very natural azure, editor translation)

23 The nephew of Nicéphore Niépce and the inventor of the albumen process; see Eder (1978, 339–42).

24 For details of their chemical procedures, see Eder (1978, 665).

most beautiful chemical colour photographs that were created on chlorinated silver plates originated from Niépce de Saint-Victor who exhibited them at the world fairs in Paris in 1862 and 1867."[25] (Eder 1978, 665) But neither Becquerel nor Niépce de Saint-Victor succeeded in fixing the chemical colour photographs. To this day, some safely protected specimens exist. They are stable in a dark environment but very quickly dissolve as soon as they are in contact with light.

These colour phenomena cannot be explained by tracing them back to physiological optics. Initially, there was no explanation for them at all. An attempt at an explanation first appeared in 1868 by a certain Wilhelm Zenker in his *Lehrbuch der Photochromie*:

> The matter looks quite different if the penetrating rays are met by rays of the same kind that issue from it. [...] Then two wave systems of the same kind meet and create a similar effect as one calls "standing waves" in water. [...] Contrary to "advancing waves" the variation of quiescent and maximum points is a peculiarity of the "standing waves" also of water. They emerge at the point at which waves break vertically on bulwarks, walls and similar structures. [...] But how do these chemical effects distribute themselves? We should not expect to find them at the quiescent points. [...] From this combination of the approaching and the reflected rays we obtain small silver points arranged in a system of layers whose reciprocal distance is half a wave length of the respective colour. This means that of all the colours that are contained in white light, the only one that remains is the one whose wave length coincides with the wave length of that light ray that has created the layers of silver points. [...] It seems to me that this is how one can very simply explain how identical colours emerge from the laws of the theory of waves. (Zenker 1868, 117–21, my translation)

Thus, Zenker is attempting to use the concept of *standing light waves*, which are produced during the exposure of Becquerel plates in the emulsion, to explain their colourful appearance. This concept originally comes from acoustics[26] (Connes 1987, 150–51) (and also hydraulics) and was transferred to optics by Wilhelm Zenker. Without the acoustic knowledge of his time, he would not have been able to change optical knowledge. Zenker's "fundamental contribution to wave optics" surfaced for the first time in 1867 "in a most unexpected place and about the most unsuitable problem" (Connes 1987, 151) when he published his idea in a text on *physiological optics* in which

25 See Becquerel (1848a, 1848b) and De Saint Victor (1855).

26 For the comparison of light and sound, see Young (1804, 12).

he attempted to explain the perception of colours by way of standing light waves in the eyeball.[27] This mingling of physiological and wave optics created problems for the propagation of the concept: "Needless to say, physiologists did not take the Zenker theory seriously, while no optician seems to have ever read his paper." (Connes 1987, 151)

The concept of standing light waves was accepted and became widely known only after Otto Wiener successfully demonstrated the standing light waves experimentally in 1889 (see Wiener 1890). In 1895, Wiener also published his essay on "Farbenphotographie durch Körperfarben und mechanische Farbanpassung in der Natur." He posed the fundamental question: "Are the colours emerging using the old procedures [i.e., Seebeck to Becquerel] pseudo-colours or body colours, i.e., do they emerge by way of interference or by absorption?" (Wiener 1895, 228, my translation). Analysing plates of the Becquerel-Niépce de Saint-Victor type and also papers à la Seebeck and Herschel, Wiener was able to find out that Zenker's explanation was only partially true for the first and not at all for the latter. While the former type of plates (Becquerel-Niépce de Saint-Victor) vary the colour slightly with the visual angle, thereby showing *iridescence* (an effect that appears both with Lippmann photography and holography), this is not the case when working with paper.[28] The colours appearing in Seebeck's and Herschel's (and partially in Becquerel's) images are not owed to the formation of the interferences of standing waves. They occur during specific chemical processes in which specific compounds of silver halide (which Carey Lea already had called "photo salts" in 1887)[29] are created in the emulsion by way of light influence. Ultimately, they are again corroded by light, but they reflect the colours contained in white light differently. On the spot where a paper of the Herschel type is exposed to a red light (for example, in a spectrum produced by a prism), only the photo salts remain, reflecting red light best. Therefore, the paper at this spot will appear as red when illuminated with white light. However, white light of course also contains light in other colours; when illuminated for a longer time, the red area will lose its colour due

27 See Zenker: "Thus we can see the elements of the retina as a system of planes on which the arriving light waves break almost vertically and from which they are also thrown back almost vertically. In doing so, standing waves must emerge, that specific form of interferences that is being produced by wave systems meeting each other." (1867, 252, my translation)

28 See Wiener (1895, 246–53, particularly 249) on the "dependence of colours from the entry angle and the necessity to be viewed in mirrored light" (my translation) in Lippmann photography.

29 See Carey Lea (1887, 352). See, for the details on the chemistry of chemical colour photography, Valenta (1894, 1–6). In this chapter's motto from Dorothy Richardson, a "violet subchloride of silver" (my translation) is mentioned – this seems to allude to the photochemical phenomena discussed here. See Eder and Pizzighelli on "violet subchloride of silver" (1881, 40, my translation)

to the chemical decomposition of the photo salt. This is also true for all other colours, and so "it seems clear that the inability to be fixated is fundamentally connected to the way in which these colours are formed since the ability to render colours is connected to their decomposability in light"[30] (Wiener 1895, 233, my translation). Thus, this media historical branch[31] of chemical colour photographs (*Photochromie*) had no future.

Thus, Zenker's application of the wave-optical concept of standing light waves was not the right explanation for Seebeck's colours or Becquerel's chemical colour photographs.[32] However, his attempt to explain them preceded Gabriel Lippmann's method by eighteen years. According to his biographers, Lippmann had already found the exact mathematical formulation of a possible use of standing waves for the reproduction of colour with black-and-white material for photography and had presented it in his lectures in 1886 (without knowledge of Zenker's approach). He began with *wave-optical theory* in an interesting contradistinction to the invention of photography as such or to chemical colour photography. A commentator called it "Mathematics conquer[s] Photography"[33] Connes (1987, 157). Ergo, Lippmann photography in the original sense of the word is a sedimentation of the wave-optical series.[34] But this sedimentation was not a simple process; Lippmann researched it for years (later also together with the Lumière brothers)[35] in order to produce the high-resolution emulsions necessary for realising the concept that had to be extremely transparent and grainless (see Connes 1987, 157). He was finally able to introduce his new procedure to the French Académie des sciences on February 2, 1891 (see Lippmann 1891). William R. Alschuler and Jens Gold in this volume chart this process's history, the international discussions about it, and its materiality.

30 See Carey Lea (1887, 349). See Geimer (2002, 320): "The appearance and the disappearance of photographic images have always been located close to each other."

31 In a current standard work on the history of photography, chemical colour photography only appears marginally; see Frizot (1998, 412–13).

32 Ives (1908, 325) is of a different opinion.

33 Therefore, Lippmann photography can be excellently represented in computer simulation; see Nareid and Pedersen (1991).

34 As if he wanted to underline Lippmann's distance from Crary, who centres on the series of physiological optics, Connes (1987, 158) remarks: "The vagaries of human vision [...] are not even considered."

35 The Lumière brothers "introduced very fine-grained silver-halide gelatin [sic!] emulsions [which] became the main recording material for Lippmann photography" (Bjelkhagen 1999a, 275); see Lumière and Lumière (1893). In 1893, they incidentally also created the first portrait on the basis of the Lippmann method. See Lehmann (1907, 1908) on later, faster emulsions.

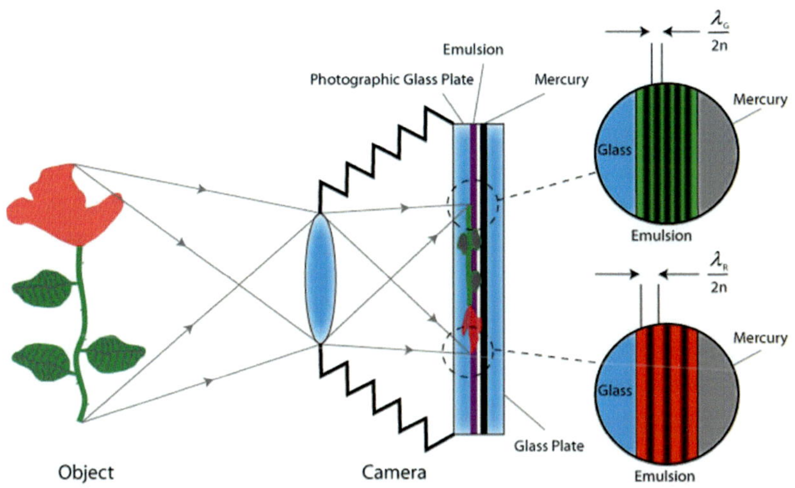

Fig. 21. Diagram of Lippmann photography, in Bjelkhagen (1999b, 56).

It works like this: The light of the object (in the diagram of a rose) is focused onto the photographic plate by means of a lens. Up to this point, this is a completely normal linear perspectival projection, in other words, it is the series of geometrical optics (that has in no way disappeared). Lippmann himself says:

> One can fixate a camera obscura image with its shapes and colours by using a permanent transparent light-sensitive coating of sufficient thickness, which during the exposure time is placed on a reflecting surface that should preferably be made of a layer of mercury. (Lippmann 1894, 97, my translation)[36]

As can be seen from the quote, the photo plate is not an ordinary one. Behind the glass plate is a thin layer[37] of special emulsions behind which

36 The original Lippmann quote in French: "On peut fixer l'image de la *chambre noire*, avec son modelé et ses couleurs, en employant une couche sensible transparente et continue, d'epaisseur suffisamment grande, adossée pendant la pose à une surface réfléchissante qu'il est commode de constituer par une couche de mercure." (Lippmann 1894, 97) See Becquerel (1848a): "De l'image photochromatique du spectre solaire, et des images colourées obtenues dans la *chambre obscure*" (483, my emphasis). This last quotation shows the ongoing usage of the *chambre obscure*.
37 The thickness of the emulsion coating depends on whether one wants to record only monochromatic (as in a sun spectrum) or polychromatic light (as in a normal scene); see Ives (1908) and Bjelkhagen, Jeong, and Ro (1998, 76).

is another strongly reflecting layer. Since silver could not be used because of possible chemical reactions, Lippmann was using a layer of mercury. Once the image of the object is projected onto the plate, the injected light is mirrored on the surface of the mercury. The injected and the reflected light interfere so that they form standing waves in the emulsion.[38] Their maxima can each be found midway on the wavelength of light – from this stems the term $\lambda/2n$, specified in illustration 21, designating λ as the wavelength of the respective light and n as the refractive index of the emulsion. At the maximum points, ultra-thin silver lamellae form within the emulsion, since the light corrodes the silver halide contained in the emulsion into silver and the halide ions absorbed by the emulsion, as in a normal black-and-white photograph.[39] When white light is directed onto the plate at a certain angle after the developing process, the original image appears in colour, since the lamellae void all colours contained in the white light that do not correspond to the original colour via destructive interference.[40] As an answer to various criticisms, Lippmann published a group of integrals in 1894 clearly describing what kind of selective reflection is being created by what frequency within the emulsion. For the extreme mathematical elegance of his method – if for nothing else – Lippmann was awarded the Nobel Prize in Physics for his invention in 1908.

Lippmann photography is a sedimentation of the wave-optical series (regarding the colour) connected to the geometrical optical series (regarding the geometrical projection of the scene).[41] According to his mathematical description of the process, which he based on Fourier, this is the truest process for representing colour physically.[42] (Cf. Nareid 1988)

38 For the development of the wave-optical concept of interference in the first half of the nineteenth century, see Buchwald (1989). It is interesting that many scholars who contributed to the research of the different photographic methods have done intensive work on wave-optical questions.

39 But Lippmann did not make his first photographs on the basis of silver halides but on that of bichromate gelatin; see Valenta (1894, 76).

40 Valenta (1894, 30–36) and Denisyuk (1984, 36–49) provide descriptive explanations. Also, by the way, the intense colours that one can see on the surface of soap bubbles emerge in a similar way; see Rood (1880, 47–50).

41 See Phillips, Heyworth, and Hare (1984, 168) who talk of a "complex mixture of geometrical and physical optics" (do not confuse "physical" with "physiological," read as wave optics). See Fournier and Burnett (1994, 508). Holograms do not resort to lenses to geometrically project a scene. Therefore, they are purely wave-optical images (regarding the laser light used, they are wave-quantum-optical hybrids).

42 In passing, Heidegger remarks in his essay "The Origin of the Work of Art": "Colour shines and wants only to shine. If we try to make it comprehensible by analysing it into numbers of oscillations it is gone. It shows itself only when it remains undisclosed and unexplained." (1993,

$$X = \frac{1}{Z}\int_A^B \int_0^Z f(\lambda)\,\varphi\left(\sin^2\frac{2\pi z}{\lambda}\right)\cos\frac{2\pi z}{\lambda'}\,d\lambda\,dz,$$

$$Y = \frac{1}{Z}\int_A^B \int_0^Z f(\lambda)\,\varphi\left(\sin^2\frac{2\pi z}{\lambda}\right)\sin\frac{2\pi z}{\lambda'}\,d\lambda\,dz.$$

Fig. 22. Lippmann's Fourier transform/integral, in Lippmann (1894, p. 104).

Despite the peerless wave-optical elegance, a "Revenge of Physiology" (Connes 1987, 163) soon took place, since Lippmann photographs did not succeed in the market. The high resolution of the almost grainless emulsions requires long exposure times that do not permit snapshots.

Every Lippmann photograph is unique and cannot be reproduced, since every reproduction or colour copy results in the disappearance of its peculiar effect of *iridescence*, as I mentioned above. On the mirroring surface, the image can only be seen from one specific angle and when seen from another converts into its complementary colours. Apparently, the relationship of the image as plane and the viewer is being reconfigured.[43] This iridescence also means that such an image can only be projected with difficulty[44] and can therefore only be seen by a few people at a time. In 1907, even before Lippmann was awarded the Nobel Prize in 1908, the first autochrome plates of the Lumière brothers, who had in the meantime discarded the commercially insecure Lippmann method, appeared commercially with sweeping success. The autochrome plate was dusted with microscopic grains of potato starch that had been dyed red, green, and blue and then covered with a thin layer of a gelatinous emulsion. The exposure took place by passing light both through the glass plate and the colour screen (Eder 1978, 662).

Obviously, the autochrome method is based on the three-colour process used to this day, "based on the physiological characteristics of human

172) Kittler (1997a, 91) already countered this by saying that, with computers and their peripheries, the frequencies could become colours again. Thanks to Lippmann's wave-optical method and his mathematical descriptions of numbers of oscillations, it was already possible to let colours shine again for roughly forty-four years before Heidegger's essay. Flusser's (2000) thoughts on colour photographs do not consider Lippmann photography insofar as colour in photography is only referred back to the "theory [...] of chemistry" (43).

43 This reconfiguration occurs in an even stronger way in three-dimensional images.

44 See Valenta (1894, 77–79) on the projection of a Lippmann photograph by Louis Lumière on August 22, 1893.

colour vision"[45] (Nareid 1988, 141), in other words, on splitting the light into the three basic colours that are then perceived again as the secondary colours.[46] Even though Lippmann photographs in principle surpass today's three-colour method, by far not only in their representation of colour but also in their durability (see Fournier and Burnett 1994) because no pigment is used, today's three-colour method prevailed in the end. It was theoretically first conceived and experimentally tested by James Clerk Maxwell in 1861.[47]

My historical reconstruction shows that, in the nineteenth century, three fundamentally different methods of colour photography existed and that only the one based on physiological optics prevailed. This confirms the assumption that the existent technologies considered "normal" today are in no way the only possible or "best" ones (whatever this may mean); they are simply those that prevailed for various historical reasons.[48] This is actually a truism and, in connection with my arguments, another point is more important: Is it possible that the implementation of the variant based on physiological optics is ultimately proof for Crary's thesis that physiological optics became dominant around 1820?

45 The context of the quote is the following: "Three-colour processes rely on a coding of the spectral distribution of the recording and reconstructing light into three primary colours. The results achieved by means of this method are, however, based on the physiological characteristics of human colour vision. There is no guarantee that the reproduction is physically accurate, even though the visual match may be excellent. There are also colour reproduction situations where a visual match cannot easily be achieved by the three-colour method, a fact of which all experienced colour printers are aware." (Nareid 1988, 141) Unlike Lippmann photographs, today's normal RGB monitors (just like TVs), which are also based on the three-colour process, cannot represent all possible colours. (See Kittler 2001, 31–32).

46 Since the artists did not use only the basic colours for the pixels of their images this physiological-optical concept *in a transformed way* is the basis for simultaneous artistic currents like pointillism; see Schug (1997).

47 See Maxwell (1861). See Evans (1961) on the fact that Maxwell was actually able to create colour photographs despite the fact that, during his time, photographic materials were mainly sensitive to blue light. On the history of three-colour photography, see Eder (1978, 639–64). It is strange that Crary (1990) does not mention Maxwell's colour photography, since it would support his thesis.

48 These reasons can be manifold. The concerted activities of different industries may establish a "worse" technology, as it happened for the VHS format, for example (see Wikipedia 2020): "VHS had been developed systematically for the private market and offered relatively low-priced and simply constructed equipment. Many see this as the decisive reason why VHS was able to win large shares in the market. Another aspect underrated by Sony's VHS competitor Betamax was the power of the porno-industry. While Sony (and thus Betamax) turned their backs on the porno-industry, the first pornographic film appeared on VHS. Shortly after, the first video rental stores opened offering exclusively pornographic material on VHS cassettes."

Even if we concede that the other series continue to exist, it might indeed be possible to imagine that at least with three-colour photography and the cinema those technologies prevailed that were based on physiological optics. However, this argument does not hold. First, black-and-white photography on paper negatives prevailed as well, even though it is not part of physiological optics (Crary's biggest problem). Second, wave optics in the twentieth century had another and much more important appearance with holography. With its unique property of three-dimensional representation, with special reference to the wave-optical representation of geometrical optical systems present in many scanners in the supermarket, in material control, in the sciences, and in safety engineering, it has become indispensable. Obviously, imaging technologies that are not physiological can also prevail (even though they are not part of mass-media phenomena that so often receive the most attention).

Disappearance and Re-Emergence

Third, for some time already, it is specifically in the area of safety engineering that precisely those particular characteristics of Lippmann photography that had originally been responsible for the fact that his method did not come out on top are appreciated again. The reproducibility that Walter Benjamin declared as the main characteristic of all technological media in his famous essay "The Work of Art in the Age of Mechanical Reproduction" (1978) (in distinction to "auratic art," as he calls painting), giving photography and the mass medium film as the only examples, is in no way the main feature of all technical media (for example, daguerreotypes are not reproducible). Reproducibility indeed is a problem for capitalist societies. It is not entirely coincidental that Benjamin starts his essay by taking recourse to Marx in order to end with the thesis "War and war only can set a goal for mass movements on the largest scale while respecting the traditional property system" (Benjamin 1978, 246). While I cannot discuss Benjamin's relationship to Marxism or the general validity of his theses here, it is a matter of his point that reproducibility not only threatens to subvert the "aura" of the artwork, but also *the copyright law.*[49] At present, the clamour about black market copies of CDs and DVDs or illegal downloads of music or movies has been overwhelming, since reproducibility has been vastly augmented

49 In the nineteenth century, particularly in France, the photograph represented a challenge for copyright laws; see Edelman (1979).

with digital media.[50] In order to preserve the commodity form, namely the "traditional property system," the reproducibility of digital media absolutely requires restrictions, either by draconian threats of punishment or by the addition and application of technological protections against copying (or both). But the restriction against copying or reproducing is not only necessary to maintain the copyright or the profit rates of the media industry, it is also necessary to secure the genuineness of certain entities, in other words, the differentiation between original and copy. Regarding these concerns about security, one must mention mainly the central medium of commodity-producing societies, such as, money (including bank and credit cards), as well as documents of governmental administration and control, like identity cards. It is in these cases that imaging technologies based on the wave-optical series are increasingly being used, among others also because the technology of copying machines as geometrical optical technologies of reproduction have continually advanced.[51]

This is precisely because using the interference of light waves creates images that cannot be copied with geometrical optical or physiological optical methods. A normal three-colour photograph can be copied very well on a good colour copying machine. With an iridescent (and, in a hologram, transplane) image this is at least difficult: "The limitations of the Lippmann technique [...] which made this type of colour photography impractical, have now become important advantages [...]."[52] (Bjelkhagen 1999b, 55) Whether an experimental method can become a useable technology then in the end depends on whether there are issues for which the method can be an answer (see Winston 2003, 5–9). Thus, Bjelkhagen (1999b, 58–60) describes that Lippmann photographs are better suited for certain security applications than holograms,[53] since it is even more difficult to copy them. The reason is that in Lippmann photographs the pattern of interference is stored three-dimensionally in the volume of the emulsion, while in many

50 Regarding these problems, Winkler observes: "Almost reminding us of Marx's contradictory terms of productive forces and relations of production, in this case the technological potential of technological reproduction and its social organisation – the copyright – exactly oppose each other." (Winkler 2004b, 29, my translation)

51 Stereoscopy was already used at the time to perform authentication (or at least there was the idea to do so); see Holmes (1861, 15), Helmholtz (1985, 304–6), Bith (1938), and Stenger (1950, 161, 1958, 202). On the use of holography on credit cards, see Colgate (1994). On the history of photocopying, see Mort (1989).

52 See the research done by Bjelkhagen (1999b), who attempts to describe possible security implementations of the Lippmann method.

53 On the alleged and real deficiencies of holograms as a security technology against copying, see Pizzanelli (1998).

types of holograms, it lies superposed flat on its surface. Therefore, since the 1980s, holograms can be copied industrially with certain methods of embossing; otherwise, they would not exist on so many credit cards or banknotes (cf. Bjelkhagen's article in this volume).[54]

However, this does not mean that the average citizen can copy these holograms (for example, with a copying machine). Reproducibility is not an attribute that either exists or does not exist; instead, it is distributed on a variety of levels and in various ways. Only for this reason, many citizens all over the world carry the so terribly marginalised hologram close to their hearts – namely, on their banknotes and the identity and credit cards in their wallets (see Schröter 2009).[55]

Conclusion

Lippmann photography is an exemplary case for the intricate problems of media historiography. On the one hand, its existence can only be accounted for in multidimensional historical models (as proposed by Schröter 2014). It does not fit into the narrative of Jonathan Crary, as was shown. On the other hand, it is an interesting case where properties that hindered its success, its non-reproducibility, became important later on. The history of media technology is full of such special and interesting cases that warrant complex and detailed historiographical discussions.

Works Cited

Apel, Friedmar. 2002. "Erobert das Unsichtbare! Für Jonathan Crary ist Aufmerksamkeit ein Mittel zur Repression." *FAZ*, August 9, 2002. p. 38.

Arago, François. 1839. *Das Daguerreotyp oder die Erfindung des Daguerre, die mittelst der Camera obscura und des Sonnenmikroskops auf Flächen dargestellten Lichtbilder zu fixieren*, Stuttgart: J. Scheibles Buchhandlung.

Atherton, Margaret. 1997. "How to Write the History of Vision: Understanding the Relationship between Berkeley and Descartes." In *Sites of Vision: The Discursive*

54 Johnston (2006a, 386–92) names techniques for copying holograms used at an early stage by counterfeiters, as well as the industry's development of increasingly refined methods against copying.

55 The security technologies used on German passports and identity documents are officially named *identigram*; see on this, the comments made by the Federal Printing Office (and its reference to holograms), to be found at Bundesdruckerei 2019.

Construction of Vision in the History of Philosophy, edited by David Michael Levin, pp. 139–165. Cambridge/MA u. a.: MIT Press.

Becquerel, Edmond. 1848. "De l'image Photochromatique du Spectre Solaire, et des Images Colourées Obtenues dans la Chambre Obscure." *Comptes Rendus Hebdomadaires des Séances de l'Academie des Sciences* 27: pp. 483.

Benjamin, Walter. 1978. "The Work of Art in the Age of Mechanical Reproduction." In *Illuminations: Essays and Reflections*, edited by Hannah Arendt, translated by Harry Zohn, pp. 217–252. 5th ed. New York: Schocken Books.

Bith, Fritz. 1938. "Banknotenfälscher und Raumbild." *Das Raumbild. Stereoskopisches Magazin für Zeit und Raum* 5 (August): p. 113.

Bjelkhagen, Hans I. 1999a. "New Optical Security Device Based on One-hundred-year-old Photographic Technique." *Optical Engineering* 38, no. 1: pp. 55–61.

Bjelkhagen, Hans I. 1999b. "Lippmann Photography: Reviving an Early Colour Process." *History of Photography* 23, no. 3: pp. 274–280.

Bjelkhagen, Hans I., Tung H Jeon, and Raymond J. Ro. 1998. "Old and Modern Lippmann-Photography." *SPIE Proceedings* 3358: pp. 72–83.

Brosch, Renate. 2000. *Krisen des Sehens. Henry James und die Veränderung der Wahrnehmung im 19. Jahrhundert*. Tübingen: Stauffenburg.

Buchwald, Jed Z. 1989. *The Rise of the Wave Theory of Light*. Chicago: University of Chicago Press.

Bundesdruckerei. 2019. "Schon immer innovativ. Bundesdruckerei-Chronik zeigt Erfindungen und technische Weiterentwicklungen." Bundesdruckerei-Chronik. Sicherheitsmerkmale im Wandel der Zeit. Last modified November 11, 2019. https://www.bundesdruckerei.de/de/Themen-Trends/Magazin/Bundesdruckerei-Chronik-Sicherheitsmerkmale-im-Wandel-der-Zeit-0.

Carey Lea, M. 1887. "On Red and Purple Chloride, Bromide and Iodide of Silver; on Heliochromy and on the Latent Photographic Image." *The American Journal of Science*, 3rd Series 33, no. 197: pp. 349–364.

Connes, P. 1987. "Silver Salts and Standing Waves: The History of Interference Colour Photography." *Journal of Optics* 18, no. 4: pp. 147–166.

Crary, Jonathan. 1985. "Notes on the Kaleidoscope and Stereoscope." *Journal. A Contemporary Arts Magazine* 5 (Autumn): pp. 38–41.

Crary, Jonathan. 1988a. "Techniques of the Observer." *October* 45 (Summer): pp. 3–35.

Crary, Jonathan. 1988b. "Modernizing Vision." In *Vision and Visuality*, edited by Hal Foster, pp. 29–49. Seattle, Washington: Bay Press.

Crary, Jonathan. 1990. *Techniques of the Observer. On Vision and Modernity in the Nineteenth Century*. Cambridge, MA et.al.: MIT Press.

Crary, Jonathan. 1999. *Suspensions of Perception: Attention, Spectacle and Modern Culture*. Cambridge, MA et.al.: MIT Press.

De Saint Victor, Niépce. 1855. "Mémoire sur la Gravure Héliographique Obtenue Directement dans la Chambre noire et sur quelques expériences scientifiques." *Comptes rendus hebdomadaires des séances de l'academie des sciences* 41: pp. 549–553.

Denisyuk, Yuri. 1984. *Fundamentals of Holography* [*1978*]. Moskau: MIR.

Edelman, Bernard. 1979. *Ownership of the Image: Elements for a Marxist Theory of Law* [*1973*]. London: Routledge & Kegan Paul.

Eder, Josef Maria and G. Pizzighelli. 1881. "Beiträge zur Photochemie des Chlorsilbers." *Monatshefte für Chemie und CVerwandte Teile anderer Wissenschaften* 2, no. 1: pp. 33–49.

Eder, Josef Maria. 1978. *History of Photography.* Translated by Edward Epstean. New York: Dover.

Evans, R. M. 1961. "Some Notes on Maxwell's Colour Photograph." *The Journal of Photographic Science* (July/August): pp. 243–246.

Flusser, Vilém. 2000. *Towards a Philosophy of Photography.* Translated by Anthony Mathews. London: Reaktion Books.

Foucault, Michel. 1966. *Les Mots et les Choses.* Paris: Gallimard.

Foucault, Michel. 1969. *L'archéologie du Savoir.* Paris: Gallimard.

Foucault, Michel. 1984. "Nietzsche, Genealogy, History." In *The Foucault Reader,* edited by Paul Rabinow, p. 76–100. New York: Pantheon Books.

Fournier, J. M. and P.L. Burnett 1994. "Colour Rendition and Archival Properties of Lippmann Photographs." *The Journal of Imaging Science and Technology* 38, no. 6: pp. 507–512.

Frizot, Michel, ed. 1998. *Neue Geschichte der Fotografie.* Cologne: Könemann.

Geimer, Peter. 2000. Review of *Eine Geschichte der optischen Medien,* by Ulrike Hick. IASLonline. Accessed October 25, 2007. http://www.iaslonline.lmu.de/index.php?vorgang_id=2084.

Geimer, Peter. 2002. "Was ist kein Bild? Zur ‚Störung der Verweisung'." In *Ordnungen der Sichtbarkeit. Fotografie in Wissenschaft, Kunst und Technologie,* edited by ibid., pp. 313–341. Frankfurt a.M.: Suhrkamp.

Goethe, Johann Wolfgang von. 1989. *Sämtliche Werke nach Epochen seines Schaffens.* Edited by Karl Richter and Peter Schmidt. Vol. 10: Zur Farbenlehre [1810]. Munich: Hanser.

Goethe, Johann Wolfgang von. 2000. *Theory of Colours.* Translated by Charles Lock Eastlake. 13th ed. Cambridge, MA and London: MIT Press.

Heidegger, Martin. 1993. "The Origin of the Work of Art." In *Basic Writings: From Being and Time (1927) to the Task of Thinking (1964)*, pp. 139–212. Rev. and expanded ed. San Francisco, CA: HarperSanFrancisco.

Helmholtz, Hermann von. 1985. *Helmholtz's Treatise on Physiological Optics. Translated from the Third German Edition.* Edited by James P. C. Southall. Vol. III. The Perception of Visions. Birmingham/AL: Leslie B. Adams.

Hentschel, Linda. 2001. *Pornotopische Techniken des Betrachters. Raumwahrnehmung und Geschlechterordnung in visuellen Apparaten der Moderne.* Marburg: Jonas.

Herschel, Sir J. F. W. 1840. "On the Chemical Action of the Rays of the Solar Spectrum on Preparations of Silver and other Substances, both metallic and non-metallic, and on some Photographic Processes." *Philosophical Transactions of the Royal Society of London* 130: pp. 1–59.

Hick, Ulrike. 1994. "Die optische Apparatur als Wirklichkeitsgarant. Beitrag zur Geschichte der medialen Wahrnehmung." *montage/av* 3, no. 1: pp. 83–96.

Hick, Ulrike. 1999. *Geschichte der optischen Medien.* Munich: Fink.

Holmes, Sir Oliver Wendell. 1861. "Sun-Painting and Sun Sculpture." *Atlantic Monthly* 8 (July): pp. 13–29.

Ives, Herbert E. 1908. "An Experimental Study of the Lippmann colour Photograph." *The Astrophysical Journal* 27: pp. 325–352.

Jay, Martin. 1988. "Scopic Regimes of Modernity." In *Vision and Visuality,* edited by Hal Foster, pp. 3–27. Seattle, WA: Bay Press.

Johnston, Sean. 2006. "Absorbing New Subjects: Holography as an Analog of Photography." *Physics in Perspective* 8, no. 2 (May): pp. 164–188.

Kittler, Friedrich. 1997. *Kunst und Technik,* Basel: Stroemfeld.

Kittler, Friedrich. 2001. "Computer Graphics: A Semi-Technical Introduction." *Grey Room* 2: pp. 30–45.

Kittler, Friedrich. 2010. *Optical Media: Berlin Lectures.* Translated by Anthony Enns. Cambridge: Polity Press.

Kosenina, Alexander. 2001. "Das bewaffnete Auge. Zur poetischen Metaphorik von Mikroskop und Guckkasten." In *Das Schöne Soll Sein. Aisthesis in der deutschen Literatur,* edited by Peter Heßelmann et al., pp. 101–127. Bielefeld: Aisthesis.

Lehmann, Hans. 1907. "Über eine neue kornlose Platte für Lippmann-Photographie." *Zeitschrift für wissenschaftliche Photographie* 5: pp. 279–283.

Lehmann, Hans. 1908. "Über die Abstimmung der Lippmannplatten." *Eder's Jahrbuch für Photographie und Reproductionstechnik* 22: pp. 157–164.

Lippmann, Gabriel. 1891. "La Photographie des Couleurs." *Comptes rendus hebdomadaires des séances de l'academie des sciences* 112 (January/June): pp. 274–275.

Lippmann, Gabriel. 1894. "Sur la Théorie de la Photographie des Couleurs Simples et Composées par la Methode Interférentielle." *Journal de physique* 3, no. 3: pp. 97–107.

Lumière, Auguste and Lumière, Louis. 1893. "Sur les Procédés Pour la Photographie des Couleurs d'Après la Methode de M. Lippmann." *Bull. Soc. Franç. Phot.* 2, no. 9: pp. 249–251.

Mann, Heinz Herbert. 2000. "Optische Instrumente." In *Erkenntnis Erfindung Konstruktion. Studien zur Bildgeschichte von Naturwissenschaften und Technik vom 16. bis zum 19. Jahrhundert,* edited by Hans Holländer, pp. 357–408. Berlin: Mann.

Maxwell, James Clerk. 1861. "On the Theory of Three Primary Colours." *Proceedings of the Royal Institution of Great Britain* 3: pp. 370–375.

Mort, Joseph. 1989. *The Anatomy of Xerography. Its Invention and Evolution.* Jefferson NC: McFarland.

Nareid, Helge and Hans M. Pedersen. 1991. "Modeling of the Lippmann colour Process." *Journal of the Optical Society of America* 8, no. 2: pp. 257–265.

Nareid, Helge. 1988. "A Review of the Lippmann Colour Process." *The Journal of Photographic Science* 36: pp. 140–147.

Nielsen, Keld. 1989. "Another Kind of Light: The Work of T. J. Seebeck and his Collaboration with Goethe. Part 1." *Historical Studies in the Physical and Biological Sciences* 20, no. 1: pp. 107–178.

Nielsen, Keld. 1990. "Another Kind of Light: The Work of T. J. Seebeck and his Collaboration with Goethe. Part 2." *Historical Studies in the Physical and Biological Sciences* 21, no. 2: pp. 317–397.

Niépce, Nicephore. 1839. "Notice sur l'heliographie. [1829]" In *Historique et description des procédés du daguerréotype et du Diorama*, edited by Louis Daguerre, pp. 39–46. Paris: Susse Frères et.al.

Paech, Joachim. 2003. "Überlegungen zum Dispositiv als Theorie medialer Topik." In *Der kinematographische Apparat. Geschichte und Gegenwart einer interdisziplinären Debatte*, edited by Robert F. Riesinger, pp. 175–194. Münster: Nodus-Publ.

Phillips, N. J., H. Heyworth, and T. Hare. 1984. "On Lippmann's Photography." *The Journal of Photographic Science* 32: pp. 158–169.

Pizzanelli, David. 1998. "Counterfeit Holograms and Simulations." *SPIE Proceedings* 3314: pp. 86–96.

Richardson, Dorothy. 1979. "The Tunnel [1919]", in *Pilgrimage* 2, pp. 9–288. London: Virago Press.

Rieger, Stefan. 2001. *Die Individualität der Medien: eine Geschichte der Wissenschaften vom Menschen.* Frankfurt a.M.: Suhrkamp.

Rieger, Stefan. 2002. *Die Ästhetik des Menschen: über das Technische in Leben und Kunst.* Frankfurt a.M.: Suhrkamp.

Rieger, Stefan. 2003. *Kybernetische Anthropologie: eine Geschichte der Virtualität.* Frankfurt a.M.: Suhrkamp.

Rood, Ogden N. 1880. *Die moderne Farbenlehre.* Leipzig: Brockhaus Weimar Universitätsbibliothek.

Schröter, Jens. 2000. "Die Form der Farbe. Über ein Parergon in Kants Kritik der Urteilskraft." In *Kants Schlüssel zur Kritik des Geschmacks*, edited by Ursula Franke, pp. 135–154. Hamburg: Felix Meiner Verlag.

Schröter, Jens. 2009. "Das holographische Wissen und die Nicht-Reproduzierbarkeit." In *Das Holographische Wissen*, edited by Stefan Rieger and Jens Schröter, pp. 77–86. Zürich: Diaphanes.

Schröter, Jens. 2014. *3D. History, Theory and Aesthetics of the Technical-transplane Image*, New York et al.: Bloomsbury Academic.

Schug, Albert. 1997. "Über die Bedeutung der neo-impressionistischen Theorie Seurats." In *Pointillismus. Auf den Spuren von Georges Seurat*, edited by Rainer Budde, pp. 10–38. Munich: Prestel.

Steigerwald, Jörn. 2000. "Das Göttliche Vergnügen des Sehens. Barthold Hinrich Brockes' Techniken des Betrachters." In *Convivium. Germanistisches Jahrbuch*, edited by Heidemann, Gudrun, Joanna Jabłkowska and Beata Mikołajczyk, pp. 9–43. Poland: Deutscher Akademischer Austauschdienst.

Stenger, Erich. 1950. *Siegeszug der Photographie in Kultur, Wissenschaft, Technik* [*1938*]. Seebruck: Heering-Verlag.

Stenger, Erich. 1958. *The March of Photography*. London and New York: Focal.

Stiegler, Bernd. 2001. *Philologie des Auges: die photographische Entdeckung der Welt im 19. Jahrhundert*. Munich: Fink.

Stiegler, Bernd. 2003. Review of *Optische Medien, Berliner Vorlesung 1999*, by Friedrich A. Kittler. *Fotogeschichte* 87: pp. 70–71.

Stiegler, Bernd. 2004. "Michel Foucault und die Fotografie." In *Foucault und die Künste*, edited by Peter Gente, pp. 277–297. Frankfurt a.M.: Suhrkamp.

Talbot, Henry Fox. 1851. "Sur la Production des Images Photographiques Instantanées." *Comptes rendus hebdomadaires des séances de l'academie des sciences* 38: pp. 623–627.

Valenta, Eduard. 1894. *Die Photographie in Natürlichen Farben unter Besonderer Berücksichtigung des Lippman'schen Verfahrens*. Halle: W. Knapp.

Waldenfels, Bernhard. 1999. *Sinnesschwellen. Studien zur Phänomenologie des Fremden Bd. 3*. Frankfurt a.M.: Suhrkamp.

Wiener, Otto. 1890. "Stehende Lichtwellen und die Schwingungsrichtung polarisirten [sic] Lichtes." *Annalen der Physik und Chemie* 40, no. 6: pp. 203–243.

Wiener, Otto. 1895. "Farbenphotographie durch Körperfarben und mechanische Farbenanpassung in der Natur." *Annalen der Physik und Chemie* 55, no. 6: pp. 225–281.

Wikipedia. 2020. "Video Home System." Last modified June 12, 2020. http://de.wikipedia.org/wiki/Video_Home _System.

Williams, Linda. 1995. "Corporealized Observers: Visual Pornographies and the 'Carnal Density of Vision'." In *Fugitive Images. From Photography to Video*, edited by Patrice Petro, pp. 3–41. Bloomington and Indianapolis: Indiana University Press.

Winkler, Hartmut. 2004. *Diskursökonomie. Versuch über die innere Ökonomie der Medien*. Frankfurt a.M.: Suhrkamp.

Wölfflin, Heinrich. 1950. *Principles of Art History: The Problem of the Development of Style in Later Art*. New York: Dover.

Young, Thomas. 1804. "Experiments and Calculations Relative to Physical Optics." *Philosophical Transactions of the Royal Society* 94: pp. 1–16.

Zenker, Wilhelm. 1867. "Versuch einer Theorie der Farben-Perception." *Archiv für mikroskopische Anatomie* 3 (1): pp. 248–261.

Zenker, Wilhelm. 1868. *Lehrbuch der Photochromie (Photographie der natürlichen Farben) nach den wichtigen Entdeckungen von E. Becquerel, Nièpce de St. Victor, Poitevin u. a.* Berlin: Self-published.

About the Author

Prof. Dr. **Jens Schröter** has been chair for media studies at the University of Bonn since 2015. He was professor of multimedial systems at the University of Siegen from 2008 to 2015. From 2008 to 2012, he was director of the graduate school *Locating Media* at the University of Siegen; he is director (together with Anna Echterhölter, Vienna; Andreas Sudmann, Bochum and Alexander Waibel, Karlsruhe) of the Volkswagen Foundation-main grant *How is Artificial Intelligence Changing Science.*

6. Richard Neuhauss's Stuffed Parrot: Interferential Colour Photography, Taxidermy, and Projection

Hanin Hannouch

Abstract
This article places doctor of medicine, anthropologist, and photographer Richard Neuhauss's (1855–1915) practice of the Lippmann process in relation to other media. It focuses on the importance of taxidermy for the choice of an adequate test object for interferential colour photography and the role the stuffed superb parrot, a colonial animal-object par excellence, played in Neuhauss's experimental practice. Moreover, this text examines the various optical media needed to present the interferential colour photograph to individual or collective spectators in Imperial Germany, especially projection, with whom it had a special relationship.

Keywords: Richard Neuhauss, taxidermy, test object, projection, interferential colour photography, optical media

Interferential colour photography emerged in relation to other media at the turn of the last century – media that were instrumental to the production and the dissemination of knowledge about it – and yet current scholarship in photographic history has rarely considered this interconnected matrix (Leonardi and Natale 2018, 2). Situating Lippmann photography within a wider constellation of technologies is essential for shedding its double historiographic isolation. First, colour photography as such is marginalised within the "general" history of photography. Second, the segregation between "direct" and "indirect" colour photographic processes has divided them into two different temporal constructions, posited as epistemologically at odds with each other. The Lippmann process is thus

Hannouch, H. (ed.), *Gabriel Lippmann's Colour Photography: Science, Media, Museums.*
Taylor & Francis Group, 2022
DOI 10.5117/9789463728553_CH06

considered a challenging and cumbersome "direct method" that produces colour photographs by resolving light waves. Within this paradigm, its position is awkward. It is considered the teleological dead-end street of colour photography, part of the "prehistory" and not the "actual" history of colour. It is also a mere "early search" for or a precursor to financially viable "indirect" methods (Isler-de-Jongh 1994, 111–12) such as three-colour photography and the autochrome, which reconstruct all shades based on three primary ones. This position has hindered the emergence of interferential colour photography as an object of historical inquiry, a narrative that urgently needs revisiting.

To overcome these conceptual and teleological limitations at the core of colour photography's historiography, it is essential to be sensitive to relationships rather than to ruptures between various colour processes. This move away from a media-specific approach to interferential photography, by which it operates only within its own expressive capacities to remain "pure," is the stepping stone for a broader transformation of how its history will be written in the future. A key attribute of the renewed history of colour photography is the complication rather than the obfuscation of the entanglements between several (colour) photographies in relation to other formative technologies, which "are related synchronically to each other in a systemic relationship or exist to develop diachronically within a specific constellation" (Schröter 2014, 11). Erkki Huhtamo suggests:

> Instead of dealing with photography in isolation from other media practices, we should embrace the connections it has with them on all possible levels. Discussing photographs as symptomatic pointers to underlying developments should be part of the endeavour but never separated from the contexts – from material to discursive – that informed their becoming and within which they radiate impulses to other media forms. (2018, 29)

My research situates Richard Neuhauss's (1855–1915) use of the Lippmann process in dialogue with a host of other media that unfolded during German imperialism and its colonial expansion. It explores Neuhauss's consistent use of a stuffed parrot as his test object and a staple of his laboratory work along with the role of taxidermy in shaping his decision. It also reveals how he propagated his colour photographs through a host of optical media that he developed and relied on, including projection. Ultimately, this approach integrates the entanglements of interferential colour photography, taxidermy, and projection within photographic science and the broader framework of media and imperial history.

The Stuffed Parrot: Taxidermy and Colour Testing

In parallel to his practice as a doctor of medicine, Neuhauss began his anthropological travels in 1884 when he visited Hawaii, Australia, and New Zealand and used a battery of research methods known to physical anthropology, such as plaster casting, determination of skin, hair, and eye colours in relation to standardised colour charts, and photography (Schindlbeck 2000, 346). In Berlin, he was known in a variety of circles: he co-founded the Free Photographic Union of Berlin (Freie Photographische Vereinigung zu Berlin) in 1889 with anthropologist Gustav Fritsch and painter Franz Goerke, and he became the editor of the journal *Photographische Rundschau* in 1894. He was also the custodian of the photographic collection of the Berlin Society for Anthropology, Ethnology and Prehistory from 1904 to 1915 (Schindlbeck 2000, 347). His theoretical and practical knowledge of photography, along with his privileged position within these networks, turned both the Vereinigung and the *Rundschau* into hubs where various colour photographies were persistently discussed and disclosed among anthropologists, scientists, and amateurs, who appropriated and transformed them within their own visual practices (Joschke 2014, 131). Around this time, the complex and challenging science of photography, understood as "the investigation of any part of the photographic emulsion in the attempt to find out just what is happening when a photograph is made (after exposure, development or fixing)" (Wilder 2009, 13), became the topic of much scrutiny during the wave of optimism about interferential colour photography's potential, a potential which remained locked behind emulsions operating according to mysterious laws.

Neuhauss knew that the interest in and development of Lippmann photography depended on "its ability to be repeated by other scientists and the replicability of its results" (Wilder 2009, 64). Results, however, are only as reliable as their testing conditions, at the centre of which lies choosing the adequate test object. The test object was not only meaningful to create a visual composition once the image is shot but to determine the quality of the photographic emulsion and to reveal the success or failure of the entire photographic process so that the scientist can chart novel territory. He considered the goals of the Lippmann process to be both the rendering of spectra, which he found relatively "easy" (*leicht*), and the expansion

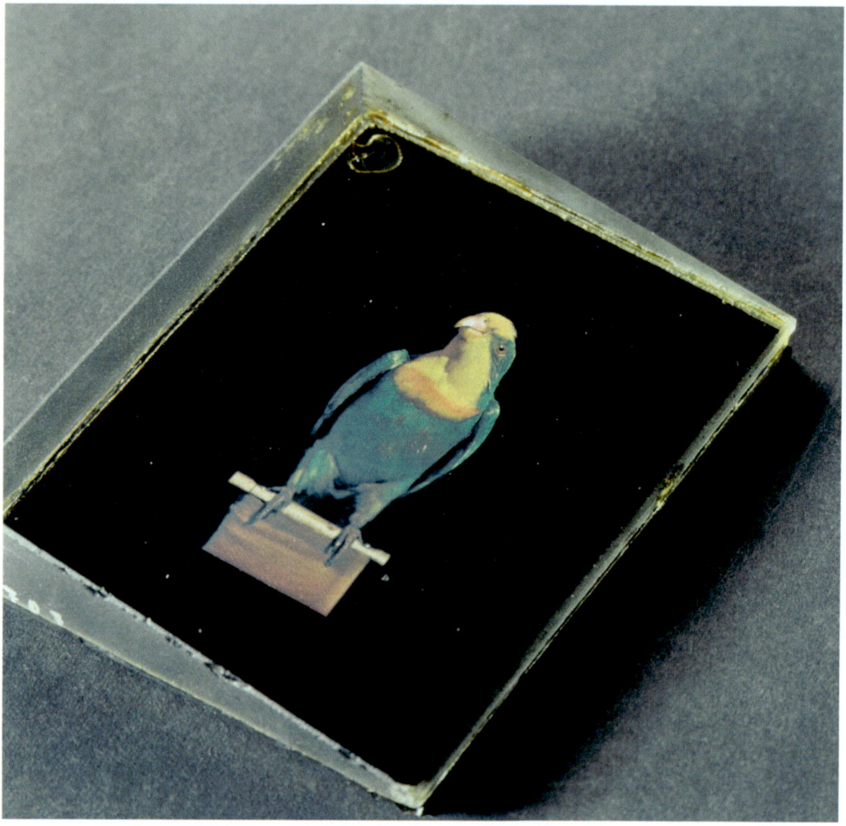

Fig. 23. Richard Neuhauss. Interferential plate with wedged prism, featuring a stuffed parrot. 1899. Inventory number 4707. Deutsches Museum (Munich).

of the medium's representational capabilities to the mixed (pigmented) colours of objects, people, and animals, etc., which he deemed to be "a diamond the size of a pigeon" (Neuhauss 1895c, 24). Nevertheless, this task presented "considerable difficulties" (*erhebliche Schwierigkeit*) (Neuhauss 1895, 20). One of the many tolls was creating an emulsion viable for the Lippmann process, which meant meeting high standards. They included transparency, a high resolution, and a grain fine enough to be able to register the physical properties of light. On the downside, such an emulsion would be insensitive to colours beyond green prior to spectral sensitisation and it would be extremely slow to record them, thereby further lengthening the exposure time of the image (cf. William R. Alschuler's article in this volume). As his laboratory notes and numerous writings reveal, Neuhauss tried several materials for Lippmann photography, such as albumen, collodion, and gelatin, in which light-sensitive silver halides can be precipitated and suspended. To ascertain how colour-sensitive they were, he depended on the stuffed parrot as a key test object – so much so that he admitted to having photographed it over three hundred times (Neuhauss 1898, 48), starting in 1898 for testing the sensitisation of the gelatin emulsion (Neuhauss 1898b, 4).

Neuhauss recognised the beauty of the *living* parrot when he saw it in 1908–10 while on an anthropological expedition sponsored by the Museum für Völkerkunde in Berlin to Papua New Guinea, an area he was also keen to map with aerial photography (Claas and Roscoe 2009, 137). Whether dead or alive, the parrot and Neuhauss's access to it, should be understood within the historical and epistemological framework of the German Empire colonising the New Guinea archipelago starting in 1884 and of the previous European economic exploitation of the area and its people. The conditions under which Neuhauss acquired the taxidermic bird remain unclear and its current location unknown. The extant plates featuring the stuffed bird depict it neither staged naturalistically in a diorama nor as part of a tableau narrative freezing it "in action to suggest both the scene's immediate past and what will immediately follow" (Henning 2007, 664). It stands, rather, in isolation on a twig, with its head slightly tilted to the right, revealing the green, yellow, and red colours of its body. Colonialism enabled taxidermy and photography's shared capacity to possess, control, and display nature. Its extractive logic brought specimens and images of rare, beautiful, and often adult male animals, which embodied the perfect expression of their type (Haraway 1989, 41), to Neuhauss's abode. This political reality accounts for how the stuffed bird, which is native to New Guinea, reached Berlin in the first place. Hence, the recurrent reference to the parrot is emblematic of Neuhauss positioning himself as the German expert on both interferential photography that was more associated with Gabriel Lippmann and France and as *the* anthropological specialist on New Guinea (Claas and Roscoe 2009, 136). The co-production of these two discourses points to an untold story of violent colonial collection practices that hangs over both the animal-object and its colourful image.

To choose an adequate test object, Neuhauss excluded others and recommended like-minded photographers to follow his lead. Although he recognised a beginners' tendency to photograph flowers as models for the Lippmann process, he quickly dismissed them as unfavourable, since, when in the field, they are easily shaken by wind and any movement; even a blossoming bud made itself "unpleasantly noticeable" (Neuhauss 1898, 47). When flowers are cut, they are quick to wilt and drop, and this subtle movement disturbs the capture of the image. Neuhauss also dismissed landscapes where the colour red (such as the roof of a house) is prominent, due to the danger it presents in contaminating the other colours unless the photographer is particularly apt (Neuhauss 1898, 47). Given the technical complexity of interferential photography, the toxicity of the needed chemicals (such as mercury in its plate cassette), its fickle emulsions (requiring repeated testing for sensitivity prior to taking the photograph), and its long exposure time (ranging from ten minutes to over three hours, depending on the image), it is best conducted in a controlled or familiar environment. These material conditions mediated its access to the world more through the still, slow, and docile, and less through the instantaneous and the moving. The ideal test object should share essential parameters with colour photography, as in it has to be as colourful as possible and it has to maintain the exact same colour over a considerable period of time. In the case of the parrot, the recourse to another medium that can stabilise its shades was necessary: taxidermy.

Neuhauss was initially reticent towards taxidermy, but in its conditioning of the expressible, the Lippmann process necessitated compromises on his behalf. He previously refused resorting to stuffed animals as models for drawing, criticising Alfred Brehm's (1829–84) widely popular thirteen-volume *Brehms Illustrirtes Thierleben* (1860) for the meagre (*dürftig*) quality of its illustrations (Neuhauss 1896a, 73). He also rejected taxidermy in his own monochromatic photography, stressing the importance of capturing living beasts (*nach dem Leben*) (Neuhauss 1896a, 72). His photographic reprints of an elephant, a rhinoceros, and monkeys held at the zoological garden of Berlin show them dwelling in enclosures. In this respect, he distinguishes himself from photographers such as J. D. Llewellyn (1850), G. C. Dawnay (1876), and H. Anderson Bryden (1893), who relied on taxidermy "in order to capture portraits of animals in a seemingly live pose and outdoor setting" (Ryan 2000, 208).

The use of taxidermy as a model for illustration and photography conflicted with Neuhauss's preferences for depicting living animals, a depiction that can, according to him, expand the knowledge of nature (*Erkenntnis der Natur*). This knowledge remains indeterminate in his article, although he considered access to it to be hampered by the unsuitable outdoor lighting conditions in zoos and by the movement of fauna (Neuhauss 1896a, 72). Although movement hinders the generation of the photograph, he thought it was an essential quality that the image should capture. Hence, photographing stuffed beasts occludes whatever information can be extrapolated from them, because the image excludes their motion and colour, leaving them stiff and grey. Neuhauss's distrust of taxidermy was a luxury he could not afford when dealing with Lippmann photography. Soon, the stuffed parrot trumped all other potential test objects.

The taxidermic bird used by Neuhauss is both an artefact of nature as well as an "experimental artefact" of colour photography (Mody and Lynch 2009, 3). As an artefact of nature, taxidermy promises the "hindering of natural decay" (Poliquin 2012, 10), thereby preserving the plumage's hues, which are what colour photographers seek and what scientists rely on to categorise beasts. As Rachel Poliquin states "whereas a black-and-white etching and a good description of any particular beast was generally sufficient to distinguish it from its close relatives, color was far more important than shape for identifying birds, and oh, the colors!" (Poliquin 2012, 10). Therefore, the stuffed parrot achieves "the semblance of nature: a pure artifactuality with the appearance of nature" (Maleuvre 1999, 215).

As an experimental artefact, the stuffed parrot is a compliant and stationary sitter with low risk of discolouration, since taxidermy preserved its shades and nullified its motion. In doing so, it prolonged the encounter between it and the photographer long enough for the exposure to actually happen and stabilised the conditions under which it can take place. Any change in its shades or position could skew the experiment's results and confound the reading of the sensitivity of the emulsion to be used in the Lippmann process. While monochrome photography flattens the shades of skin that taxidermy preserves best, interferential photography outbids it by conveying the stuffed parrot's most distinguishable attribute: its vivid plumage. At best, the inscription of the stuffed bird's colours on the photographic plate with maximal brilliance and accurate colour rendition is indicative of correct sensitisation of the emulsion and of the overall success of the steps needed to achieve the results. At worst, the lack of translation of its colours meant that the photographic process had yet to be optimised on either one or several levels. Overall, the stuffed parrot's role as the "Shirley card" of the Lippmann process is foundational in its operation as the chromatic standard against which Neuhauss gauged the success of his colour experiments.[1] In guaranteeing the appearance of nature and in stabilising the parrot's colours, taxidermy allowed the stuffed parrot as test object to safeguard the stability of the testing conditions and ultimately the veracity of the results. The still and immutable properties enabled by this technology of preservation turned the parrot into a staple of the material culture of laboratory science as the test object of interferential colour photography.

1 Cf. Lorna Roth, 2013, "The Fade-Out of Shirley, a Once-Ultimate Norm: Colour Balance, Image Technologies, and Cognitive Equity", *Canadian Journal of Communication* 34, No.1, 111–136. and Tanya Sheehan, 2020, "Color matters: rethinking photography and race," *The Colors of Photography*, edited by Bettina Gockel, in Collaboration with Nadine Jirka and Stella Jungmann, 55–71. Berlin: De Gruyter.

The durability of colour promised by taxidermy also allowed the test object the advantage of being readily available for contrasting with its photographic image for a long period of time. Also, both taxidermy and colour photography "peel a layer from the world which they then present as truth" (Hauser 1998, 8), thus maintaining their relationship to their referent, be it skin or object (Henning 2006, 138). This relationship to the referent the stuffed bird has could help validate its image in the eyes of doubtful critics. Neuhauss noted how "the strict adherence to a good, durable test object may be a bit monotonous, but among other advantages, the original is always available for comparison with the pictures" (Neuhauss 1898, 48). This comparative methodology aimed at ascertaining interferential colour photography's "ability to tell the truth about nature" (Wilder 2020, 309) and, with it, colour photography's evidentiary power as well as the scientist's credibility. Often, the results of the comparison of the photograph with the test object it depicts had "reflexive implications" (Mody and Lynch 2009, 3), as in these results also tested the photographer's commitment to their work.

For instance, in 1892, Gabriel Lippmann projected interferential colour photographs of a group of flags, a plate of oranges, a four-coloured stained-glass window (red, green, blue, and yellow) and a stuffed bird at a combined meeting of all the photographic societies in France during the Champ de Mars exhibition in Paris. A British observer, Cameron Swan, lamented that the original subjects were not brought to the exhibition to be shown for comparison. He damagingly concluded that the problem of colour photography had not yet been solved (Mitchell 2010, 328), whereas out of all objects depicted, only the flags and the stuffed bird would have maintained their colours long enough to have been able to be transported to the exhibition space to begin with.

In the German Empire, Hermann Wilhelm Vogel, lithographer Emil Ulrich, and Vogel's son Ernst created a three-colour photomechanical printing technique whose paragon, also from 1892, features an "oil painting and natural butterflies" depicting South American butterflies mounted around the painting of a woman working a needle. Neuhauss compared the print with the actual mounted insects themselves. He repeated the claim of an unnamed expert stating that the colour of the butterflies' antenna is not always blue in real life whereas this is the only shade the photograph conveys. This contention led to further scepticism vis-à-vis Vogel's printing method, which cannot guarantee a perfect overlap of the three printing matrixes needed to generate the colour image and sometimes can only convey one shade in some of the prints' finer details like the delicate antennae, thus becoming "inadequate" (*mangelhaft*) (Neuhauss 1896c, 111). Hence, Neuhauss joined the genealogy of three-colour printing objectionists, such as Edmond Becquerel who dismissed Ducos Du Hauron's three-colour prints on the grounds of them being less the imprint of light and more the fruit of "arbitrary artificial dyes and chromatic manipulation" (Boulouch 2001, 54) that alter the shades of nature at their will. Some years later, when Neuhauss sent Vogel a Lippmann plate featuring the stuffed parrot to demonstrate the success of his process, Vogel proved less than enthusiastic. He judged the parrot's colours as similar to nature (*Naturähnlich*) but not corresponding (*entsprechen*) to nature (Vogel 1898, 300). Vogel understood this correspondence to nature as the primary task of his three-colour technology and by extension of other photographic processes, basing his argument against Neuhauss's in the surface reflection of the plate. Although Neuhauss succeeded in photographing the shades *of* the stuffed bird and wove a visual similarity between the colours of its plumage and those in the photograph, he failed to reconstruct the act of looking *upon* the colours of the taxidermic bird. The stuffed parrot as an object can be seen just fine, but its colours in the photograph change (and at times disappear altogether) depending on the angle through which light strikes their glass surface, to Vogel's great displeasure (Vogel 1898, 300). The image's colours are too shifty. In this divergence between colour and light angle lies the failure of the Lippmann process in conveying any kind of "truth" (*Wahrheit*) about the world (Vogel 1898, 301). Three-colour photography, however, from Vogel's viewpoint, distinguishes itself from this plasticity by reproducing the overall instantaneousness and the enduring quality of colour.

Inversely, in Neuhauss's book relaying his trip to Papua New Guinea, *Deutsch Neu-Guinea* (1911), he saw the living parrot he often photographed stuffed. He called it "a blob of colour" (*Farbenkleckser*), due to the brightness of its adjoining shades, while the bird of paradise was to him a "colour artist" (*Farbenkünstler*). Together, they constitute a "paradisiac perfection" (*paradiesischer Vollkommenheit*) in their uniqueness and in their deviation from the archetype (*Urform*). When describing the bird of paradise's shimmering colours (*Schillerfarben*), he compares them to the shades of a Lippmann plate. The iridescent shimmer of both bird and photograph gives them maximum liveliness (*höchste Lebendigkeit*) (Neuhauss 1911, 487). For Neuhauss, the Lippmann process not only records nature's colours but, through its iridescence, reproduces the shimmering and lively quality of the bird's feathers that result from the phenomenon of light waves interference – the very principle of the Lippmann's process. Vogel's rendition of the mounted butterflies' colours is, however, impure, because it falsifies the insect's shades due to the use of dyes. It increases the gap between colour photography and nature.

Neuhauss's use of the stuffed parrot as test object epitomises the entanglement of taxidermy and colour photography both politically and technologically, since the depiction of the former was essential to testing the cultural legitimacy of the latter. Neuhauss's and Vogel's perceived level of success in representing the parrot or the butterflies, respectively, was couched in terms such as "correspondence" or "proximity," which operated as a metonymy for the colour technology's ability to render nature as such. The assumed quality of the relationship between these two media gave colour photographers enough ammunition to downplay each other's achievements.

The stuffed parrot as test object solidified its status as a staple of Neuhauss's laboratory also by having reflexive implications for both process and scientist. By comparing the parrot as animal-object to its photographic representation, Neuhauss tracked his own progress in creating and refining the emulsions needed to generate the image. Yet, irrelevant of the stability afforded by the (seemingly) unchanging stuffed parrot, Neuhauss was not entirely shielded from the arbitrariness of his photographic experiments. "Since the days of the earliest photographic processes, photographers had been dealing with substances whose activities were neither predictable nor completely under their control." (Geimer 2018, 42) Like them, he was often incapable of determining the exact nature of the phenomena precluding his success. He once recounted how he ran out of the gelatin needed for an emulsion, just as soon as he began achieving repeatable results. After purchasing a second batch from the same provider, photographing the parrot failed for the next three years, despite numerous retrials. His colleague Dr. Cowl from the Physiological Institute of Berlin later gave him the gelatin he used to generate bacterial culture, and Neuhauss could finally reproduce anew the parrot's colours in their splendour (*Farbenpracht*) and truth to nature (*Naturtreu*), making this breakthrough his thirty-fourth trial at emulsion sensitivity (Neuhauss 1898b, 4–5). Still, he remained oblivious to the exact cause of his initial failure, since the substances used for the Lippmann process (including gelatin) were quite unreliable, as his answer to a reader of the *Rundschau* in 1895 reveals. The reader heeded the advice provided by the main scholars writing about the Lippmann process all throughout winter, and still, after repeated trials, no colour was in sight, so they asked Neuhauss for help. The only guidance Neuhauss could provide was that practising the Lippmann process during the winter season is not conducive to good results for causes that remain unclear. The solution is simply to try again in the summer (Neuhauss 1895a, 159).

Given the little control scientists had over interferential photography, the reliable test object proved essential in anchoring the photographer amid the confusion surrounding its science. Its colours were enduring and its motion lacking. It also laid bare the failure of the trial when its colours could not be rendered, and in doing so, it helped generate new knowledge about emulsion sensitivity, optimising the process, and ultimately examining Neuhauss's virtue and persistence, helping him increase his expertise. The test object's resistance to being photographed also tests the photographer, since only through diligent replication can they reach mastery and hope to be credible under scrutiny; otherwise, success is written off as chance. Neuhass declares:

> The author did not rest until all these colours [of the parrot] were perfectly reproduced in the picture. In the end, the photo was not only taken once, which one could describe as a particularly lucky coincidence, but a whole series of images of the same quality was produced. (Neuhauss 1898, 47)

The fortunate depiction of the stuffed bird synthesised in a single image: Neuhauss's years of experimental and theoretical research and his painfully acquired "great confidence" (Neuhauss 1898, 48) in himself and his work. The parrot's image has reached such an iconic status as a "carrier of the great weight of evidence" (Wilder 2009, 58) of the successful rendition of mixed colours to the point that Neuhauss sent it to Gabriel Lippmann himself as proof (cf. Pauline Martin and Carole Sandrin's article in this volume) and to the Deutsches Museum in 1906 at the behest of its founding director Oskar von Miller (1855–1934), as a gift and exemplary sample of the process.[2]

Despite colour photographers' reliance on stuffed animals as test objects, taxidermy's relationship to the Lippmann process unfolded in a laboratory environment. To locally disseminate his success and his knowledge of the science of colour, Neuhauss relied on a variety of optical media, including projection, which were tailored to individual as well as collective spectatorship.

Disclosing Colour: Projection and Other Optical Media

The technical genesis of the Lippmann plate is such that surface reflection, irreproducibility, and fragility are integral to its materiality, making it more geared towards an individual observer. The primary issue that all

2 Correspondence from Oskar von Miller to Otto Wiener from February 17, 1906. Object Identity. 605332. Inventory Number 7160. Deutsches Museum, Munich.

Lippmann photographers encountered and sought to resolve was the surface reflection of the image. It is the unwanted light from the top surface of the emulsion that dilutes the interference colours with white light, causing colours to change depending on the viewing angle. Hence, the interferential photograph refutes the "ease of decipherment" (Handy 2019, 28), as in the observer's ability to behold what is depicted in a split-second moment often associated with photography.

The photograph's visual elements demand a slower consumption, since they unfold only when it is carefully tilted, shedding light at times on one coloured part of the parrot and not another. A lot of awkward hand movement and bodily engagement are necessary for the calibration and the alignment of plate, hand, and eye in order to find the right angle from which the colours will reveal themselves and to unlock the elusive depiction. Beholding the plate forces the slow attention to the somatic aspect and the quality of one's own vision, since its direct visual impact remains locked in the observer's own bodily and spatial coordination.

Furthermore, in the 1900 *Handbook of Photography in Colours* published in London, the authors distinguish between the possibility of reproducing the plates in theory and their impossibility in practice. Reproduction could work by using parallel light on a backed interferential image to make a transparent positive. Yet, the colours are too faint to be transmitted and risk shifting in the process (Bolas, Tallent, and Senior 1900, 59). One exception is German physicist and engineer Hans Lehmann succeeded by using the three-colour halftone process to reproduce a Lippmann plate in colour depicting an oil painting in 1907 (Eder 1932, 974). Today:

> using the interreference of light waves creates images that cannot be copied with geometrical optical and/or physiological optical methods. A normal three-colour photograph can be copied very well on a very good colour copying machine. With an iridescent [...] image this is at least difficult. (Schröter 2014, 158)

Fully aware of these limitations, Neuhauss relied on optical devices to disclose Lippmann plates, such as the wedge prism, the *Schaukasten*, and the projector. Each device has its own unique way of negotiating with the image's materiality, generating at times a double feedback loop between the two of them. Far from seeing optical media as mere prosthetic accompaniment, or as expressions of a teleological quest for mimesis within photography, or as heralding the arrival of cinema, I contend that they function, rather, as "archives on a different level: that of modes of perception and systems of representation" (Turquety 2019, 53). Like the image whose success carries the great weight of evidence, its various optical media are also historically relevant. They are so both epistemologically and archaeologically. Epistemologically, they are representative of the viewing device's general abstract principle and the very "concrete detail that ensures its cohesion, properties (technical, economic, aesthetic), and singularity" (Turquety 2019, 34). Hence, they clarify how the Lippmann plate's materiality shaped the solutions to its difficult visibility. Archaeologically, they are machines that can historicise and materialise the perception, performance, and production of images (Turquety 2019, 25). Hence, they construct the "multisensory nature of the photographic experience" (Batchen and Gitelman 2018, 208) that single or collective spectators had of the Lippmann process. By focusing on the various problems they posed and solved, the optical media elucidate the spectatorial positions of Lippmann photography and explore the quality and the conditions of the observation they afforded.

Although surface reflection can be momentarily countered by manually handling the plate for as long as one is looking at it, Neuhauss strongly advised against it, writing "never handle it directly by hand" (*niemals direkt in der Hand*) (Neuhauss 1898, 57). Tactile engagement with the photograph was dangerous, since the examiner's fingers could leave print marks, and if they were to speak in the act of looking, their saliva stains would be too laborious to remove. Moreover, he knew that, for a long time, the plate could neither be protected through a varnish because the latter risked obfuscating its colours and reducing their lustre, nor could it be directly covered with glass, as was the case with projection slides, because it made it look "dull and powerless" (*flau und kraftlos*) (Neuhauss 1898b, 142). Also, to get the most out of its shades, the operator must use a

> diffuse unfocused light, and preferably natural skylight entering an otherwise darkened room, rather than by the impoverished yellow-tinted light of gas or electric lamps. The plate must also be lit and viewed nearly face-on so that the optical interference in the emulsion yields the correct colour (viewing at an angle shifts the colours towards the blue end of the spectrum). (Johnston 2016, 25)

In a laboratory environment, Neuhauss relied on physicist Otto Wiener's "container method" to assuage the surface reflection. In 1895, he reviewed Wiener's *Neue Untersuchungen über Farbenphotographie* (Neuhauss 1895b, 258), in which the author suggested placing the plate in a container filled with benzene or xylene. This way, the colours unfold through the careful tilting of the bowl enough for the liquid to form a small angle of refraction over the plate, thus preventing the light from being reflected on the surface (Wiener 1899, 489). The plate's surface reflection decreases while its iridescence increases and, with it, the contrast of its colours, without the spectator having to touch the photograph much. The perceptual experience this solution enables is that of beholding the image "under water" while gently rocking its container. The advantage this method presents is that, since the liquid shares the same refraction index with glass (n = 1.5), it gives the photographer a sense of what the plate would look like once attached to a glass prism. Although Neuhauss tried this method several times, it was, however, less adept to uninformed users in a more relaxed setting.[3]

For both casual individual use and for collective experience, a wedge prism was often needed to safely bring the plate out of the laboratory and into the world. Usually, the most iconic and successful of all plates, such as those of the stuffed parrot, were protected with the wedge prism.[4] Its implementation was consistent across numerous countries.[5] Specifically in the German Empire, companies such as Picht & Co. in Rathenow and Dr. A. Hesekiel and G. Braun in Berlin manufactured prisms tailored to the size of the interferential photograph (Neuhauss 1900b, 51). Neuhauss cemented the prism, as was common, onto its emulsion side using Canada balsam, while he painted the back side of the plate black, and glued a paper onto it, where he signed his name, wrote the name of the photographic process, and dated the image. Once put together, the prism can no longer be removed without annihilating the emulsion, thus fusing the image and optical device into a single hybrid. In relation to the world, the prism hermetically seals and protects the colours from pollutants and humidity variations (Bjelkhagen n.d). In relation to the viewer, it is a permanent separation between image and observer, locking the image away from idle hands that are now confronted with its "objectness" and increased weight. It suppresses the surface reflection as much as possible and increases the iridescence of the plate, thereby contributing to its status as a photograph worth keeping.

3 Richard Neuhauss' Journals 1894–1900. Entry number 82: 19–10, 1899, The Lippmann Color Process. Prof. Dr. Albert Narath Archives, Preus Museum, Norway.

4 Lippmann plates made by Neuhauss and featuring the stuffed parrot with a wedge prism can be found at Deutsches Museum, the Victoria and Albert Museum, and Lette Verein, among others.

5 Several practitioners of the Lippmann process used it independently of each other: Lippmann, Ives, I. F. Usagin, and Neuhauss all mention it around the years 1895–96.

For the individual or (a limited number of) spectator(s) to "calmly" (*in Ruhe*) (Neuhauss 1894b, 363) observe the colours in a more intimate setting and at greater ease, the plate along with the prism could be inserted inside of a viewer called *Schaukasten*, which Neuhauss initially designed in 1894. It was endowed with a convex lens (*Sammellinse*) and unlike most viewing devices at the time, it was also uniquely furnished with a flexible ball-and-socket joint (*Kugelgelenk*), controlling the plate so that the user can determine the optimal position for the iridescent image to reveal its colours, especially when placed near a window to get daylight or in the vicinity of an opal globe. Subsequent users need only look through the box to see them, with no further manipulation needed. Its proximity to the spectator's face and its capacity for somewhat enlarging the image intensified the experience of colour and made it more intimate. Neuhauss presented one interferential plate with the *Schaukasten* at the world exhibition in Sydenham near London in 1898 and six plates with the same device at the Paris world fair two years later (Neuhauss 1900a, 126). In the latter case, he also turned the viewing device into a vehicle for seriality, as in "the accompaniment of one photograph by another, and their arrangement in sequence" (Ruchatz 1998, 31). Its curatorial use in such an important setting implies that Neuhauss considered the *Schaukasten* to be a legitimate solution to the material vulnerability and visibility of the Lippmann plate.

But this optical solution engendered new problems. Neuhauss reconsidered his design in 1898, because showing several plates one after the other mechanically damaged them. The second version of the optical tool consists of a wooden base board over which two transparent mirror glass panes are attached like a roof, slightly magnifying the image. The wider glass pane absorbs daylight, while the other pane is the one through which the observing subject looks at the image. The contraption is shielded by matte covers held together by two rubber cords to make sure that only a certain amount of light enters. Spectators could use the *Schaukasten*'s long handle to hold it to their faces or as a prop to place the device on a table and look into it. Neuhauss considered this hands-free position "optimal" (*günstigste Lage*) to gaze upon colours (Neuhauss 1898b, 141). Overall, both models of the *Schaukasten* rendered the experience of the colour image intermedial thanks to their seriality and magnification of the image – key attributes they share with projection.

By the time Gabriel Lippmann first revealed his colourful plate to the French Académie des sciences in 1891, William and Frederik Langenheim had been projecting photographic glass slides since 1845. Thus, projecting interferential colour can and should be inscribed in the tradition of disclosing photographs via projection. Projection transformed colour photography from a visual experience enjoyed alone or in small company into a "multimedia spectacle" for large audiences in both scientific and amateur circles, one that involves seriality, magnification, and oral performance by the speaker. In turn, interferential colour photography propelled the "rationalisation" of the projector, turning it increasingly into a scientific instrument.

For a projection session to be successful, it necessitated a photographic society that could pay for an instrument as expensive as an optical lantern, an agenda compelling enough to attract the crowd, and a prestigious place to hold the sessions. Having co-founded the Vereinigung, Neuhauss found that projection events were significant for financing the camera club and ensuring its continuous success (Neuhauss 1908, 135). They were also a platform in Berlin for the regular presentation of photographs of all parts of the world made by club members. Thus, the Vereinigung's ambition to feature documentations of the world through a financially viable spectacle should be understood as a product of the German Empire's colonial governance and as a result of anthropologists's use of photography as a method to construct otherness and to sell it as education. These historical conditions also account for the choice of host institution for these events, the Museum für Völkerkunde, following a long negotiation between Franz Goerke from the Vereinigung and the museum's founder and first director Adolf Bastian (1826–1905).

The projection of interferential colour photographs took place at a time when the instrument was being "rationalised." The rationalisation of the projector is both a discursive and a material shift by which the eighteenth-century magic lantern, with its phantasmagorical ties, changed into the "optical lantern," a scientific and widely accepted educational instrument in the mid-nineteenth century (Ruchatz 2000, 43, Eisenhauer 2006, 204). Neuhauss's knowledge of projection exemplifies the interdisciplinarity of the instrument: his earliest texts focus on the different projectors needed for presentations of all things microscopic, such as photomicrographs, drawings, or biological-cellular samples in scientific studies (Neuhauss 1894, 41). Also, the Lucaesian apparatus, which functioned as both drawing machine and geometric projection device and was designed by German anthropologist Gustav Lucae, was extensively used to facilitate craniological measurements in anthropology – Neuhauss's second discipline (Zimmermann 2001, 99). Projection catered to several disciplines, ranging from art history to the natural sciences, and especially in the "material practices firmly grounded in experiential knowledge" (Schmidgen 2011, 46). Its popularity created a "boom in lantern construction towards the end of this period when German intellectuals suddenly connected to modern methods of visual education that had long been used elsewhere" (Rossell 1999, 20). Manufacturers answered these demands by building versatile and user-friendly devices capable of accommodating heterogeneous objects and disciplines.

The change in the instrument was also material. The optical lantern distinguished itself from its predecessor by a stronger light source with a reflector, an achromatic lens, a condenser that guided and intensified the light hitting the projection slide, and "a more austere look than the ones used by street lanternists, or those made as children's toys. Made of black, varnished metal or covered in wood" (Sandrin 2017, 52). Moreover, both the accuracy of the photographic slide and the reliance of the projector on electricity propelled it more in the realm of science, making it adequate for the projection of Lippmann plates. It is in fact only *after* the projection of Lippmann photography that projection as such would become "the preferred means of visualising these new types of colour images such as the autochrome and three-colour photography" (Boulouch and Gisinger 1999, 45). Neuhauss considered the display of Lippmann plates as a "special purpose" (*besondere Zwecke*) of projection (Neuhauss 1908, 95), since they are part of the category of "opaque objects" (*undurchsichtiger Gegenstände*) (Neuhauss 1908, 102). The plates are distinct from the traditional glass slide positive that refers to "a transparency developed out of an initial negative with the ability to be traversed by light" (Boulouch 2014, 3). Their opaqueness is the result of its encasement in the glass prism and of the numerous silver layers in them, which "reflect the light rays like a mirror in a certain direction, pulling them one way" (Neuhauss 1908, 96). So this needed an optical lantern capable of megascopic projection of opaque objects. At the Museum für Völkerkunde's assembly hall, the Vereinigung was allowed to permanently place a lantern it purchased in 1894 from Schmidt & Haensch, which, as was typical for its day, was entirely funded by the club members' donations. It was endowed with a strong light source, three condensers of 16 cm diameter each, which could accommodate the Lippmann plate's various formats: 9 x 12 cm, 6 x 6 cm, 6 x 9 cm, (Lavédrine 2009, 69) and 3 x 11 cm in case of spectrographic images made by photographer Hermann Krone.

Projection solved the problem of the colour shift of the plate and best helped the image reach total collective visibility, since it implies increasing its size. Unlike colourised slides, the Lippmann plate had a lot to gain and little to fear from magnification. While researching stereoscopic photography, scientist Sir David Brewster considered the mediocre quality of hand-painted lantern slides as the chief culprit for the reticence towards the optical lantern in the early nineteenth century (Ruchatz 2000, 45). Photographic slides greatly improved the image's quality and consistency, thereby "rationalising" the projector. They freed it "from its entanglements with the perceived imperfection of the rendering hand, elevating its production to an entirely technicologically-mediated correspondence between one technology of vision, the camera, to another, the projected image" (Eisenhauer 2006, 201). Hand-painted slides, photographic or not, reveal all the imperfections of the painter's technique once they are magnified – and disappoint the audience. Magnifying the Lippmann plate, however, proved the *mechanical* inscription of colour and disproved colourisation. It ascertained the distance between it and any manual application of pigments unto the surface of a photograph associated with the magic lantern (Eisenhauer 2006, 200). It also asserted the difference between its own high photographic resolution and unparalleled accuracy and that of the monochrome camera slide. It greatly exceeded the photographic precision offered by any material at the time, rationalising the projector even more. Magnification also allowed the Lippmann plate to overcome its small size and to have its colours astonish the senses, making them "truer to life beyond their original stature" (*viel lebenswahrer als die kleine Originalaufnahme*) (Neuhauss 1894c, 177).

Often, several photographs, both colour and monochrome, were projected and magnified, one after the other. In 1896, at the Gesellschaft zur Förderung der Amateur-Photographie in Hamburg's projection session, Neuhauss revealed his Lippmann plates, along with his slides depicting engineer Otto Lilienthal's attempted flights, the aforementioned photographs from the zoological garden of Berlin, among others, all under the umbrella term of "projection lecture" (*Projektionsvortrag*) (*Vereinsnachrichten* 1896, 38). Also, Neuhauss's Lippmann plates were showcased in October 1894 in Halle (Saale) after Charles Scolik's presentation of the Oriental Society's slides of Palestine and Egypt, and were followed by Gustav Droysen's slides featuring modern painting, Dr. Schmidt's depiction of illness, and Franz Goercke's images of the Baltic sea (Photographische Gesellschaft zu Halle 1895, 13). Seriality, along with the narrative embedded in the selection and in the order of the photographs, loosens the boundary between still and moving images. It testifies to projection's ability to unite the various photographic formats and depictions in one homogeneous visual and auditory presentation. It situates Neuhauss's research on Lippmann within the complete panorama of his overall photographic work while juxtaposing pervasive anthropological, colonial, and orientalist images projected alongside Lippmann plates culturally legitimises the colonial gaze as scientific. Also, Neuhauss demonstrated the various steps needed to master the Lippmann process and the theoretical basis of his work in 1895 at the Physikalische Gesellschaft zu Berlin (Berlin Physical Society). During this session, he projected several spectrographic and ten mixed-colour plates, augmenting them to two by two metres (Neuhauss 1895, 21), in the presence of knowledgeable colleagues in order to convince them of the stability of his results. The serial and magnified projection gave credibility to the Lippmann process thanks to its display of the repeatability of Neuhauss's experiments that engendered, not one, but several pristine plates. Projecting interferential photography confirmed its ability to hold up to collective scrutiny and suspicion of colourisation, as it astonished the audience.

Furthermore, the Lippmann plate required electricity for it to reach its full visual potential, as in an even more "rationalised" scientific optical lantern. In projection, light does not simply "illuminate existing scenes; it creates them" (Schivelbusch 1988, 220) thanks to "the new and enhanced brightness of the lantern's illuminants" (Vogl-Bienek 2015, 217) operating in parallel to the darkening of the auditorium. This fact, along with the reliance on a white screen, which is the best possible reflective material (Neuhauss 1908, 97), was a shared feature between photographic and cinematographic projections. As an article in the *Rundschau* confirms, the darker the room, the more pictorial (*plastischer*) the enlarged photograph will appear (Rieck 1888, 11). This play between light and shadow sought to articulate "a direct communication between the spectator and what was being presented, to the exclusion of all distracting, external factors" (Schivelbusch 1988, 206). These distractions came in many forms and forced the speaker presenting the images to be attentive to their performance to maintain the audience's attention, such as the stench of kerosene light, which must be replaced by stage lighting (*Kalklicht*) and by arc lamps (*Bogenlicht*) (Neuhauss 1908, 99). Neuhauss also stressed the importance of the speaker's or photographer's own performance during the projection session. He asserted that people often waited anxiously for the appearance of the first slide and should be given the satisfaction of beholding it without first being exhausted by a long speech. He also recommends a short break between the slides so that the image sinks into the spectator's psyche. Ensuring the synchronicity between the image on the screen and the voice of the speaker means forming an "invisible union" (*unsichtbarem Zusammenhang*) between them (Neuhauss 1908, 132) that goes hand in hand with the erasure of reflexivity. Neuhauss recommends that the speaker stand next to the projector, directing their voice to the ceiling so that the audience in front and behind the machine can hear well. They should also adjust their speaking pace to the rhythm of the slides, improvising just enough to make the session "lively" (*lebendig*) (Neuhauss 1908, 132). A performer standing at the front can be seen by the audience giving an auditory or optical signal to the person changing the slides, thus thwarting the public's excitement and surprise for the upcoming image (Neuhauss 1908, 132). Successful projection means carefully orchestrating speech, image, and gesture to create a heightened sensual and intermedial experience of the image, one that produces "a bodily impression" on the audience (*Eindruck des Körperlichen*) (Neuhauss 1894c, 177).

The Lippmann plate helped rationalise the optical lantern by severing its discursive ties with the magic lantern's hand-coloured slides. In turn, the projector helped sensualise interferential photography by transmitting its magnified mechanical colours. It also disclosed the success of the Lippmann process's test results, which engendered flawless tints.

Conclusion

The Lippmann process was excluded from the "general" history of photography and was marginalised in the history of colour photography for numerous reasons. The conception of it in isolation from other media and coeval photographic practices, added to its challenging materiality (surface reflection and irreproducibility), have contributed to its historiographic erasure. Hence, connecting the Lippmann plate to the various media that shaped the science behind it, such as taxidermy, as well as its reception, such as optical devices, is foundational for updating its history. In the case of Neuhauss, his calibration of colour-sensitive emulsions took place in relation to the colours of a colonial animal-object par excellence – a stuffed bird whose very presence in the laboratory was enabled by another technology of empire: taxidermy. The relationship between both is such that taxidermy gave colour photography what it coveted best: immutably coloured beautiful skin or plumage that can be used as tools to baseline emulsion sensitivity and an animal that is docile enough to be artistically depicted or scientifically tested as long as it takes to reach the ideal calibration. Colour photography gave taxidermy the only mechanical image that is truly in its own likeness, one that best encapsulates what taxidermy preserves best: skin. The reference to stuffed beasts around 1900 and the complication of their representation remained central to the triangle of colour, nature, and vision. Other than taxidermy, the Lippmann process can only be understood in relation to the various optical media that allowed for its disclosure. The wedge glass prism turned it into a hybrid, safe from pollutants and easier to behold, especially when inserted into a viewing box *Schaukasten*, which Neuhauss tailored to individual spectators looking to experience his Lippmann plates in series.

Projection is a foundational medium with which interferential photography had a special and long-lasting relationship. Interferential photography helped further rationalise the projector by offering it an accurate, high-quality, and mechanical colour rendition, unmatched by any other technology at the time. In turn, projection introduced it to entertainment and scientific circles, allowing for a collective and intermedial experience of its magnified shades, transforming the mere act of looking upon a coloured image into a multimedia spectacle. The focus on the Lippmann process's ties to taxidermy and projection connects media history to colour photography and brings it out of its historic vacuum, as it carves a new space for it in the current scholarship.

Acknowledgment

I warmly thank Darran P.M. Green and William R. Alschuler for our inspirational discussions and fruitful exchanges.

Works Cited

"Vereinsnachrichten".1894. *Photographische Rundschau.* 8 (1).1.

"Vereinsnachrichten".1896. *Photographische Rundschau.* 10 (3). 38.

Andrew Zimmerman. 2001. *Anthropology and Antihumanism in Imperial Germany.* Chicago: University of Chicago Press.

Batchen, Geoffrey and Lisa Gitelman.2018. "Afterword: Media History and History of Photography in Parallel Lines". In *Photography and Other Media,* edited by Nicoletta Leonardi and Simone Natale. Pennsylvania: The Pennsylvania State University Press.

Bjelkhagen, Hans. n.da. "The True Colours of Photography". Accessed February 22, 2021. https://www.academia.edu/23030595/The_True_Colour_of_Photography

Bolas, Thomas, Alexander A. K. Tallent, and Edgar Senior. 1900. *A Handbook of Photography in colors.* New York: E. & H.T. Anthony & Co. https://archive.org/details/handbookofphotog00unse.

Boulouch, Nathalie, and Arno Gisinger. 1999. "Der Große Erfolg der Autochrome-Platten liegt in ihrer Projektion": das projizierte Bild als privilegierte Präsentationsform früher Farbfotografie". *Fotogeschichte.* vol.74, 45-58.

Boulouch, Nathalie. 2001. "Peindre avec le soleil ?" *Études Photographiques.* vol. 10. 50–75.

Boulouch, Nathalie. 2014. "Photographie Illégitime, Cinéma du Pauvre : le Destin Impossible de la Diapositive". *Intermédialités.* (24–25): 24–25.

Claas, Ulrike, and Paul Roscoe. 2009. "Hot Air and the Colonialist 'Other': the 'German-English-Dutch airship expedition' to New Guinea". *The Journal of the Royal Anthropological Institute, Incorporating Man.* 15 (1): 131–150.

Correspondence between Franz Goerke of the *Freie Photographische Vereinigung zu Berlin* and Museum für Völkerkunde in 1893. Reference Number597, MV, IVa, 1893/02020–1920/1043. File Number: 202/93. Zentralarchiv, Staatliche Museen zu Berlin-Stiftung Preußischer Kulturbesitz.

Correspondence from Oskar von Miller to Otto Wiener from 17.02.1906. Object Identity. 605332. Inventory Number 7160. Deutsches Museum, Munich.

Diapositive: Histoire de la Photographie Projetée : [exposition, Lausanne, Musée de l'Elysée, du 1er juin au 24 septembre 2017]. Edited by Lacoste, Anne, Boulouch, Nathalie, Lugon, Olivier and Carole Sandrin[Lausanne]: Les Ed. Noir sur blanc. Exhibition catalogue.

Eder, Josef Maria. 1932. *Geschichte der Photographie 2*. Halle (Saale): Knapp.

Eisenhauer, Jennifer F. 2006. "Next Slide Please: The Magical, Scientific, and Corporate Discourses of Visual Projection Technologies". *Studies in Art Education.* 47 (3): 198–214.

Erkki Huhtamo. 2018. " Elephans Photographicus: Media Archaeology and the History of Photography". In *Photography and Other Media,* edited by Nicoletta Leonardi and Simone Natale. Pennsylvania: The Pennsylvania State University Press.

Geimer, Peter. 2018. Inadvertent Images: A History of Photographic Apparitions. Trans. By Gerrit Jackson. Chicago: University of Chicago Press.

Handy, Ellen. 2019. "Dancing with Images: Embodied Photographic Viewing". *Open Arts Journal* issue 7, 27-39.

Hauser, Kitty. 1998. "Coming Apart at the Seams: Taxidermy and Contemporary Photography". MAKE: The Magazine of Women's Art (82): 8-11.

Henning, Michelle. 2006. "Skins of the Real: Taxidermy and Photography". In *Nanoq: Flat Out and Bluesome: a Cultural life of Polar Bears,* edited by Snæbjörnsdottir, Bryndís, and Mark Wilson. London: Black Dog Publishing.

Henning, Michelle. 2007. "Anthropomorphic Taxidermy and the Death of Nature: The Curious Art of Hermann Ploucquet, Walter Potter, and Charles Waterton." *Victorian Literature and Culture.* 35 (2): 663–678.

James R. Ryan. 2000. "'Hunting with the Camera': Photography, Wildlife and Colonialism in Africa". In *Animal Spaces, Beastly Places New Geographies of Human-Animal Relations,* edited by Philo, Chris, and Chris Wilbert. London: Routledge.

Johnston, Sean. 2016. *Holograms: A Cultural History.* Oxford: Oxford University Press.

Joschke, Christian. 2014. *Les yeux de la nation: Photographie Amateur et Société dans l'Allemagne de Guillaume II,* 1888–1914. Dijon: Presses du réel.

Lavédrine, Bertrand. 2009. *Photographs of the Past: Process and Preservation.* Los Angeles: Getty Conservation Institute.

Leonardi, Nicoletta and Simone Natale. 2018. "Introduction". In *Photography and Other Media,* edited by Nicoletta Leonardi and Simone Natale. Pennsylvania: The Pennsylvania State University Press.

Ludwig Maria Vogl-Bienek. 2015."Screening Sensations and Live Performance: the Creative Blending of Traditional and New Projected Media at the Start of the Twentieth Century Screening Sensations and Live Performance". In *Performing New Media, 1890–1915,* edited by Yumibe, Joshua, Williams, Tami, Pelletier, Louis, Gray, Frank, Curtis, Scott and Kaveh Askari. Bloomington: Indiana University Press.

Maleuvre, Didier. 1999. *Museum Memories: History, Technology, Art.* Stanford (Cal.): Stanford University Press.

Mitchell, Daniel Jon. 2010. "Reflecting Nature: Chemistry and Comprehensibility in Gabriel Lippmann's "Physical" Method of Photographing Colours". *Notes and Records of the Royal Society of London.* 319 - 337.

Mody, Cyrus C.M., and Michael Lynch. 2010. "Test Objects and Other Epistemic Things: a History of a Nanoscale Object". *The British Journal for the History of Science.* 43 (3): 423–458.

Neuhauss, R.1895. "Ueber die Photographie in in natürlichen Farben".*Verhandlungen der Physikalischen Gesellschaft zu Berlin.* 13. 17–24.

Neuhauss, Richard. 1894a. *Die Mikrophotographie und die Projection.* Halle a.S.: Knapp.

Neuhauss, Richard. 1894b. "Die Photographie in natürlichen Farben". *Photographische Rundschau.* 8 (11). 327–329. 359–364.

Neuhauss, Richard. 1894c. "Ueber den Bau der Projektions-Apparate". *Photographische Rundschau.* 8 (6). 177–180.

Neuhauss, Richard. 1895a. "Fragekasten". *Photographische Rundschau.* 9 (5): 159.

Neuhauss, Richard. 1895b."O.Wiener's neue Untersuchungen über Farbenphotographie". *Photographische Rundschau.* 9 (9): 257–260.

Neuhauss, Richard. 1895c. "Photographie in Natürlichen Farben". *Photographische Rundschau.* 9 (1): 24–25.

Neuhauss, Richard. 1896a. "Das Photographiren in Zoologischen Gärten." *Photographische Rundschau.* 10 (3): 71–75.

Neuhauss, Richard. 1896b. "Die Abtheilung für Wissenschaftliche Photographie."*Photographische Rundschau.* 10 (3): 297–299.

Neuhauss, Richard. 1896c. "Selle's Farbige Aufnahmen". *Photographische Rundschau.* 10 (4): 110–112.

Neuhauss, Richard. 1898a. *Die Farbenphotographie nach Lippmann's Verfahren neue Untersuchungen und Ergebnisse.* Halle a. S: Knapp http://digital.slub-dresden.de/id1733654569.

Neuhauss, Richard. 1898b."Weitere Untersuchungen über Photographie in natürlichen Farben". *Photographische Rundschau.* 12 (1): 1–9. 140–146.

Neuhauss, Richard. 1900a."Die Photographie auf der Pariser Weltausstellung". *Photographische Rundschau.*14. 125–132.

Neuhauss, Richard. 1900b. "Weitere Untersuchungen über Lippmanns Farbverfahren". *Photographische Rundschau.*14. 48–52.

Neuhauss, Richard. 1908. *Lehrbuch der Projektion.* Halle a.S.: Knapp.

Neuhauss, Richard. 1911. *Deutsch Neu-Guinea. Band I.* Berlin: D. Reimer.

*Photographische Gesellschaft zu Halle a.S. XXVII. Sitzung Montag, den 8. October 1894, Abends 8 Uhr im Hotel "zum goldenen Ring".*1895. *Photographische Rundschau.*9 (2): 12–14.

Poliquin, Rachel. 2012. *The Breathless Zoo: Taxidermy and Cultures of Longing.* Pennsylvania: Pennsylvania State University Press.

Richard Neuhauss· Journals 1894–1900. The Lippmann Color Process. Prof. Dr. Albert Narath Archives, Preus Museum: The National Museum of Photography, Norway. https://holowiki.org/data/lippmann/Neuhauss_Journals.pdf

Rieck, Ernst. 1888. "Werner's Vergrösserungs- und Nebelbilder-Apparat (Verbessertes Scioptikon)". *Photographische Rundschau.* 2 (1): 7–11.

Rossell, Deac. 1999. "Überblendungen: Aspekte der Laterna Magica-Projektion in Deutschland". *Fotogeschichte.* Vol. 74, p.13-23.

Ruchatz, Jens. 1998. "Ein Foto Kommt Selten Allein. Serielle Aspekte der Fotografie im 19. Jahrhundert". In *Fotogeschichte*, Nr. 68, p. 31-46.

Schindlbeck, Markus. 2000. "Vulkane und Leprakranke: Richard Neuhauss und Seine Photographien von Hawaii". *Baessler-Archiv* XLVIII. 345–378.

Schivelbusch, Wolfgang. 1988. *Disenchanted Night: The Industrialization of Light in the Nineteenth Century.* California: The University of California Press.

Schlesische Gesellschaft von Freunden der Photographie in Breslau. Sitzung vom Freitag, den 15. März 1895, Abends 8$^{1/2}$ Uhr im Höcherl-Bräu."1895. *Photographische Rundschau.*9 (6): 8–9.

Schmidgen, Henning. 2011. "1900–-The Spectatorium: On Biology's Audiovisual Archive". *Grey Room.* 1 (43): 42–65.

Schröter, Jens. 2014. *3D: History, Theory, and Aesthetics of the Transplane Image.* New York: Bloomsbury.

Schütt, F. *Freie Photographische Vereinigung zu Berlin. Deutsche Gesellschaft von Freunden der Photographie in Berlin. Gemeinsame Sitzung am 16. November 1894.*1895. *Photographische Rundschau.*9 (1): 4–6.

Timby, Kim. 2015. *3D and Animated Lenticular Photography. Between Utopia and Entertainment.* Berlin: Walter de Gruyter & Co.

Turquety, Benoît, and Timothy Barnard. 2019. *Inventing Cinema: Machines, Gestures and Media History.* Amsterdam: Amsterdam University Press.

Vogel, H.W, 1898. " Beobachtungen an farbigen Interferenz-Photographien". *Photographische Mitteilungen.* 34 (19). 300–301.

Wiener, Otto Heinrich. 1899. *Ursache und Beseitigung Eines Fehlers bei der Lippmann'schen Farbenphotographie, Zugleich ein Beitrag zu Ihrer Theorie.* Leipzig: J.A. Barth.

Wilder, Kelley. 2009. *Photography and Science.* London: Reaktion Books.

Wilder, Kelley. 2020. "Science, Art, and the Business of Color". *The Colors of Photography / Edited by Bettina Gockel ; in Collaboration with Nadine Jirka and Stella Jungmann.*

About the Author

Dr. **Hanin Hannouch** is curator for analog and digital media at the Weltmuseum Wien (Vienna). She has guest-curated the exhibition "Slow Color Photography" about interferential colour photography at Preus Museum: Norway's National Museum for Photography and is the guest-editor of the journal PhotoResearcher Nr. 37 "Three-Colour Photography around 1900: Technologies, Expeditions, Empires". She was post-doctoral researcher at the Kunsthistorisches Institut in Florenz, Max-Planck Institut and at the Ethnologisches Museum in Berlin. She is currently writing her monograph about the history of colour photography in Imperial Germany.

7. Brilliant! Enthusiasm for the Aesthetic Qualities of Lippmann's Interferential Photography

Elizabeth Cronin

Abstract

In 1908 Edward Steichen wrote in *Camera Work*: "... we must not forget Professor Lippmann, who gave us what is undoubtedly the most wonderful process of colour photography." Comparing the striking colour to that of a good Renoir, Steichen placed aesthetic value on the photographs and yet, the Lippmann method is remembered primarily for its scientific value and for its failure as a stunning but impractical process. This chapter resituates the Lippmann process as a visually compelling object by examining what its practitioners thought about it, where they exhibited the plates and how they were received. Lippmann photographs represent more than difficult science. They were the product of a diverse web of scientists, artists, and photographers all seeking to advance colour photography.

Keywords: Gabriel Lippmann, Edward Steichen, Josef-Maria Eder, art history, aesthetics, interferential colour photography

For the first time, in 1908, the high-quality photographic art journal *Camera Work* dedicated its pages to colour photography. It was not the only journal to do so. *The Studio*, a British art photography journal, also published a special summer issue celebrating colour photography. These two issues showcased the autochrome, a colour process that relied on dyed grains of potato starch for its vivid colour. Within a year of its marketing, artistic photographers such as Heinrich Kühn (1866–1944), Baron de Meyer (1868–1946), and Alvin Langdon Coburn (1882–1966) were making their own autochromes, which were then reproduced in these volumes via a four-colour halftone process.

Hannouch, H. (ed.), *Gabriel Lippmann's Colour Photography: Science, Media, Museums.*
Taylor & Francis Group, 2022
DOI 10.5117/9789463728553_CH07

The autochrome, however, was not the first colour method to be reproduced, nor was it the first to have commercial application. Among the many different solutions for colour photography, Gabriel Lippmann's interferential colour process had already emerged triumphantly, as it dazzled the eye and caught the imagination of scientists and viewers alike. It was introduced into the market as a viable process and yet its success, if one can even call it that, was short-lived.

Only referencing Lippmann's interferential photography in passing, both *Camera Work* and *The Studio* introduced the nascent history of colour photography by explaining the myriad attempts to reproduce colour. Writing for *Camera Work*, Edward Steichen declared, "[...] we must not forget Professor Lippmann, who gave us what is undoubtedly the most wonderful process of color photography" (Steichen 1908, 14). In contrast, Dixon Scott's commentary in *The Studio* was less complimentary, complaining that, "it would take a full volume to explain the useless process of Lippmann" (Scott 1908, 3). (At the time, there were in fact already many full volumes explaining the process and none of them described it as useless.) While most photographers and scientists had accepted that no widespread commercialisation would come of Lippmann's method, they felt the need to mention it and some practitioners, such as the American Herbert Eugene Ives (1882–1953), even continued to improve upon it.[1]

Throughout its short history, interferential photography was simultaneously lauded for its absolute brilliance and disparaged for its dismal complexity. Either way, it remained intriguing for much longer than other failed methods. Why? As is often explained in the literature, the physicist Gabriel Lippmann substantiated a scientific theory with his process, for which he won the Nobel Prize in 1908, although a variety of political reasons also played a role (Mitchell 2010, Connes 1987). Lippmann's process, however, also produced incredibly beautiful and eye-catching photochromes, a fact that has so far eluded scholarship on the process. Adhered to sheets of glass, the radiant visual colour of the interferential photographs is distinctive, and it made a great impression on those who witnessed it. It is therefore not surprising that scientists, photographers, and critics almost always described interferential photochromes as "brilliant," "luminous," "beautiful," or a combination thereof. This article re-evaluates interferential photography based on its production of visually compelling objects. It contends that the aesthetic value and attraction of Lippmann plates contributed greatly

1 Ives succeeded in creating interferential photographs without using mercury but used celluloid as a mirror instead.

to the enthusiasm for the process, which in turn helped solidify its brief commercialisation, exhibition, and ultimately, its memorialisation. In expanding the engagement with interferential photography, locating it between scientific circles and artistic ones, the restrictive boundaries between "art" and "science" can be loosened and a better understanding for the process, its importance, and its historical sway can be gained.

Discovered and announced in 1891, Gabriel Lippmann's process was unlike any of the previous "solutions" to colour photography, for it relied solely on the physical phenomenon of an interference between incidental and reflected light waves for its colour. Interference happens when reflected light waves directly interfere with incident light waves, causing the formation of stationary waves, which produce layers of colour. The most common example is found on a soap bubble or an oily film on water. Lippmann's new technique instantly attracted and mesmerised scientist photographers, and they worked ardently to improve its ease, complexity, and material peculiarities. Even Lippmann desired to perfect his own method, which required a transparent emulsion, and he continued to make refinements and exposures. The year following his announcement, he presented four additional colour photochromes, each with a permanently fixed image, to the Paris Académie des sciences (Valenta 1893, 18–19; Valenta 1912, 58). The subjects he photographed aimed to show the potential applications of his method, because they demonstrated a variety of colours and subjects: a stained-glass window, a plate of oranges with a red poppy flower, a bunch of flags, and a stuffed bird. In October of the same year, Lippmann refined his process even further to produce plates without using silver salts (Eder 1894, 448).

As Lippmann's adjustments continued, journals throughout Europe and the United States reported on his discovery, excitedly characterising the genius of his method and his images. Such was the enthusiasm for Lippmann's method that in 1893 both German photochemist Hermann Wilhelm Vogel (1834–1898) and Viennese chemist Eduard Valenta (1857–1937) compared Lippmann's interferential photographs to the state of daguerreotypes in their early years (Vogel 1893, 285–86; Valenta 1893, 22). The comparison is apt – sort of. The two media were physically similar, they were both produced on hard substrates, and, with their reflective surfaces, they both had to be held at a specific angle to the light source for the image to be seen clearly. They both required long exposures, were not reproducible, and the image was often permanently fixed and stable. Daguerreotypes, however, proliferated quickly and their production was not as complicated as that of Lippmann's process. Already in the early 1840s, hundreds of daguerreotypists could be found throughout the world. In this light, Vogel's comparison seems overly optimistic. Nevertheless, other scientists who diligently investigated and worked with Lippmann's process must have shared Vogel's sentiment that the process would become widespread, despite its real and obvious complexities. Valenta, for example, produced hundreds of spectra as he investigated the science behind colour photography. Highly concerned with the physical properties of natural sunlight and artificial light, he examined the reaction of various coloured light waves to different chemicals as well as the colour gradations of spectrographs. In 1894, he published *Photography in Natural Colours*, elucidating the technicalities of the interference method. A valuable and highly detailed resource, Valenta's text was careful to refer to other scientists who worked with the method and their contributions towards improving it.

Around the same time that Lippmann's process and other contrasting methods like three-colour photography were burgeoning, photography was starting to be recognised as a serious art form. This timing led to an interesting intersection of science and art, promoting debates about what constituted art photography and what the application of colour photography might mean for the medium's artistic aspirations. In 1905, for example, Edmund von Sallwürk, a literary professor, wrote an article titled "Farbenphotographie und Kunst" for a journal dedicated to artistic photography. Sallwürk believed art photography required a skilled artist who conveyed aesthetic individuality. In contrast, he thought colour photography could be learned by anyone (Sallwürk 1905, 244). Referring to the additive three-colour process, he wrote that colour photography could not communicate any personal touch because it was highly unpredictable in its results. From his perspective, which other critics shared, colour photography simply could not have the same status as an artwork due to the scientific parameters that underpinned it and its all too truthful depiction of life. In short, he found it too technical to become a creative process. In 1907, photographer Heinrich Bachmann similarly criticised the technical process of colour photography. Speaking on the occasion of an exhibit of art photography, he advised amateur photographers to use some colour (in the form of gum prints) to create a mood or add effects, but in no way would he recommend trying to exactly reproduce nature (Bachmann 1907, 229–234). Bachmann reasoned that a painter would want to leave a personal mark and would not dare to literally copy the colours of nature. Art photographers, he wrote, must similarly produce creative work and not exact copies.

Neither Sallwürk nor Bachmann wrote specifically about the Lippmann process, but it can be assumed that they would have been equally dismissive of its use as a form of art. The lack of reference to it points to the limited application and use of Lippmann's process, more than a decade after its announcement. Amateur photographers in clubs were presumably neither exhibiting nor practising interferential photography. Instead, they were experimenting with more widespread and straightforward indirect colour processes, like gum and additive three-colour prints.

While cultural critics derided what the "art" they believed to be as purely scientific pursuits, scientists interested in the Lippmann process were equally attuned to and critical of these indirect methods of colour photography favoured by artists and amateurs. In 1898, Richard Neuhauss, who was one of the most prolific and successful interferential photographers, published his own book on the Lippmann process. In it, he stated that Lippmann's method produced the only true colour photography and complained that obtaining colour photographs by means of three-colour printing left the choice of colours completely under the control of the photographer (Neuhauss 1898, foreword). His comment reflects the common idea that manipulated photographs were more artistic and those that, like the Lippmann process, archived light without an alleged human intervention were more scientific and truer to nature. Years later, an exhibition organised by the scientific and technical department prompted Neuhauss's assessment of the state of colour photography in the art photography annual *Photographische Kunst*.[2] The artistic works included in the exhibition led Neuhauss to question the types of colour photography that can be artistic. He believed colour photography in art exhibitions, however impressive to the layperson, were not true representations of natural colour (Neuhauss 1907, 12). In describing the various kinds of direct and indirect colour photography, he declared that most artists lacked the technical know-how and equipment to produce a three-colour photograph. Even if they could do it correctly, he believed, then the photographs would still be too "natural" and not artistic enough (Neuhauss 1907, 24). With his slightly condescending tone, Neuhauss reinforced the idea that art photography should not be realistic, but rather should show the creative hand of the artist. In other words, Neuhauss thought colour photography's ability to mimetically render the world discounted it as an expression of individuality. He seemed to advocate for a separation of art from science.

For Neuhauss and other practitioners, interference photography was science. They were not interested in consciously and artistically altering

2 The exhibition was held in the house of representatives, and the scientific and technical department most likely refers to the Technical University in Berlin.

photochromes to create a desired mood. They wanted to perfect their process and have a realistic and accurate colour representation of whatever they were photographing. Still, the realms of art and science overlapped in a variety of ways and were not as distinct as Neuhauss professed. While the points of view seem to suggest a defined and accepted division between art and science, it was not so clear-cut in practice. Photographer Hans Lehmann, who was extremely adept at interferential photography, briefly mentioned the "artistic effect" of the plates in his theoretical and practical primer on the process (Lehmann 1906, 44). Lehmann did not expound on this characterisation, but he proudly displayed his Lippmann photochromes by hanging them on the wall of his home. He felt their luminous and rich colours contributed to their practical value, because they also provided viewers with aesthetic pleasure (Lehmann 1906, 45).

The British novelist and philosopher Edward Douglas Fawcett was one such viewer. He delighted in seeing Lippmann's plates so much that he began to produce his own and wrote detailed instructions as to how to make them easily. His article, which was published in *The Photographic News*, sheds light on how taken he was with the imagery. He portrayed them as "the finest colour miniatures," which were "superior to what the painter's brush could effect" (Fawcett 1900, 248). He furthermore described them as brilliant and lively with a "gorgeous show of colours." With these descriptions and commentary, Fawcett treats the plates as if they were art objects. He clearly valued the uniqueness of the images and preciousness of the objects. The French photographer Gaston-Henri Niewenglowski similarly prized the Lippmann plates for their singularity when he cited Louis Ducos du Hauron's statement regarding multiplicity and rarity, the latter of which, he commented, can be likened to great art and has more prestige (Niewenglowski 1899, 520). Lippmann plates were indeed rare because they were not reproducible, and their luminescence made them even more treasured objects.

Importantly, the photographic realism of the Lippmann plates was a non-issue for Fawcett. He recognised that the main objective of photography was realism, and in regard to Lippmann plates, he found it was "well catered for" (Fawcett 1900, 248). For him, there existed no conflict between their aesthetic value and their realism. Realism was, of course, a chief concern for the scientists testing and working with the process: a more realistic result meant that the process was better.

Examining the extant interferential photochromes provides evidence as to how the photographers worked and what motivated their choices. The subject matter and compositions were rooted in scientific experimentation that sought to reproduce real colour, but they also reflected an artist-like practice in their desire to show off visual variety and skill. Still-life motifs, for example, may at first glance appear to be no more than a rather insignificant selection of test objects that were conveniently nearby. Indeed, there is some truth to this – Neuhauss's stuffed green parrot permanently sat on his balcony (cf. Hanin Hannouch's article in this volume). But while the interferential photographers were scientists and mainly concerned with cause and effect, they were also invested in a result with practical and aesthetic allure, and at times, they seem to have made decisions that are difficult to argue from a strict scientific viewpoint. The choices, such as the iridescent plumage of a bird, seem more related to the expressive capacities of interferential photography and popular appeal.

Landscapes are desirable, beloved subjects, and photography was an ideal way to showcase one's travels and foreign geographies. Determined to use his process and prove its practicality, Lippmann travelled throughout Switzerland and to the Savoy region of France and Venice, taking interference photochromes (Girardin 2000). Neuhauss does not seem to have travelled for the same purpose, but he was likewise resolute when he repeatedly photographed his house in Berlin, which featured green trees, red window frames, and white curtains – an altogether difficult juxtaposition of colours to render in interferential photography. His extant plates of this motif deviate in colour and demonstrate his struggle in obtaining accurate results. The process was fallible, a fact that detracted its scientific value, though not its aesthetic.

If the Lippmann plates were to have broad application and attraction, then they would need to be used for all genres and be made familiar to a far-reaching audience. Photographic exhibitions and world fairs provided an avenue for the interference method to be seen. They also enhanced the information exchange between international photographers. The 1892 Exposition Universelle in Paris, for example, showed the photochromes Lippmann had presented to the Academy earlier that year. The following year, at the International Photography Exhibition in Geneva, Auguste and Louis Lumière exhibited a landscape interference photochrome of a red-tiled country home with a colourful flower garden. Valenta admired its representation of red, green, and blue when the same photograph travelled to the Vienna Photographic Society (Eder 1894, 447). The Lumière brothers also exhibited their portrait of a woman with her head resting on a table. Set with fruits and bottles, the table, in combination with the cast shadows on the sitter's forehead and chequered dress, clearly establishes a depth of space. The composition is extraordinarily dense, which makes it all the more remarkable as an interference photochrome. It was even labelled the first interference colour portrait (Eder 1894, 447).

The Lumière brothers clearly understood the importance of portraiture for commercial success. In 1894, at the Salon du Photo-Club in Paris, they exhibited four colour portraits: Louis's son in Far Eastern dress, Louis's son in his laboratory, an army officer, and a small girl in a garden (Girardin 2000).[3] These successful interference plates travelled to other exhibitions in Paris, Chicago, Geneva, and Vienna. Neuhauss, however, disparaged the exhibited Lumière plates because they were always the same, reasoning that the Lumière brothers did not have many good examples of the interference photochromes with a variety of mixed colours (Neuhauss 1894, 327). Indeed, other spectators criticised some of the duller hues apparent in the landscapes when the photochromes toured the Photographic Congress and Camera Club in London (Ives 1894, 18–19). Nevertheless, these international and local events testify to a circulation, however limited, among photographic audiences as well as their acceptance in these circles.

News of the process and plates also appeared frequently within contemporary photographic literature. The 1893 depiction of the woman resting on the table made its way to Eder's *Jahrbuch* as a black-and-white reproduction. Eder praised the photograph for its depiction of a living person whose complexion seemed quite real, even though, as he mentioned, the glass of wine appeared green instead of red (Eder 1894, 447). His remark points to the difficulty of attaining a perfect copy of all the colours. But, more importantly, it suggests that the excitement for the process often eclipsed the reality of its inaccurate representation. For Eder, and presumably other enthusiasts, a significant colour deviation did not seem to matter because the composition was otherwise rendered successfully. He furthermore felt the results were beneficial and highly meaningful to science because all the details were clearly visible (Eder 1894, 447). His feelings suggest a genuine passion for the process that was shared among other scientists and practitioners.

When the *American Journal of Photography* reported on Lippmann's photography in 1895, it admitted that the colour values are not always rendered accurately, but countered that the colours were reproduced exactly and the results were wonderful (Lippmann 1895, 108). The underlying tension in these proclaimed successful scientific developments was a paradox. As a natural recreation of true colour, the Lippmann interferential process was

3 Girardin describes this event as showing the first interference portraits, but they were most likely the same as the portraits exhibited in Geneva in 1893.

oftentimes at odds with correctly reflecting the colours of the photographed subject, such as the wine. This problem of course meant the process was not completely under the control of the photographer and it was often lamented that the colours were rarely ever correct (Porter 1908, 613). And yet, the scientist, much like the artist, desired to regulate the outcome of the process. Many portraits with simpler compositions, however, illustrated colour rendition to near perfection. In Neuhauss's self-portrait from 1901, he appears in his dark suit, draped in a bright red tartan shawl. The photograph is as much about the reproduction of the red colours of the cloth as it is about representing Neuhauss. Based on his own experience, he recommended using black velvet for backgrounds to prevent a loss of plasticity in the subject matter (Neuhauss 1898, 37). Lippmann's earlier 1892 self-portrait, with his head turned slightly in a three-quarter profile, is also a realistic portrayal, although, ironically, it does not emphasise colour. It attests to the struggle he had in representing mixed colours (resulting from pigments in objects and in people etc., as opposed to spectral colours emanating from a ray of light split through a prism), which were said to appear matt and washed out in interferential photographs. The chemist Leopold Pfaundler concluded that the problem of mixed colours would ensure that the process would only remain an exciting and highly interesting physical experiment and nothing more (Pfaundler 1904, 233). As mixed colours were not accurately reflected by the Lippmann process, they, too, challenged the presumptive truthfulness of its representation.

Despite such setbacks, the interest in colour and interferential photography grew around the turn of the century. Lippmann's and Neuhauss's photochromes were again displayed at the 1900 world fair, but exactly how much attention they received is hard to gauge. Critic Paul Neven's report from the photographic sections of the fair only mentions other processes: photogravures, three-colour transparencies, and three-colour halftones, etc. (Neven 1900, 104–5). At the Royal Photographic Society's exhibition in 1900, interferential photographs by Neuhauss and Fawcett were also displayed, but a reviewer complained that the exhibition room was not well lit, and therefore, he could not see the interferential plates correctly (Royal 1900, 686). Consequently, he was not impressed. It is likely that inadequate lighting disadvantaged the plates from gaining more attention and plagued other exhibitions of the photochromes. Only when the light is reflected at the right angle does the interferential photograph appear in its brilliant colours; otherwise, it is a rather undefined and unimpressive image. Nevertheless, there were enough good displays, because a sizeable cohort of Lippmann fans and practitioners existed at the time. In 1903, for example, the Société

Française de Photographie even gave a prize for the best series of Lippmann plates, and in the following year, the first international exhibition for colour photography took place in Paris (Koshofer 1981, 18, 75). These events (and no doubt others) prove that practitioners of the Lippmann process were not locked in their laboratories, but rather, they were engaged and promoting their work constantly.

As the inventor of the process, Lippmann was especially committed to showing off his photochromes and he did not just entertain scientists. Photographic artists, novelists, and intellectuals visited him. From Lippmann, Fawcett must have understood that the common perception regarding the process was that it was too complicated. Fawcett, therefore, set out to right this misconception. In his article, he explains that many improvements had been made and states that interferential photography "is now a cheap, perfectly practicable method of obtaining results superior to those obtainable in three-colour work, and, of course, permanent" (Fawcett 1900, 248). Fawcett was even given a plate of Versailles, complete with an attached viewing prism. The gift would allow for knowledge of the process to spread, as others in Fawcett's circle could now be similarly mesmerised with the brilliance and realism of the plate.

Apart from the viewing prism, which aided viewers in witnessing interference and the colour, Lippmann plates could be projected via a specially designed lantern. The Lumière brothers and Lippmann often showed their photochromes in this manner. Competitions on specifically interference photochromy projection were held in both France and Belgium. Photographer Frederic Eugene Ives contended that projecting the images was "the only really satisfactory way" of seeing the brilliancy of the photochromes (Ives 1894, 18). The projection of the slides was also what had impressed Steichen when he visited Lippmann. He found the slides "simply dazzling" and argues that "one would have to go to a good Renoir to find its equal in colour luminosity" (Steichen 1908, 14). Without projection, the Lippmann plate could only be viewed at a specific angle and that made it less than ideal. One critic of the process predicted that the curious will only regard the photochrome with short-lived interest, and after that, they will tire of its novelty (Jones 1899/1900, 152). Lehmann brushed this kind of critique aside when he agreed that the viewer could only see the image from a certain angle but added that at least the image was still clearly visible from a distance of a few metres (Lehmann 1906, 56). And yet, the photochromes, when not projected, are best held up-close, in the hand. They are small precious objects, jewel-like in their iridescence and also captivating because of this private engagement.

Almost two decades after its discovery, the Lippmann process was still only being practised by a dozen or so advocates. Two of its early practitioners, the Lumière brothers, had sealed the fate of the Lippmann process in 1904 when they developed the autochrome, which they then began marketing in 1907. Like interferential photography, the autochrome was incapable of being duplicated and it also needed viewing devices, but its easier production process made it an instant alternative. While the colours of the autochrome were not as brilliant in their luminosity, they were vibrant and presented fewer complications. Steichen was among the many art photographers who made autochromes and credited the Lumière brothers for the "first practical solution of color photography" (Steichen 1908, 14). Nevertheless, Steichen could not forget the "undeniably real color photography" of the Lippmann process, and he was not alone. Only a year before, Lehmann, for example, had used tri-colour printing to reproduce an interferential colour photograph of a painting as he tried to increase the practical application of the method (Eder 1932, 974). Others continued to experiment with the process and write of its importance. In 1912, Valenta came out with a second edition of his book on direct colour photography, which is interferential, and even planned to do a third edition in 1927. All of these efforts speak to the attraction of the plates on viewers and attest that they could not be dismissed so easily.

The shining intensity of colour produced by interference remains inimitable and should, if nothing else, be appreciated simply for this aesthetic reason. Lippmann's photography, when successful, accurately recorded and recreated interference. It was thus construed as "direct," based on "natural colours," rather than on a combination of dyes or the superimposition of colours, as was often associated with "indirect colour" photography. Its physical phenomena and scientific underpinnings are what encouraged its dismissal as an art form, rather than the quality of the images it allowed. Vibrant in their saturation, the interferential photographs contributed to the tenuous relationship colour photography was to have with art photography. For artists and critics, a purely objective practice such as interferential photography, a practice that did not allow for the artist's hand to alter the recorded image, was in no way artistic. And yet, as we have seen, its luminescence made the Lippmann plates attractive to the eye, and the scientists themselves invested in making their images aesthetically pleasing, popular, and practical. Steichen was not alone when he granted these photographs aesthetic value and saw the importance of safeguarding it in the medium's history. Even though the Lippmann method never gained any traction as a viable means of producing artistic photography, its successes and failures contributed to a complex web of efforts by scientists, artists, and photographers to the understanding and appreciation of colour. The aesthetic and visual attractiveness of interferential photography complicated and bridged the supposed interstice between art and science. And this makes it worthy of our continued interest.

Works Cited

Bachmann, H[einrich]. 1907. "Die Farbe in der Photographie." *Der Amateur: Illustrierte Monatschrift für Photographie und Projektion* 6 (June): 229–234.

Eder, Josef Maria. 1894. "Photographie in Natürlichen Farben." In *Jahrbuch für Photographie und Reproductionstechnik für das Jahr 1894* 8: 446–456.

Eder, Josef Maria. 1932. *Geschichte der Photographie*, 4th, rev. ed. Halle a.S.: Wilhelm Knapp.

Fawcett, E. Douglas "Amateurs and the Lippmann Process" *The Photographic News* (April 20, 1900): 248– 249, 266–267, 279.

Girardin, Daniel. 2000. "Gabriel Lippmann" *Du* 708 (July-August): 30.

Ives, F[rederic]. E[ugene]. 1894 "The Lumière-Lippmann Color Photography" *Journal of Franklin Institute* 137, no.1 (January): 16–21.

Js[Jones]., C[hapeman]. "Wood's Colour Photography" Camera Obscura 1899/1900: 152.

Koshofer, Gert. 1981. "Geschichte der Farbphotographie in der Pionierzeit." In *Farbe im Photo: Die Geschichte der Farbphotographie von 1861 bis 1981*, 71–77. Cologne: Josef-Haubrich-Kunsthalle.

Lehmann, Hans. 1906. *Beiträge zur Theorie und Praxis der direkten Farbenphotographie mittels stehender Lichtwellen nach Lippmanns Methode.* Freiburg i. Br.: C. Trömer's Universitäts-Buchhandlung.

Neuhauss, Richard. 1894. "Die Photographie in natürlichen Farben" *Photographische Rundschau* 8, no.10: 295–302; no. 11: 327–329; no. 12: 359–364.

Neuhauss, Richard. 1898. *Die Farbenphotographie nach Lippmann's Verfahren: Neue Untersuchungen und Ergebnisse.* Halle a.S.: Wilhelm Knapp.

Neuhauss, Richard.1907. "Ueber den Gegenwärtigen Stand der Farbenphotographie" In *Die Photographische Kunst im Jahre 1907*, ed. F. Matthies-Masuren (Halle a.S.: Wilhelm Knapp, 1907) 12–24.

Neven, Paul. "A Walk Round the Photographic Sections at the Paris Exhibition, 1900" Penrose's Pictorial Annual 6 (1900): 104–105.

Niewenglowski, M. G[aston]. H[enri]. "Color-Photography Up to Date" *The Literary Digest* 18, no. 18 (May 6, 1899): 520–521.

Pfaundler, L[eopold]. 1904. "Ueber die Zenkerschen Streifen in nach Lippmanns Verfahren hergestellten Photographieen sich überdeckender Spektra und ihre Bedeutung für die Leistungsfähigkeit dieses Verfahrens bei gemischten Farben." *Jahrbuch für Photographie und Reproductionstechnik für das Jahr 1904* 18: 229–233.

"Royal Photographic Society Exhibition" *The Photographic News* 44 (Oct. 19, 1900): 686.

von Sallwürk, Edmund. 1905. "Farbenphotographie und Kunst" In *Photographische Kunst: Halb-Monatschrift für künstlerische Fach-Photographie*, ed. G.H. Emmerich, 243–248. vol. 4. Munich: Süddeutsche Verlagsanstalt.

Scott, Dixon. "Colour Photography" *The Studio: and Other Recent Developments of the Art of the Camera*

Steichen, Edward. "Color Photography" *Camera Work* 22 (April 1908): 14.

Valenta, Eduard. 1893. "Über die Photographie in natürlichen Farben" In *Vorträge des Vereines zur Verbreitung naturwissenschaftlicher Kenntnisse in Wien*. Vol. 33, no. 13. Vienna: Verein zur Verbreitung naturwissenschaftlicher Kenntnisse.

Valenta, Eduard. 1894. *Die Photographie in natürlichen Farben mit besonderer Berücksichtigung des Lippmann'schen Verfahrens sowie jener Methoden, welche bei einmaliger Belichtung ein Bild in Farben liefern*. Halle a.S.: Wilhelm Knapp. (2nd revised ed. 1912)

Vogel, Hermann Wilhelm. 1893. "Ueber Photographie in natürlicher Farben" *Jahrbuch für Photographie und Reproductionstechnik für das Jahr 1893* 7: 285–287.

About the Author

Dr. **Elizabeth Cronin** is the Robert B. Menschel Curator of Photography at the New York Public Library and author of *Heimat Photography in Austria: A Politicized Vision of Peasants and Skiers*. She has curated and co-curated such exhibitions as *Viewpoints: Latin America in Photographs* and *Public Eye: 175 Years of Sharing Photography*. She holds a PhD in art history from the Graduate Center, City University of New York, an MA in art history from the University of Minnesota, and an MA in German studies from the Technical University of Dresden.

8. From Maxwell to Mercury: The Cultural History of Gabriel Lippmann's Colour Photography

Rolf Sachsse

Abstract

This article presents discussions of the curatorial team of the "Farbe im Photo" exhibition (Cologne 1981) had when dealing with Lippmann photography as part of the history of its modern reception. It uses these reflections to re-integrate Gabriel Lippmann in his cultural milieu and to complicate his relationship to it in several ways. His knowledge of the importance of colour to the arts ca. 1900 and his implicit understanding of the link between liquid mercury to colour. As such, the Lippmann process can be understood in all its nuances; both scientific and cultural.

Keywords: museum exhibition, mercury, interferential colour photography, aesthetics, cultural history

While preparing *Farbe im Photo*, one of the first exhibitions in the German-speaking world on the history of colour photography in 1981, the curatorial team I was part of – together with Siegfried Gohr, Gert Koshofer, and Klaus op ten Höfel – stumbled upon a small glass plate that looked like a cassette in the possession of the Deutsches Museum in Munich, with the photograph of a stuffed parrot that struck its viewers with a brightness of colours unparalleled to any other specimen in this field (Gohr 1981, 72). We learned that the photograph was made by Richard Neuhauss, a renowned anthropologist and frequent author in photographic magazines at the time, after a photographic process designed by Gabriel Lippmann who had received the Nobel Prize for this invention in 1908, the first in the field of physics with the help of photography (Lippmann 1894). Of course, we traced some

Hannouch, H. (ed.), *Gabriel Lippmann's Colour Photography: Science, Media, Museums.*
Taylor & Francis Group, 2022
DOI 10.5117/9789463728553_CH08

material from and about Lippmann in museums worldwide. We reserved for the exhibition a number of colour photographs on paper from several imaging technologies as well as one or the other portrait of the scientist in his lab at the Sorbonne University in Paris (Gohr 1981, 284, 4). Fairly soon, the interest in Lippmann and his process among us diminished into the corner of scientifically interesting but medially – or even artistically – unsatisfying displays. Intellectually, Lippmann's colour photography was absolutely fascinating, and its results were astonishingly bright, but they did not function as photography in the way the medium was understood at the time: neither could they be produced in whatever size – the mercury cassette that had to be installed in the back of the camera as well as in the display case had to be small to be manually operable, nor could the images be duplicated or even printed in larger numbers. Photographing the plate for our exhibition became a complex process, since we had to commit to the view from the polarising angle and the overall iridescence. The Lippmann photographs simply were tiny and unique pieces for the connoisseur of unsurpassed bright photographic colours – nothing less, nothing more. This essay examines the difficult reception of the Lippmann process in the late 1970s and early 1980s at the exhibition *Farbe im Photo.* It traces the emergence of Lippmann's colour photography in the cultural context of its time, positioning it at the threshold between scientific experimentation and artistic knowledge of colour.

Throughout the 1970s, the emerging interest in photography as a fine art had a strong impact on placing the medium within the art market. Starting with the documentary use of the medium by several artists, commonly summarised under the term *conceptual art*, photographers who started their artistic practice with long-term projects in the preservation of buildings – like Hilla and Bernd Becher (Sachsse 1999) – were represented by avant-garde galleries invested in industrial photography (Meyer 1972, XI–XII). From there, a common interest in the history of photography arose in conjunction to photographers who had been involved in their own long-term documentary projects, like August Sander or Walker Evans (cf. Hinson 1996, 1–19, Heiting 2009, 14–15). Throughout the late 1970s, photography was increasingly accepted by galleries, museums, and art magazines, and there were early attempts of discussing the field as an integral part of the visual arts (Goldberg 1981). The exhibition *Farbe im Photo* is inscribed within this tradition of documentary photography. It marks the border between the latter's way of looking at photography from a technical viewpoint and the new interests in its aesthetic values with a link to the visual arts. Half of its curators came from an industry-bound historiography of technical developments, the other half from art history. Within this context, the basic quality of Lippmann's colour process was seen as the intellectual game of a versatile scientist – and consequently, it was not regarded as an important contribution to the understanding of photographic imagery but a result of chemical and physical inquiries. The Lippmann process was positioned in the historic moment of the birth of colour photography, but with no continuity within the technical history of the medium (Koshofer 1981, 71–72). Both Lippmann and scientist James Clerk Maxwell (1831–1879) fell awkwardly in between these fields – the technical and artistic history of photography – and they shared the same fate at the exhibition.

Although *Farbe im Photo* opened on the 120th anniversary of Maxwell's presentation of the three-colour tartan ribbon to the Royal Institution on May 17, 1861, we could only show a simulation of this projection in minor quality. Similar to the later exhibition on the history of colour photography in Rochester, New York, 1982, we had no possibility of reproducing this image in the catalogue (Sobieszek 1982, 2). Both Maxwell and Lippmann were trapped in a historical oxymoron: until the late 1970s, the history of photography had been written mostly in technical terms, even if it was presented by art historians, and the next generation of photographic historians concentrated on iconography and style but – except for market terms like *vintage print* – did not care about the visual and material qualities of the pictures presented in exhibitions and books (cf. Baier 1964, Wagner 2008, 14–26). There was little room for the relationship between photography and science, especially if the images could not be reproduced. The prolongation of this oxymoron into the media debates of the 1990s and 2000s is discussed at length by Jens Schröter and other contemporaries (Schröter 2013, 235–63). Although Schröter mentions the irreproducibility of photographs after Lippmann's process, he does not discuss former attempts to do so, as practised by the Deutsches Museum in preparation for the exhibition in 1981 (Gohr 1981, 72). Supposedly, our discontinued interest was due to the subject of the image: a stuffed parrot depicted alone standing on a twig is neither a family photograph (which was the use discussed by Schröter as a basic function of imagery in colour), nor is it a subject of wider interest in iconography, since it is not a part of the still-life genre that art historians are susceptible to. It is in fact a test object for emulsion sensitivity that, due to its irreproducibility, cannot be part of the documentary, technical, or artistic history of colour photography. Consequently, in the 1980s, the Lippmann process and the stuffed parrot plate were in a historical and curatorial vacuum (cf. Hannouch in this volume).

The subject of James Clerk Maxwell's experimental projection at the Royal Institution in 1861 suffered from a similar problem in its iconography:

it showed a narrow view of the Clerk family's tartan ribbon with a knot placed on a velvet ground (Clerk Maxwell 1858–63, 370–74). The depicted motif is neither the reproduction of artworks through the novel medium of three-colour photography, nor is it a worthy visualisation of precious objects (such as jewels or fashion) to be preserved or distributed by the new technology for commercial purposes, despite the qualities of colour reproduction. The projection of the ribbon merely marks a step of cultural improvement, compared to the lectures Maxwell had previously given on the nature of colour and light in which he had used children's toys to explain the mixture of the three basic colours (Clerk Maxwell 1860, 57–84). His novel projection session was a huge step in understanding the nature of light and colour – even with all the embarrassments and obstacles put forth by the optical lantern and the colour-blind emulsions he had to use. He could show his basic idea of mixing any given colour photographically into a trustworthy emanation of reality. As his contemporaries, Hermann Helmholtz and Edmond Becquerel, Maxwell was not interested in the subject of a photograph but in the description of light this image technology can provide. And Gabriel Lippmann continued in this line of understanding the materiality of colours as a part of light, not as substance, a notion that is foundational to the wave theory upon which his photography resides. A generation younger than his tutors and those who inspired him, Lippmann's approach to these matters is slightly different, for both personal and cultural reasons.

When Lippmann finally announced his process of colour photography in 1891, he claimed to have worked on it for more than a decade, but there is no indication as to the exact beginning of his research, and there is no sign why he went to the public exactly in the beginning of the 1890s when a number of colour photographic processes were on the edge of being announced and when photographic printing in three or four colours was already heralded (The Photographic Times, 1893, 82). In 1882, Lippmann published an overview of the studies he conducted between 1872 and 1882, and there is no sign of developing a photographic process, but there are notes on the question of polarised reflections on metallic surfaces – a problem that had been treated before by Edmond Becquerel and, to a certain extent, by James Clerk Maxwell (Lippmann 1882, 6).

I claim that it is particularly fruitful to understand the Lippmann process at the juncture of considerations about both science and art, because, given the interest that Gabriel Lippmann had in the arts of his time, there are one or two trajectories to be followed in finding out about his fundamental curiosity about colour and photography *before* 1891 (Fechete 2016, 903–8).

First, there is the materialisation of his ideas in physics with the help of mercury: Lippmann knew about Maxwell's experiments as he researched electrocapillarity, especially in the times of his German postgraduate studies, which he spent in Heidelberg and Berlin, namely with Hermann von Helmholtz (Pier Massad 1997, 3). His Paris doctorate from 1875 includes the invention of a capillary electrometer soon to be named after him, and this electrometer already contains a thin glass cassette partly filled with mercury that helps visualise the results of energetic processes. But, as with Maxwell, his main interest at that time seems to have been the mathematics of his findings, resulting in numbers of equations on electrocapillarity and similar problems.

Moreover, when Edmond Becquerel had started his depiction of the visible spectrum of light on the basis of the same highly polished silver that had been used in the production of daguerreotypes, Lippmann – as in the words of Herbert Ives in his dissertation on Lippmann's photography – simply used thin glass cassettes filled with mercury instead. He also employed thinner and more sensitive emulsions of the late 1880s, which enabled him to produce the standing waves responsible for the bright physical colours of his process (Ives 1908, 325–51). Mercury as a material, however, was commonly used in the textile industry, and Lippmann was a descendant of a family of *tanniers* – or, according to other sources, *gants*, makers of gloves (Wagener 2016, 38–40) – a business where mercury was needed for leather bleaching and for the de-colouring processes (Bintz 1997, 5–9). According to some of his biographers, Lippmann's father seemed to have been so successful that he could follow the bourgeois habits of separating the production from the sale: the family moved to Paris while most of the production stayed in the districts of Alsace-Lorraine where it had been before (Pier Massad 1997, 82, Lebon 1911). As part of the textile and fashion scene there, Lippmann could not escape the many debates on systematising colours for industrial use. Most of the colour systems at the time were ordered in forms of wheels or triangles. The most common in French industries was the delicately labelled circle designed by Eugène Chevreul in 1839, whereas German and English physicists followed the trichromatic triangle after Tobias Mayer, James Clerk Maxwell, and Hermann von Helmholtz. With their inscription of a curve according to thermal energy, they designed the initial colour scheme that was to be defined in 1932 as the final diagram of light and energy by the Commission Internationale d'Eclairage as the so-called CIE triangle (Sachsse 2003, 26). When Lippmann studied with Helmholtz, the latter had just finished his research on the nature of light and started to combine his findings with assumptions on human recognition, reaching into the fields of psychology (Conrat 1903). By that time, mercury was also a steady part of Lippmann's scientific interests, as his CV from 1882 exemplifies, and so was his implicit knowledge of the relationship between mercury and colour (Lippmann 1882, 5–6).

Second, Gabriel Lippmann knew of the debate between Charles Cros and Louis Ducos du Hauron in relation to three-colour photography in the years of 1869 and 1870, as posited by Ariane Isler-de Jongh (1994, 116). The discussion about the systematisation of both their findings in colour photographic processes was rewarmed from time to time in the 1870s and the 1880s. Although Ducos du Hauron had published his findings partly in 1878, Isler-de Jongh considers their knowledge of each other's work to be a "pure conjecture" – despite them both being well-known visitors to the salons of Paris. With this, I contend that she underestimates the influence of one of the most important art critics at the time, Charles Blanc (Pietsch 2004). Blanc distinguished with precision between art and industry in his attempt to establish a *l'art pour l'art* theory within his critical practice. What is often overlooked, however, is that Blanc employed several terms and discussed ideas that would account for modernism in the arts. In one of his lectures "Des Expressions de la lumière," published in *Le Temps*, a key newspaper of the time, he stresses the importance of light in painting as its main subject, thus making a turn from his *Grammaire des arts du dessin* (Pietsch 2004, 272, Weibel 2006, 95). In contrast to his earlier writings, which gave the line in drawing the supremacy over the constitutive idea behind any artwork, Blanc emphasises in "Expressions de la lumière" the importance of light as the generator of colour so much that Vincent van Gogh used his work as a source of inspiration in exactly this aspect (Gethöffer 2018, 311–14).

The importance of both debates is discussed by Laura Anne Kalba by introducing a comparison between fashion and art (Kalba 2017, 74–75). On one hand, there is Charles Blanc with his emphasis on colour and light; on the other hand, there is the eminently influential director of the fashion magazine *La Monde Illustrée*, Emiline Raymond. Raymond's attempts to colourise the Parisian fashion are firmly grounded on Chevreul's circle and its relevance for the dyeing of textile fabrics, but even if both were "circulating in different social circles" (Kalba 2017, 74) they were convinced that colour was an important addition to social and cultural progress. In this climate, Lippmann may have recognised the need for further research. By 1881, he seems to have finished his inquiries on the reverse piezo effect and handed over further explorations in this field to his assistant, Pierre Curie. He shifted his interest to another field: the representation of colour in a painting or in photographic processes at a time when its centrality dominated aesthetic criticism (Lippmann 1882, 8–10). Gabriel Lippmann's vicinity to the Salons of Paris strengthened throughout the 1880s. In 1888, he married Laurence Cherbuliez (1857–1934), the daughter of the renowned novelist Charles Victor Cherbuliez, a central figure to the debates on the importance of the arts to society (Encyclopaedia Britannica 1911). Looking at the history of colour in painting, the 1880s were the times of a diminishing influence of Michel-Eugène Chevreul on the academic painting and the growing importance of Helmholtz's as well as Maxwell's theories for the artists later labelled as Neo-Impressionists (Gage 1999, 196–227). Although he did not utter a single word about the theoretical base of his painting, Georges Seurat started mixing hues directly on his canvas by putting dots of primary colours together; a large number of younger painters followed him, irrelevant of where they positioned themselves in relation to the ideas of Charles Blanc. Throughout the 1870s, photography had improved from a craft into an industry of its own by the introduction of the dry plate and by the growing sensitivity of the emulsions; the early 1880s saw the introduction of pre-print and printing technologies for photographs, which instigated the advent of a new image journalism (Hannavy 2008, 692–703). The installation of photography as a medium happened with an interview that the photographer Nadar led with Michel-Èugene Chevreul on the occasion of his hundredth birthday; it is known as the first fully printed photographic interview (Nadar 1886, 284–85). Only seven years later, this interview could have been printed in colour (The Photographic Times 1893, 82).

Little is known about the years of preparation before Gabriel Lippmann announced his *Présentation du procédé interfériel de la photographie en cuoleurs* at the Académie des sciences in Paris on February 2, 1891 (Pier and Massad 1997, 3). Following the description given to this presentation, he revealed an image of the sun spectrum, which seems to be one of very few images taken by Gabriel Lippmann himself (Gohr 1981, 285, A106). His subsequent official lectures and publications were illustrated with several photographs relying on his method. These photographs had been taken by the brothers Auguste and Louis Lumière from Lyon who had opened in 1882 an extremely successful fabrication of dry plates and were looking for any expansion their business could provide (Hannavy 2008, 875–77). There is no document explaining how and where the Lumière brothers met Lippmann, but there is a link between the two in the person of the entrepreneur and philanthropist Albert Kahn (Franziska-Maria Scheuer, personal communication to the author, 2020). Lippmann, who always had foreign students – as he and his father-in-law had been studying in Germany in the 1860s and 1870s – was invited to Albert Kahn's committee Bourses des voyages autour du monde, which he had set up after moving to Boulogne-Billancourt, where he also hosted Henri Bergson. This society – one of many instigated by Kahn – financed foreign studies for students and postgraduates and, upon their return, were frequently invited to Kahn's house to present their research and discuss it with Bergson and other members of the society. Lippmann had taken part in a number of meetings at this place as did the Lumière brothers (Beausoleil 1995, 196, 241). The motifs provided by the Lumière brothers for Lippmann fit into the scheme later seen in the variety of their short films: a lesson in reality with a feeling for both tourism and didactics (Allan 2008: 159–70). Under this aspect, Lippmann's colour photography should be understood intermedially in relation to both the Lumières' autochrome projection and their work on early cinema: nobody could have predicted how a medium like the projection of still and moving images would change the visual world (Burgin 2018, 217–30). But given the delicate process and materials, as well as the low sensitivity of Lippmann's emulsions it is no wonder that the set of images the Lumières provided for the 1893 and 1894 lectures by the inventor were of lesser interest to photographic critics and historians – despite the beauty of their colours, there is nothing much interesting to see in them: still lifes with flowers, sunlit landscapes, and one single portrait of the professor himself, fixed into the chair as in the early portrait photography of the 1850s. As is the case with all new media, colour photography had to settle in society by using stereotypical motifs and compositions first, and only in a second step could it lead its spectators with its own unique aesthetic choices (Gohr 1981, 117–19).

Shortly after Lippmann's announcements, several attempts were starting to present an executable method of colour photography. John Joly began producing his plates in 1895, and others followed him. When Gabriel Lippmann was awarded the Nobel Prize in 1908, the Lumière brothers had already successfully introduced their autochrome process, and the whole complex of colour photography had moved from the experimental state to a state of practicability for specialists (Gohr 1981, 117–32). Therefore, it seems important to notice that most sources referring to the subject of Lippmann's Nobel Prize give a concise explanation of the process: Ernest Lebon writes he received the Prize "for his method based on the phenomenon of interference, allowing for the photographic reproduction of colour "my translation" (*pour sa méthode fondée sur le phénomène de l'interférence, permettant de reproduire photographiquement les coleurs*) (Lebon 1911, VII). And later descriptions go even further: "In 1908, he received the Nobel Prize for Physics for having elaborated his undulatory wave theory of light, from which the interferential photograph is the practical result." (Nobel

Foundation 2011) Here, the photograph is the technical result of a process, nothing more and nothing less. For good reason: the prize was submitted for the detection of a physical discovery with its mathematical backgrounds that instigated a number of inventions to follow – until today – but it was not given for producing aesthetically satisfying photographs.

Around the time of the exhibition *Farbe im Photo* in the 1980s, the Lippmann plate of a stuffed parrot produced by Neuhauss was an unlikely candidate for an exhibition. Its distance from aesthetics, owing to the fact that it can neither be reproduced, nor admired at ease, trumped the sophistication of the science behind it for the curatorial team. Also, the Lippmann process was an ill fit with the documentary focus of historical narratives prevailing when we had not yet considered the importance of scientific photography. When reconsidering Gabriel Lippmann's research on colour today, especially his "silence" about it before 1891, it is essential to integrate him in his cultural milieu and to complicate his relationship to it. This article has done so by unpacking his understanding of the rapport of mercury to colour,as well as his reception of the importance of colour to the arts. As such, the Lippmann process can be understood in all its nuances – both scientific and cultural.

Works Cited

Allan, Michael. 2008. "Deserted Histories: The Lumière Brothers, the Pyramids and Early Film Form." *Early Popular Visual Culture* 6, no. 2: 159–70.

Baier, Wolfgang. 1964. *Quellendarstellungen zur Geschichte der Fotografie*. Halle (Saale): VEB Fotokinoveri.

Beausoleil, Jeanne et al. (eds.). 1995. *Réalités d'une utopie, Albert Kahn 1860–1940*Boulogne: Musée Albert-Kahn. Exhibition catalogue.

Bintz, Jacques. 1997. "Gabriel Lippmann 1845–1921. " In Pier, Jean-Paul, Massard, Joseph A. (eds.): *Gabriel Lippmann (1845 – 1921), Commémoration par la section des sciences naturelles, physiques, et mathématiques de l'institute grand-ducal de Luxembourg du 150e anniversaire du savant né au Luxembourg, lauréat du prix Nobel au 1908*, edited by Jean-Paul Puer and Joseph A. Massard, 5–9. Luxembourg: Imprimerie de la Cour Victor Buck.

Burgin, Victor. 2018. "The Film Essay and the Crisis of Naming [2012]." In *The Camera: Essence and Apparatus*, 217–30. London: Mack.

Clerk Maxwell, James. 1860. "On the Theory of Compound Colours and the Relations of the Colours of the Spectrum." *Philosophical Transactions of the Royal Society of London* 150: 57–84. Cit after: Isler de Jongh, Ariane. 1994. "The Origins of

Colour Photography. Scientific, Technical and Artistic Interactions." *History of Photography* 18, no. 2: 111–19.

Clerk Maxwell, James. 1861. "On the Theory of Three Primary Colours." *Proceedings of the Royal Institution of Great Britain* 3 (63): 370–74. Cit. after: Isler-de Jongh, Ariane. 1994. "The Origins of Colour Photography. Scientific, Technical and Artistic Interactions." *History of Photography* 18, no. 2: 111–19.

Conrat, Friedrich. 1903. "Helmholtz' Verhältnis zur Psychologie." PhD diss., Friedrich-Wilhelms-Universitat zu Bonn.

Encyclopaedia Britannica. "1911 Encyclopaedia Britannica/ Cherbuliez, Charles Victor." Last edited June 26, 2016. https://en.wikisource.org/wiki/1911_Encyclop%C3%A6dia_Britannica/Cherbuliez,_Charles_Victor.

Fechete, Ioana. 2016. "Jonas Ferdinand Gabriel Lippmann: The Pioneer of Color Photography or Primus Inter Pares." *C. R. Chimie* 19: 903–8.

https://cyberleninka.org/article/n/1418928/viewer.

https://cyberleninka.org/article/n/1418928/viewer [2021–01–26].

Gage, John. 1999. *Color and Meaning. Art, Science, and Symbolism*. Berkeley: University of California Press.

Gethöffer, Afra. 2018. "Ein Temperament, gesehen durch die Natur. Emotion in den Landschaftsdarstellungen von Vincent van Gogh." PhD diss., Ludwig-Maximilians-Universität München..

Gohr, Siegfried, ed. 1981. *Farbe im Photo*, Cologne: Josef-Haubrich-Kunsthalle. Exhibition catalogue.

Goldberg, Vicky, ed. 1981. *Photography in Print. Writings from 1816 to the Present.* New York: Simon & Schuster.

Hannavy, John, ed. 2008. *Encyclopedia of Nineteenth-Century Photography*. London: Routledge.

Heiting, Manfred, ed. 2009. *At the Still Point, Photographs from the Manfred Heiting Collection*, vol. III, 1969–2000. Amsterdam: Cinubia.

Hinson, Tom E. Hinson, ed. 1996. *Catalogue of Photography*. Cleveland: Cleveland Museum of Art.

Isler-de Jongh, Ariane. 1994. "The Origins of Colour Photography. Scientific, Technical and Artistic Interactions." *History of Photography* 18, no. 2: 111–19.

Ives, Herbert E. 1908. "An Experimental Study of the Lippmann colour Photograph." *The Astrophysical Journal* 27: 325–52.

Kalba, Laura Anne. 2017. *Color in the Age of Impressionism: Commerce, Technology, and Art*. University Park: Penn State University Press.

Lebon, Ernest. 1911. *Savants du jour: Gabriel Lippmann. Biographie, bibliographie, analytique des écrits*. Paris: Gauthier-Villars.

Lippmann, Gabriel. 1882. *Notes sur les travaux scientifiques 1872–1882*. Paris.

Lippmann, Gabriel. 1894. "Sur la théorie de la photographie des couleurs simples et composées par la méthode interférentielle." *Journal Physique* 3, no. 3: 97–107. https://www.nobelprize.org/prizes/themes/lippmanns-and-gabors-revolutionary-approach-to-imaging/. [2021–01–26].

Meyer, Ursula. 1972. *Conceptual Art*, New York: E.P. Dutton.

Nadar, L. 1886. "L'art de vivre cent ans, Trois entretiens avec Monsieur Chevreul." *Le Journal illustré* 23, no. 36 (September): 284–285.

N.N.: Printing in Colors, in: *The Photographic Times*, Vol.23 1893, 82.

Nobel Foundation: *Entry on Gabriel Lippmann*, https://web.archive.org/web/20111123112555/http://dahl.lippmann.lu/index.php?id=37&L=2 [2021–01–28].

Pier, J. P, and J. A. Massard, eds. 1997. *Gabriel Lippmann, Commémoration par la section des sciences naturelles, physiaues et mathématiques de l'Institut grand ducal de Luxembourg du 150e anniversaire du savant né au Luxembourg lauréat du prix Nobel en 1908*. Luxembourg: Imprimerie de la Cour Victor Buck.

Pietsch, Kristiane. 2004. "Charles Blanc (1813–1882) Der Kunstkritiker und Publizist." PhD diss., Heinrich-Heine Universität Düsseldorf.

Sachsse, Rolf. 1999. *Hilla und Bernhard Becher. Silo für Kokskohle, Zeche Hannibal, Bochum-Hofstede, 1967. Das Anonyme und das Plastische der Industriephotographie, kunststück*. Frankfurt: Fischer Taschenbuch.

Sachsse, Rolf Weissbunt. 2003. "Das Gehirn der Welt." *Wilhelm Ostwald: Farbsysteme. Das Gehirn der Welt*, 65–88, edited by Peter Weibel. Karlsruhe: Zentrum für Kunst und Medientechnologie.

Scheuer, Franziska Maria. 2020. *Bilder für den Frieden. Gestaltung und historischer Gebrauch der Autochrome des Multimedia-Ensembles Les Archives de la planète (1908–1931)*, PhD diss., Philipps-Universität Marburg.

Schröter, Jens. 2013. "Von der Farbe zur Nicht-Reproduzierbarkeit." *Akteur-Medien-Theorie*, edited by Tristan Thielmann and Erhard Schüttpelz, 235–63. Bielefeld: Transcript.

Sobieszek, Robert, ed. 1982. *Color As Form, A History of Color Photography*. Rochester: George Eastman House. Exhibition catalogue.

Wagener, Renée. 2016. "Gabriel Lippmann, ein Moderner Wissenschaftler." *Ons Stad* 113: 38–40.

Wagner, Anselm. 2008. "Integrating Photography into the History of Art. Remarks on the Life and Scientific Estate of Heinrich Schwarz (1894–1974)." *Photoresearcher* 32, no. 11: 14–21.

Weibel, Peter. 2006. "The Development of Light Art." *Light art from artificial light, lichtkunst aus kunstlicht*, edited by Peter Weibel et al., 86–223. Ostfildern: Hatje Cantz. Exhibition catalogue.

About the Author

Rolf Sachsse was trained by Karl-Hugo Schmoelz and Walde Huth as a photographer with a specialisation in advertising, architectural, and industrial photography. He studied art history, communications research, and German literature in Munich and Bonn. He completed his PhD on the history of architectural photography in Germany (1900–1975) in 1983. Until 2017, he was professor for design history and design theory at the Hochschule der Bildenden Künste Saar in Saarbrücken. Since 2018, he has been a lecturer at the University of Bonn.

Part III

Contemporary Reception and Future Trajectories

9. Materiality, Identification, and Conservation of Lippmann Plates

Jens Gold

Abstract
Few institutions and collectors worldwide have interferential colour images in their possession. This fact has less to do with the fragility of the object and more with its practical use at the time of its emergence. There is a clear need for additional awareness about this medium in the field of photograph conservation, especially the long-term preservation of this photographic material. The purpose of this paper is to present the twelve Lippmann colour plates and other holdings at Preus Museum – the National Museum of Photography in Norway. After sharing the principle of the Lippmann technique, describing the physical and optical properties of the plates, the chapter will discuss typical damages, deterioration issues, and challenges when using and handling these types of objects, both in exhibitions and for long-term preservation.

Keywords: Richard Neuhauss, Hans Lehmann, museum collection, conservation, interferential colour photography, photographic technique, Lippmann plate

Background

The Lippmann process is one of the most interesting examples of early colour photograph technology. Unlike most others, but no less complex than other colour techniques around 1900, it does not use dyes or pigments to refer to the colours of an object, but still delivers real and very permanent recording of light and its "true" colour. This is interesting not only for the expert in the field of early colour photography but also for the general lover of early photography and photo history. Today, only a few institutions and collectors worldwide have original examples of these rare colour images in their possession. This fact has not so much to do with the fragility of the object as with its practical use and application at the time of its origin. However, because of the special properties in terms of the technique itself, its presentation form, and its requirements for viewing, there is a clear need for additional awareness about this medium in the field of both history of

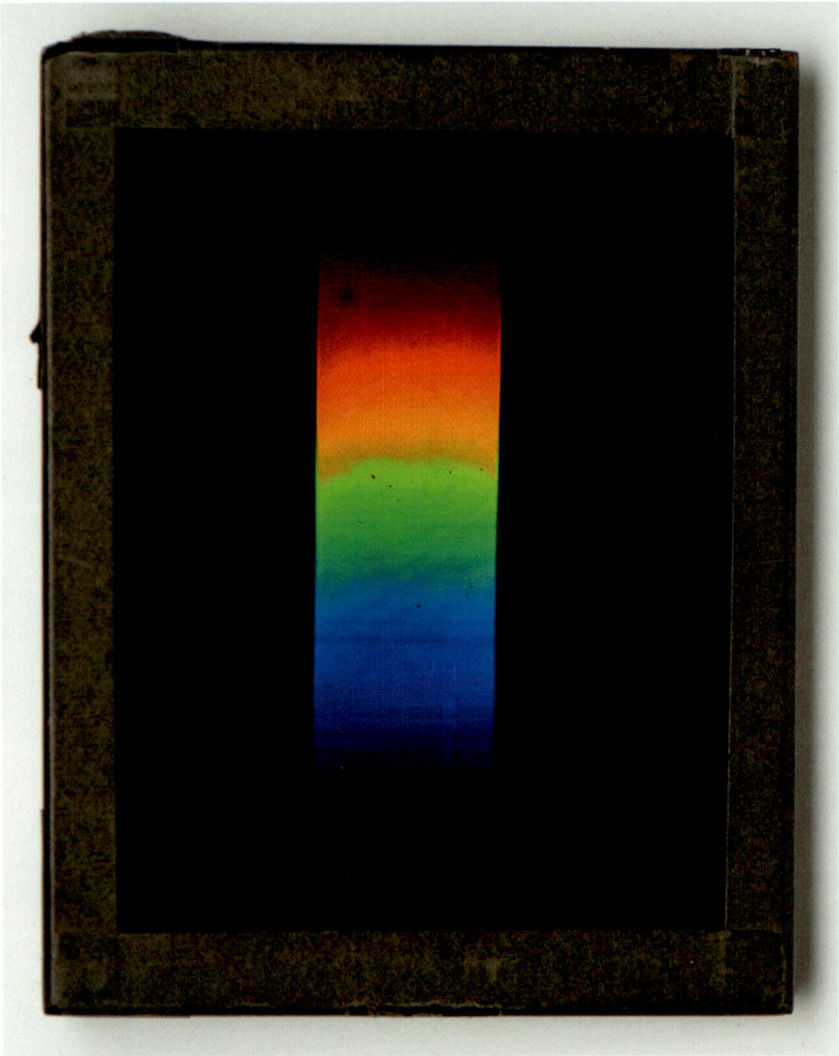

Fig. 24. Solar spectrum from 1901, plate made by R.G. Neuhauss.

photography and photograph conservation, with special attention to the long-term preservation of this photographic material. Preus Museum – the National Museum of Photography in Norway – has a collection of twelve Lippmann colour plates, together with objects and papers related to these photographs. The purpose of this paper is to provide information about the museum's Lippmann collection, share the principle of the Lippmann technique, and describe the physical and optical properties of the plates. In addition, the chapter will discuss typical damages, deterioration issues,

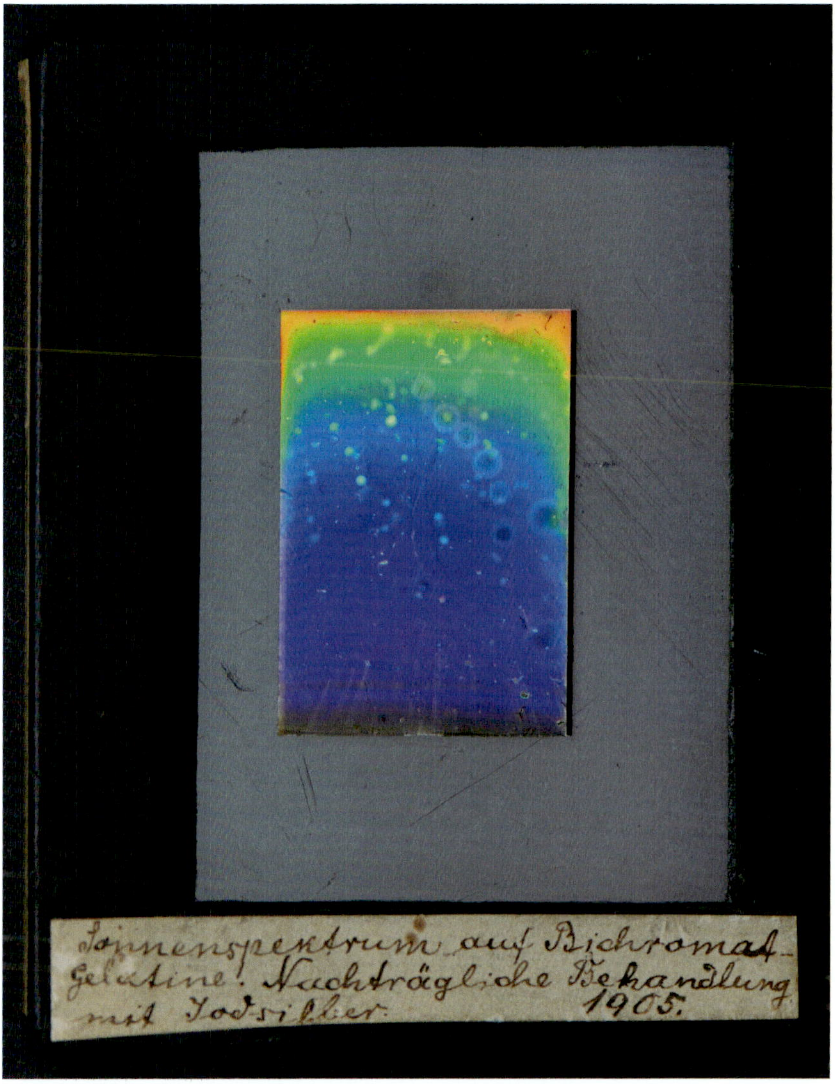

Fig. 25. Solar spectrum, gelatin bichromate plate, made by R.G. Neuhauss.

and challenges when using and handling these types of objects, both in exhibitions and for long-term preservation.

Lippmann Plates at the Preus Museum

The Lippmann plate collection of early colour photography at Preus Museum is one of the most important of its kind. It came to the museum in 1974

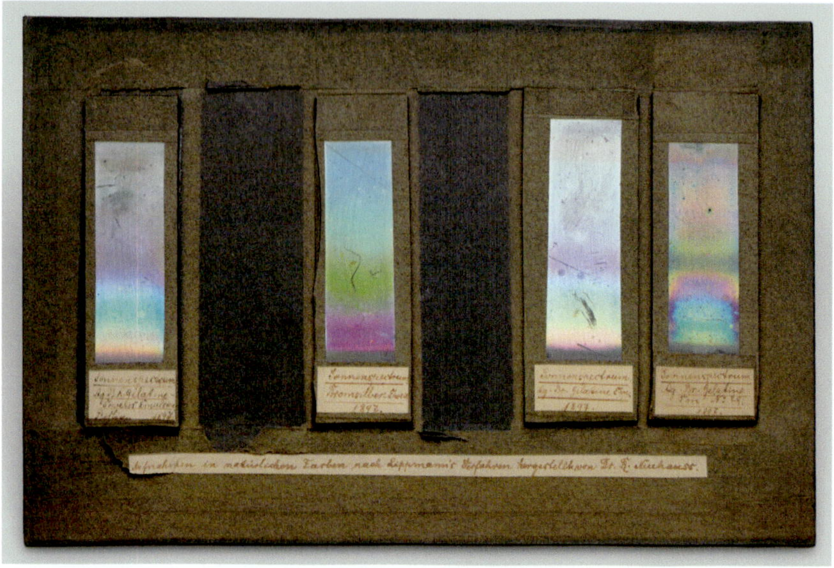

Fig. 26. Solar spectrum of the light experiments, plates made by R.G. Neuhauss.

with the acquisition of the research archive of Professor Dr. Albert Narath (1900–1974). This archive served originally for research and educational purposes at the Institute for Applied Photography and Motion Picture Technology at the Technische Universität Berlin (TU – Berlin). It consists of papers, books, objects, and, above all, photographic and graphic prints and samples mostly of a scientific character. It is one of a kind and comparable to collections like the Collection of the Austrian Federal Education and Research Institute for Graphics in Vienna (Sammlungen der Höheren Graphischen Lehr- und Versuchsanstalt – Wien), collections at institutions like the George Eastman Museum in Rochester, New York, or the National Museum of Photography, Film and Television in Bradford, England.

The Lippmann collection at Preus Museum is known and mentioned in the international literature regarding early colour photography (Coote 1993, 13; Bjelkhagen 2014, 22). Since its arrival at the museum in 1974, it has only occasionally been shown to the audience or visiting researchers. Most of the Lippmann colour plates at the museum were produced by Richard Gustav Neuhauss, a German physician, anthropologist, and photographer dedicated especially to scientific photography. His contributions are among the greatest when it comes to successfully producing colour plates, improvements, and publications concerning the Lippmann colour photography. One of the plates (Figure 27) was made by Hans Lehmann, a photochemist and physicist working for Zeiss in Jena and the Ernemann Company in Dresden.

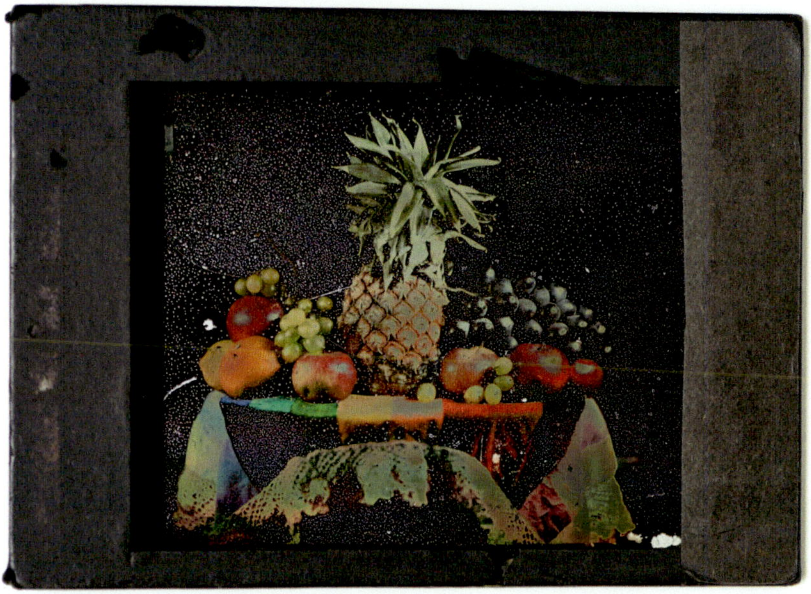

Fig. 27. Still-life, plate made by H. Lehmann.

His work was dedicated to cinematography but also many other specialties like interference photography (Eder 1932, 973).

The Lippmann plates in the Preus Museum collection feature different subjects:

- various spectra of the light in different qualities, two of which are cemented to a glass prism (by Neuhauss, Figures 24–26),
- close-up images of fabrics and various coloured patterns, (Figures 28 & 44)
- views of the Neuhauss garden house in Berlin – Grosslichterfelde, (Figures 29 & 47) and
- a still life with fruits (by Lehmann, Figure 27).

Several of the plates have paper labels (Figure 30) with handwritten information. In addition to the processed plates, the collection contains a package of unexposed 9 cm x 12 cm Lippmann plates manufactured by VEB Filmfabrik AGFA – Wolfen after the Second World War (Figure 31).

In addition to the plates, the collection contains special Lippmann plate holders with a mercury chamber in the standard formats 9 cm x 12 cm and 13 cm x 18 cm. The 9 cm x 12 cm holder has been attributed with high certainty to the fabrication of the German manufacturer Emil Busch AG-Rathenow (Figure 32 a & b) (stamp on the plate holder lock E.B. 211) (Hannouch, personal communication, 2020). The 13 cm x 18 cm plate holder

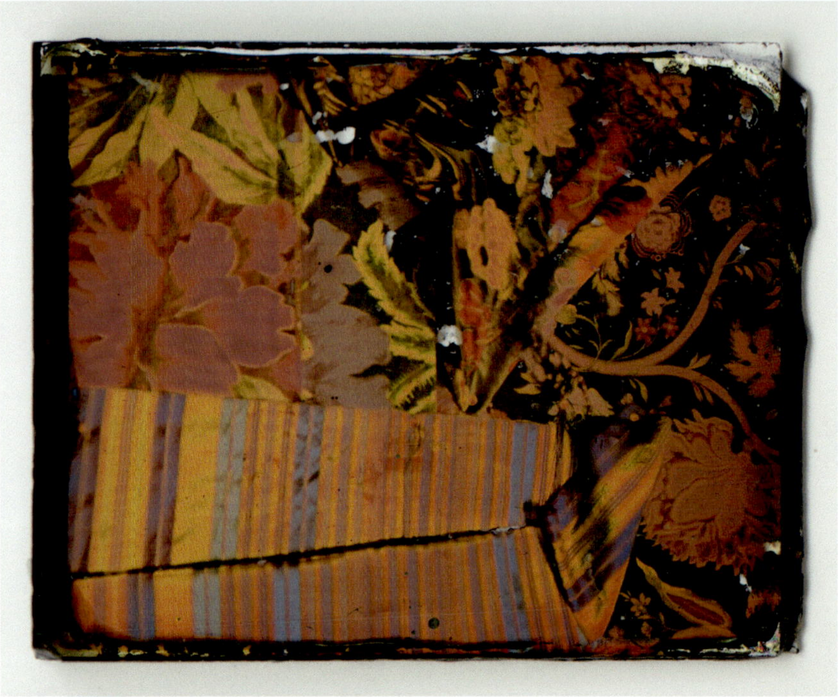

Fig. 28. Fabrics closeup, plate made by R.G. Neuhauss.

has a factory label from the French manufacturer of photographic equipment L. & A. Boulade Frères and can clearly be attributed here (Figures 33 a & b). Furthermore, the museum owns original papers by Neuhauss concerning the Lippmann technique as well as several original book publications from the time related to the topic, making it a privileged institution for the study of the Lippmann process history and materiality.

Lippmann Colour Photography History in Short

Observations and experiments related to the record of coloured light on light-sensitive materials, like silver halogens, are as old and older as photography itself. Key names and years in this regard are, for example, Senebier 1782, Seebeck 1810, Herschel 1840, Becquerel 1847, and Niépce de Saint-Victor 1851 and 1866. Becquerel and Niépce de Saint-Victor managed to obtain images but without notable success in regard to image permanence and colour reproduction (Eder 1932, 961, Koshofer 1981, 341). In 1873, Hermann W. Vogel discovered the sensitising dyes for silver halogen-based light-sensitive

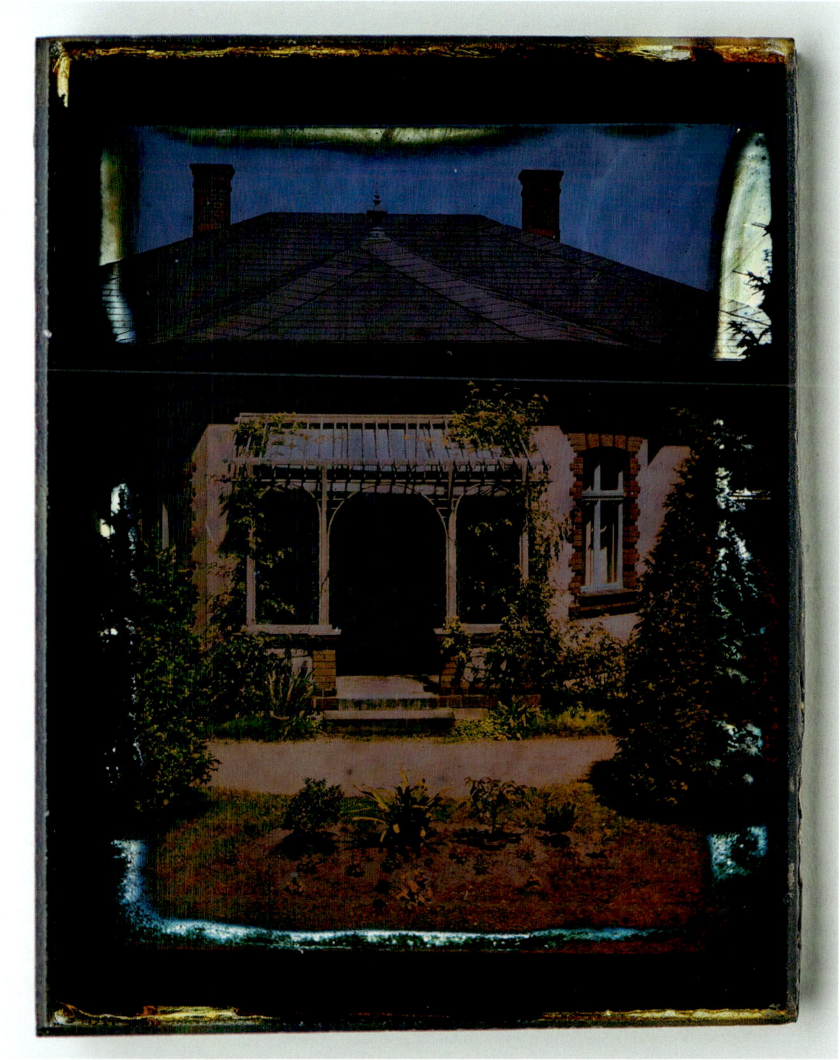

Fig. 29. View of the Neuhauss garden house in Berlin – Grosslichterfelde, plate made by R.G. Neuhauss.

emulsions, a discovery that made these materials sensitive to a broader part of the light spectrum (Peres 2007, 64). Gabriel Lippmann was aware of earlier attempts and discoveries while working on his own research using dye sensitising to prepare his experimental work. In 1891, Lippmann announced and presented at the Académie des Sciences in Paris the first samples of his invention, named after him but also known as interference heliochromy (Senior 1900, 211–26, Grebe, Hübl, and Wall 1929, 219–36). Two years later, he

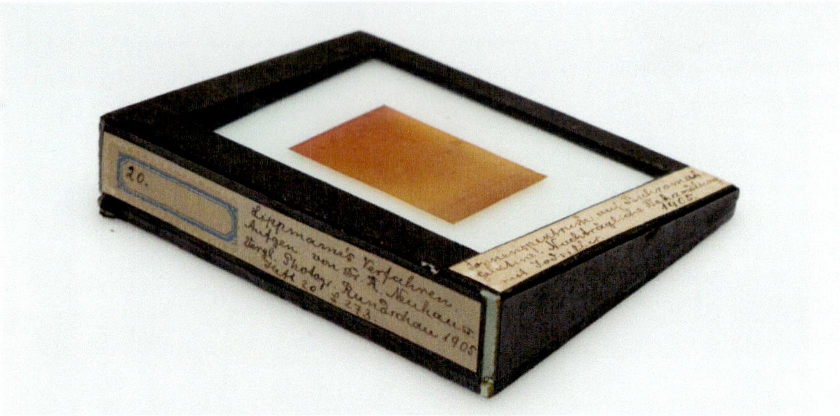

Fig. 30. Plate with paper labels, by R.G. Neuhauss.

presented another set of colour plates to the Académie, this time of higher quality. These plates were taken by Auguste and Louis Lumière to show the technique's practicability and potentially high quality. Eventually, the complete theory of his technique and discoveries was published in 1894 (Lippmann 1908, lecture, Bjelkhagen 2008, 862).

The fascinating process produces real and permanent colour images on an isochromatic or panchromatic black-and-white photographic emulsion by recording an interference pattern of the different wavelengths of coloured light.[1] The recorded and processed silver image reflects a colour image, in the same wavelength, when seen in white light from a certain angle or with the aid of a viewing device (a prism or viewing apparatus). For the design of his wonderful invention, Lippmann received the Nobel Prize in Physics in 1908 (Eder 1932, 970–71; Mutter 1967, 111–12; Pénichon 2013, 17–18). The invention was a promising sensation in early colour photography. But the technique required special skill, equipment, and photographic materials, as well as long exposure times, up to ten minutes, in sunlight (Koshofer 1981, 72), limiting the Lippmann technology to reach mostly scientific interest together with special attention from the early colour photography pioneers, like for example Valenta, Miethe, Krone, Neuhauss, Lehmann, Cajal and Lumière... . Other techniques like the autochrome plate, which did not even match the colour-recording quality and resolution of the Lippmann plate, were more practical and economical in use and therefore became the

1 Depending on the type of colour sensitisation, many photographic materials at the end of the nineteenth century and the beginning of the twentieth century were still not sensible to the whole spectrum of light (panchromatic).

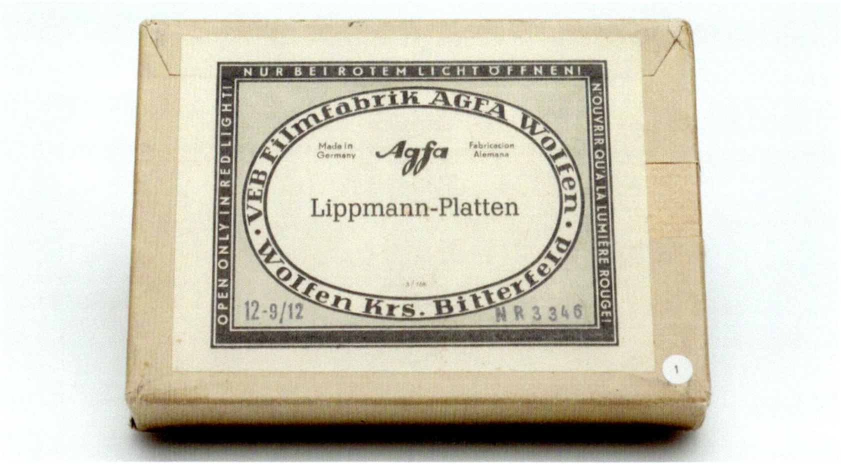

Fig. 31. AGFA – Wolfen – Lippmann plates – approximately 1950s.

milestone, starting the age of commercial colour photography. Nevertheless, the Lippmann colour photography is probably one of the most exceptional and beautiful techniques ever invented in the field of colour photography, due not in the least to its genuine technology, which allowed real colours in the correct, original wavelength of light reflected from the recorded object, as well as its diamond-like appearance when mounted correctly to a viewing prism. In many cases, it also has extraordinary permanence compared to most other colour photographic techniques up to the present day.

Short description of the Lippmann process

Lippmann himself described the principle of the image formation during his Nobel Lecture on December 14, 1908:

> The method is very simple. A plate is covered with a sensitive transparent photographic layer that is even and grainless. This is placed in a holder containing mercury. During the take, the mercury touches the sensitive layer and forms a mirror. After exposure, the plate is developed by ordinary processes. After drying the colours appear, visible by reflection and now fixed. This result is due to a phenomenon of interference which occurs within the sensitive layer. During exposure, interference takes place between the incident rays and those reflected by the mirror, with the formation of interference fringes half a wavelength distant from each other. The fringes imprint photographically through the whole

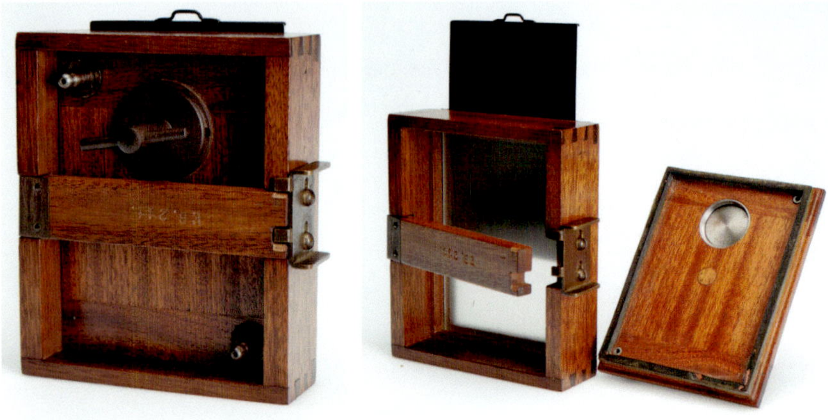

Fig. 32 a & b. Plate holder with a mercury chamber in the standard formats 9 cm x 12 cm by German manufacturer Emil Busch AG-Rathenow.

> thickness of the film and form a casting for the light rays. When the shot is afterwards subjected to white light, colour appears because of selective reflection. The plate at each point only sends back to the eye the simple colour imprinted. The other colours are destroyed by interference. The eye thus perceives at each point the constituent colour of the image. This is no more than a phenomenon of selective reflection as in the case of the soap bubble or mother-of-pearl. The print in itself is formed of colourless matter like that of mother-of-pearl or soap film. (Lippmann 1908, lecture)

Lippmann's description makes the technique to obtain a real colour image seem like an easy, straightforward operation. In reality, this was not the case, even when the skilled photographer used equipment and standardised plate material from the era's leading companies and manufacturers (e.g., Lumière, Carl Zeiss Jena, Jahr, L. & A. Boulade Frères, Braun, Stegemann, Emil Busch AG Rathnow, Watson & Sons, Penrose & Co., and Kranseder & Cie. AG München) (Eder 1932, 972–74, Koshofer 1981, 18–19, Koshofer 1981, 77, Bjelkhagen and Green n.d) (Hannouch, personal communication, 2020) . As already mentioned, several researchers and pioneers in the field of colour photography actively practised and tried to improve the technique. Lippmann himself used dye-sensitised albumen plates as an image carrier for his first experiments (Eder 1932, 966). The Preus Museum owns Lippmann plates prepared and exposed by Richard Neuhauss and Hans Lehmann, who both used self-manufactured plates based on albumen and gelatin (Neuhauss), as well as premanufactured gelatin-based plate types (Lehmann). Lehmann had a cooperation with the dry plate manufacturers

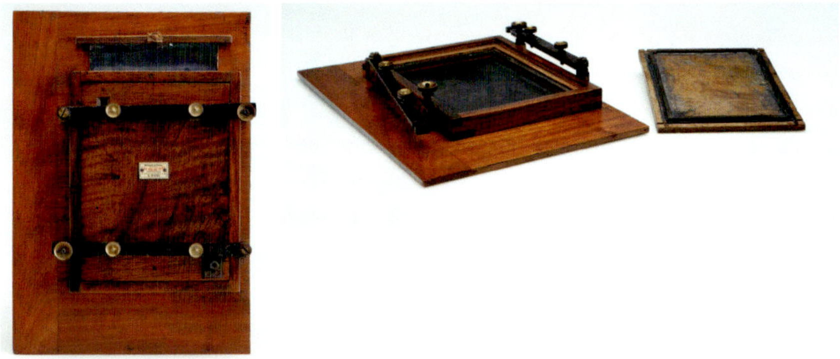

Fig. 33 a & b. Plate holder with a mercury chamber in the standard format 13 cm x 18 cm by the French manufacturer of photographic equipment L. & A. Boulade Frères.

Trockenplattenfabrik Kranseder in Munich and Trockenplattenfabrikant Jahr in Dresden. These plates are described as some of the best of their time (Eder 1932, 974, Koshofer 1981, 19). Like many other formulas of manufactured photographic materials, the way of fabrication of these plates were kept a secret for a long time.

Neuhauss also experimented with Lippmann's technique on the basis of gelatin bichromate hardening (Figure 25) (Valenta 1912, 105–7). His book *Die Farbenphotographie nach Lippmanns Verfahren*, however, describes in detail the fabrication and use of albumen, albumen/gelatin, and gelatin emulsions. These descriptions are made from his experiences in his laboratory. They give a good indication of what making these plates was like and of the struggles a "Lippmann photographer" had to overcome. The Preus Museum's collection consists mostly of Neuhauss plates. In the following, I will therefore present the basic technique in ten steps on the example of making an albumen-based plate, in line with Neuhauss's own description from 1898 (Neuhauss 1898, 16–28). This may give the reader a better idea of the labours the technique entailed, but also an indication of what to look for when identifying and inspecting a Lippmann plate.

Step 1: Preparation of the albumen emulsion

Albumen is a part of the egg white with exceptional properties for making photographic emulsions. By beating and settling the egg white, albumen is separated as a clear substance from the other parts of the egg white. During this step, halogen salts are added. Neuhauss suggests potassium bromide or potassium iodide but not a mixture of both. After the liquid has settled, the clear albumen is decanted and often also filtered before use. A key factor in order to succeed, Neuhauss points out, is the quality of all the raw materials

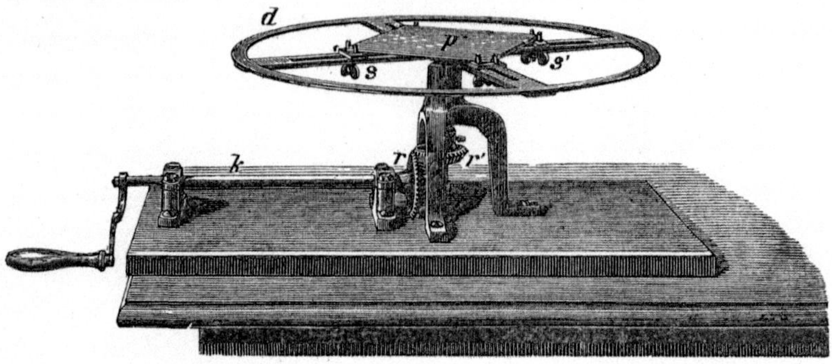

Fig. 34. Photographic plate-centrifuge.

(especially of the egg) but also the absolute purity of the solution and the accuracy of the photographer's work.

Step 2: Cleaning the carrier glass

To secure a good adhesion of the emulsion on the carrier glass, it is necessary to remove any dirt, grease, fingerprints, or glass corrosion products from the glass surface. Neuhauss suggests a twelve-hour bath in nitric acid, followed by a wash with water, and a final polishing with the aid of linen cloth, ammonia, and ethanol.

Step 3: Coating the glass plate

For coating the plate, it is best to hold it horizontally between two fingers of the left hand. An appropriate amount of albumen is poured on the centre of the plate. Air bubbles should be avoided. A clean glass stick can be used to disperse the liquid evenly or remove air bubbles. Neuhauss also suggests spinning the plate slowly on a plate centrifuge (Figure 34) to make the albumen layer even more uniform. The coated plate is trayed in a dust-free area on a water-levelled support. The whole operation can be done in daylight, since the albumen emulsion is not light-sensitive yet (unlike gelatin plates, where the emulsion is light-sensitive already during coating). When dry, the plates have a very long permanence and may be stored and used over a long time.

Step 4: Sensitising the emulsion with silver nitrate

During this operation, light-sensitive silver halogens are formed in the photographic emulsion. Working in a darkroom is therefore a requirement.

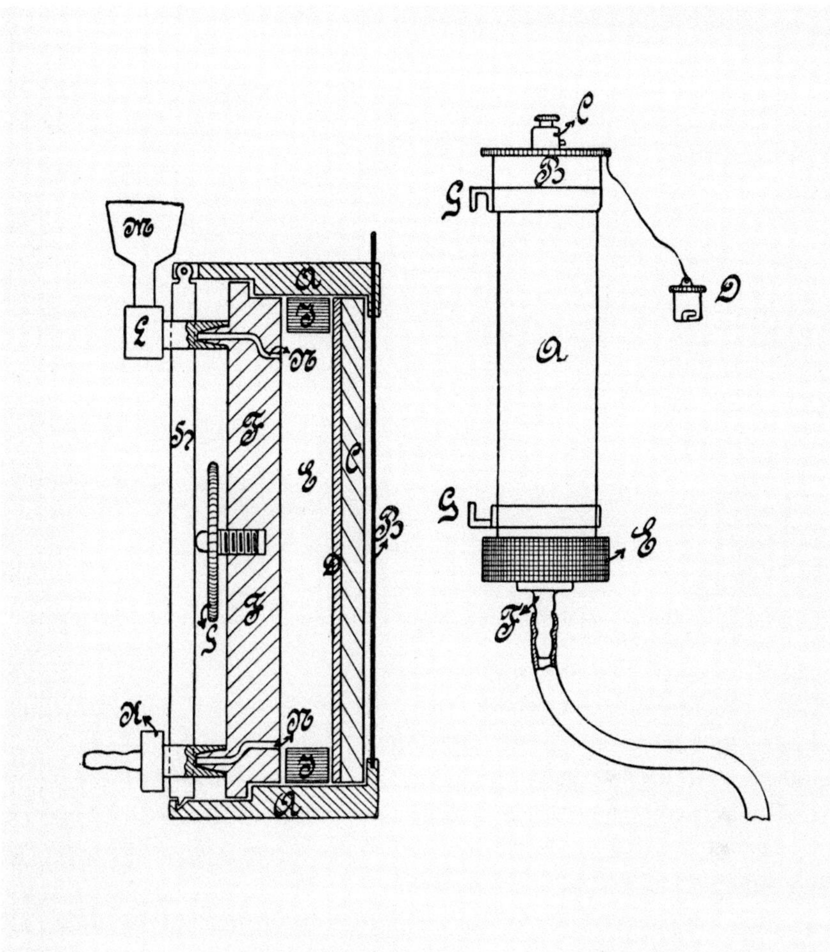

Fig. 35. Photographic plate holder with mercury chamber and reservoir bottle.

The ambition is to achieve an almost “grainless” and transparent photographic emulsion. In this operation, the albumen plate is placed for two minutes in a bath containing silver nitrate (10 g), distilled water (100 ml), and vinegar (10 ml). After two minutes, the plate is removed and placed in a water bath, which should be refreshed from time to time during the next fifteen minutes. The plate is now light-sensitive to the blue and ultraviolet part of the light spectrum.

Step 5: Colour sensitising the emulsion with dyestuff

Neuhauss recommended a two-bath operation with cyanine dye (bath no. 1) and erythrosine dye (bath no. 2). The dyes are diluted 1:500 in ethanol, and these solutions are mixed for each dye bath with 200 ml water. The plates, already

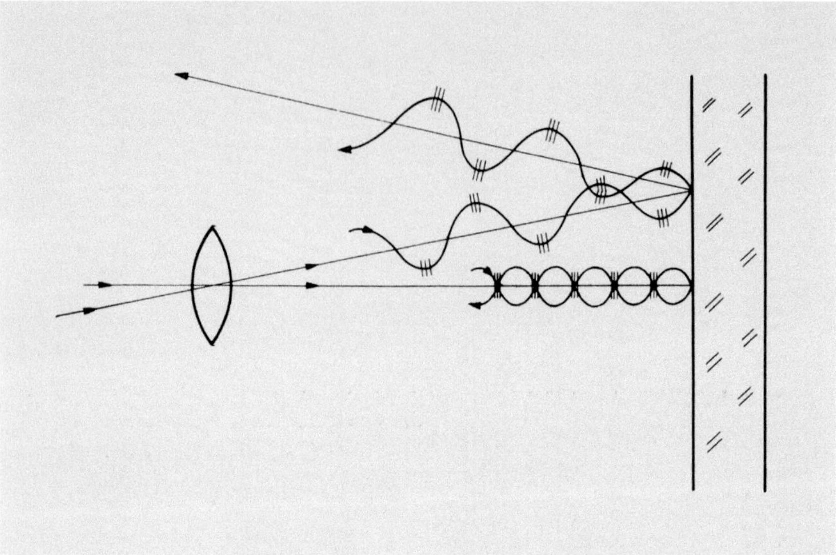

Fig. 36. Photographic exposure in the Lippmann emulsion. The different wavelength of light in the Lippmann emulsion bouncing back from the mercury mirror (on the right) through the emulsion again and herby creating standing waves, registered as exposure in the overlapping areas.

washed after the silver nitrate bath, are now placed first in bath no. 1. After three minutes, they are moved to bath no 2. Finally, the excess of the dye solution is removed with the aid of a plate centrifuge. After drying in the dark, the plates are useable for about two weeks before fogging, with non-image silver, occurs.

Step 6: Setting up the camera and plate holder

The camera is placed and focused on the subject to be photographed. Bright daylight – but not too bright – was recommended by most of the photographers. The key tool in terms of equipment for the Lippmann technique is the "mercury" plate holder (Figures 32, 33, & 35), allowing the use of liquid mercury as an efficient mirror substance. To operate, the photographic plate is placed in the plate holder with the emulsion side facing the mercury chamber on the back of the holder. This closed and tightened chamber is filled with liquid mercury right before exposure. The operation of the camera and plate holder would have been as follows (Valenta 1912, 64–65): the plate holder, loaded with the plate facing the uncoated side to the lens, was placed in the camera. Now, the mercury was filled into the plate holder with the aid of a reservoir bottle, which is connected to the plate holder by a rubber hose (Figure 35). A valve on the bottle is opened, the bottle is lifted, the mercury runs in the space in the back of the plate holder, touching the surface of the photographic emulsion. The filling is completed when mercury is visible on a control opening at the top of the holder. When this occurs, a valve in the reservoir is closed, and the

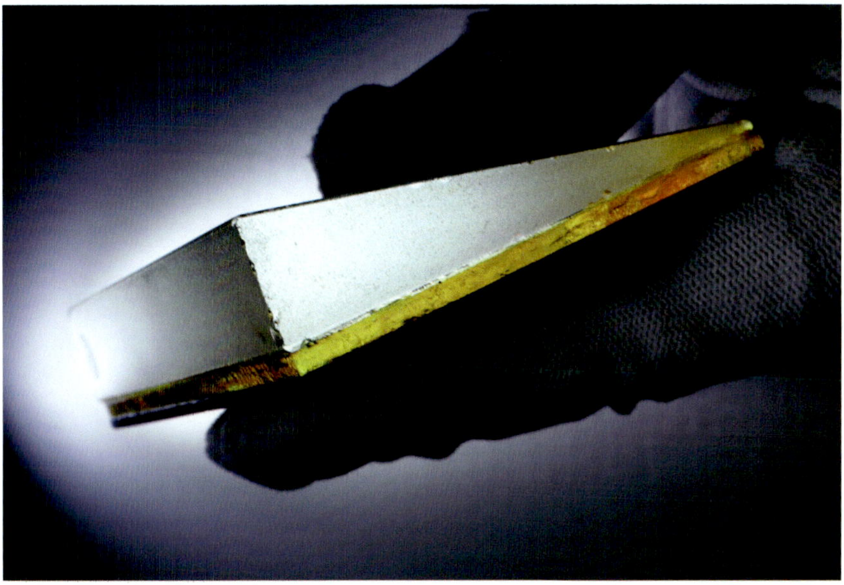

Fig. 37. Lippmann plate with 10° prism.

plate is now ready for exposure (The mercury free method, using the light reflection phenomenon on the air/emulsion interface was not recommended by Neuhauss or Lehmann and is therefore not discussed here.).

Step 7: Exposure

The exposure times for colour plates after Lippmann are very long (several minutes or up to hours depending on the subject). There are several reasons for this, but the major one is that the silver halogen crystals in the emulsion need to be extremely small to record the very small interference pattern of the standing light waves of the colours of light (Figure 36). The disadvantage is that small silver halogen crystals catch less light, resulting in a low photographic speed of the photographic material. Lippmann photography requires, therefore, without exception, the use of a tripod.

Step 8: Removing the plate from the plate holder

After exposure, the mercury is removed from the plate holder by opening the valve at the reservoir bottle, lowering the bottle, and letting the mercury flow out of the plate holder and into the reservoir. Now the valve at the bottle is closed and the connecting hose is removed. Simultaneously, the dark slide of the plate holder is closed. After that, the plate holder can be removed from the camera and moved to the darkroom, or a changing bag for removal of the plate. The plate is now ready for development.

Step 9: Processing the plate (without additional treatments like intensification and toning)

As previously mentioned, to record the interference patterns of standing light waves in the black-and-white emulsion, it is essential that the silver halogen crystals – and finally the developed image silver particles – are of an extremely small size. This can be achieved by avoiding a ripening process of the emulsion, for example, by making and using the emulsion in lower temperatures and by using appropriate developers. The recommended developers to do so were pyro- or ferrioxalate developers or pyro-ammonia-potassium-bromide developers (Neuhauss 1898, 12–14, Valenta 1912, 27, Eder 1932, 967). They secured not only a small grain size but also the whitish character of the silver for maximum light reflection of the recorded image. After the development is completed, the image is fixed in a solution of sodium thiosulphate / hypo (Valenta 1912, 28, Neuhauss 1989, 13). This is followed by the final wash, before the plate is dried in a dust-free environment. At this point, the colour image is visible when held against a black background and viewed from an angle, depending on the position and type of the light source. Still, reflections from the verso of the carrier glass and surface of the emulsion are rather disturbing (Wiener 1899, 524–25, Zenker 1900, 149–57, Valenta 1912, 82–83). The easiest way to examine the colour plates after processing, is the method recommended by Otto Wiener. He suggests inserting the plate in a benzol-filled glass cuvette at a 10–degree angle to the glass wall and holding this against a black background. The refraction index of benzol is close to glass and enables the viewer to examine the plate in an improved way by removing disturbing reflections from the carrier glass and emulsion surface. This method is appropriate for a first quality inspection but may not be practical for long-term display and use (Zenker 1900, 149–57, Wiener 1899, 524–25, Valenta 1912, 83, Mutter 1967, 117).

Step 10: Mounting

The processed plates show in transmitted light a rather flat yellowish, almost colourless black-and-white image of the recorded subject. Some authors mention that complementary colours could be seen, although this has not been fully confirmed (Mutter 1967, 117). However, it gives the photographer a first impression of the quality of the exposure, but no indication of the recorded colour image. The developed Lippmann plate needs some finishing to show the recorded colour image in the best possible way. Lippmann colour plates are viewed in reflected light. In an optimal situation of the developed image, each small silver grain has an almost perfectly round metallic structure, reflecting light in all directions, making the image material appear whitish (Eder 1932, 967). To some degree, similar structures are seen in other early photographic techniques like the daguerreotype or the wet collodion positive, even though the silver grains in a Lippmann plate are much smaller. When seen under enlargement in reflected light, however, the grains themselves appear also whitish (Gold 2018, 16–18 and 83, Valenta 1912, 27). These silver particles build in their sum the interference patterns of the recorded wavelengths throughout the whole emulsion (Figure 36). When seen in white light, these small patterns reflect light waves only according to the recorded wavelength or colour, and all other unrecorded light waves will be absorbed. To see the colour image, however, the light has to hit the plate surface at a certain angle and the viewer has to look at the plate slightly from the side.

Step 11: Finishing steps

To improve the viewing of the colour plates, several final finishing steps have been performed:

1) To limit the reflections from the back side of the carrier glass, the verso of the glass plates was painted with a black varnish made of asphalt (a black semi-solid substance made of bitumen and inert mineral, such as pigments) or black varnish based on pigmented cellulose nitrate (Zapon lacquer) (Neuhauss 1989, 49) (Figure 42 a). Alternatively, practitioners were recommended to mount a black glass sheet on the verso of the plate (Figure 42 b). In the Preus Museum collection, the plates with varnished verso are often also sealed with paper (Valenta 1912, 83, Senior 1900, 226).
2) To remove or limit disturbing reflections from the image surface, the application of a transparent lacquer like collodion or Zapon lacquer was recommended. The cementing of another glass sheet with the aid of Canada balsam (an oil resin consisting of a viscous yellowish to greenish liquid exuded by the balsam fir of North America) was suggested alternatively (Zenker 1900, 156–57, Valenta 1912, 83).
3) An even better method to limit the disturbing surface reflections and additionally improve the viewing angle was to mount a ten-degree prism (Figure 37) on the surface of the image (Valenta 1912, 83, Zenker 1900, 156). For this purpose, Canada balsam is poured on the centre of the image surface. The prism of appropriate size is then placed carefully on top of the photographic plate. This was done in a manner that prevented the formation of air bubbles. As this step is completed, the sandwich is carefully pressed together, and the excess of Canada balsam is removed from the edges. Finally, the whole sandwich is sealed with a paper seal. The use of Canada balsam is a method known from the optic industry for lens assembly, or used by nineteenth-century photographers in the production of ambrotypes. The material has a similar refraction index as glass, but unfortunately because of that, tends to optically interact

with the colour image on the plate surface. To avoid these interactions, Richard Neuhauss mentions in his notes a coating of the image material before mounting the prism, with, for example, gelatin or Zapon lacquer (Neuhauss 1894–1900, 31–32). (Gold 2018, 34–35).
All three steps improve the image quality significantly, making the colour image appear clearer and more brilliant.

Viewing / Presentation: Light and Light Sources

The quality, quantity, and direction of light is very important in order to see colour images properly. By the end of the nineteenth century, when the Lippmann plate was first presented, most people were used to looking at photographs in daylight to be able to see details in the image. Sunlight coming from one direction, or diffuse light in the shadow or a cloudy sky, would be the light source. Daylight has the best range of the light spectrum and the best direction, and it is intense; no other light source can match this when looking at Lippmann plates. Artificial light was available from a variety of different light sources, but as a strong, bright-white light source it was luxurious.

Early colour photographers keenly described viewing conditions and light, and their descriptions were according to the possibilities at the time (Valenta 1912, 82–88, Neuhauss 1898, 57–59). Artificial light produced by for example gaslight, or still very frequently petroleum light, was common when daylight was not available. Electrical arc light was available as a very strong and intense light source and mostly used for projection, but it was not very practical to use for standard illuminations in rooms. The incandescent light bulb was about twenty years on the market and its qualities were impressive, especially when it came to intensity and flexibility in use. Artificial light sources had the advantage that they could be positioned and adjusted from the viewer according to the needs of direction and intensity, whether for close inspection by one or a few persons or as an illumination source for projection.

Viewing and Viewing Devices

One of the things that enhanced the presentation of Lippmann plates was the use of viewing devices and projectors. The presentation of Lippmann plates is not uncomplicated – for the viewing it is essential to have a white light source and the right viewing angle. The removal of surface reflections

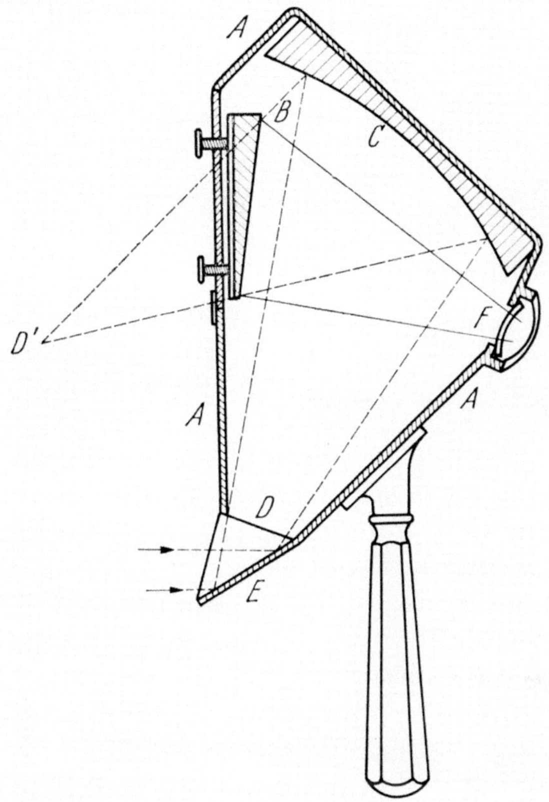

Fig. 38. Katopitsches Betrachtungsgerät made by Carl Zeiss Jena.

is also vital. The right positioning of the light source, with or without a mounted prism, should be close to the optical axis of the human eye or the projector lens. This is easy to achieve under normal daylight conditions; with artificial light, photographers are advised to cover the light source with a diffusing screen, paper, or opal glass (Senior 1900, 226). Viewing devices, manufactured for example by Carl Zeiss Jena, could be used with artificial light or daylight (Mutter 1967, 117) (Figure 38). With an insight arrangement of mirrors and lenses, they secured the optimal positioning of light and of the viewer's eye (Valenta 1912, 84–85, Neuhauss 1898, 57–59).

The projection of Lippmann plates to reach a greater audience was not at all an easy endeavour since the plates needed to be projected with reflected light. Particularly the projection of non-translucent objects was challenging

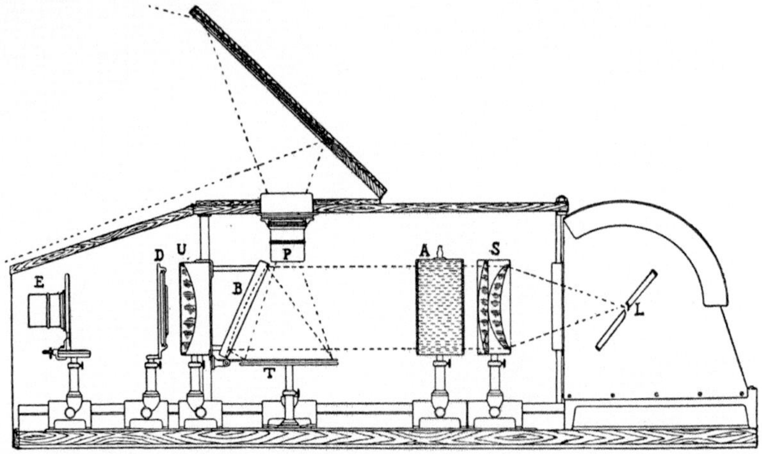

Fig. 39. Projector for episcope and standard projection

(Liesegang 1903, 140–144, Neuhauss 1908, 95–103) but to do so, an episcope projector type or additional episcope equipment to a standard projector with a very strong and bright light source was required. The projection light would travel from the light source reflected and enhanced by a concave mirror and/or by a condenser lens, from there to a mirror and then to the photographic image to be projected (Figure 39). Since light sources at that time produced a significant amount of heat radiation, a heat absorbing cooling device was often installed between light source and mirror (Neuhauss 1908, 21–24, Hassack and Rosenberg 1907, 13–16) (Figures 39 & 40).

A classic one was a water-filled glass tank / cuvette installed in the light path of the projection unit (Figure 40). However, the light reflected by the Lippmann plate would travel through a projection lens, again hit a mirror, and ultimately be projected on a screen. There were several additional obstacles to overcome when projecting Lippmann plates. First, the plates reflected only a little part of the light, namely the recorded colour image. Most of the light was absorbed by the object materials. Cooling devices / chambers, mirrors, condensers, and lenses also absorbed a considerable part of the light used in this kind of projection. Often very strong light sources, like arc lights or other electrical lights, were used. Historical records mention that the size of the final projected image was limited. Valenta describes an example where the Lumières in 1893 used screen sizes of about 40 cm x 70 cm to project Lippmann plates (Valenta 1912, 86–88). Other sources advise against enlarging the projected image more than one meter at the maxium in order to have a bright enough colour image on the screen (Mutter 1967, 118).

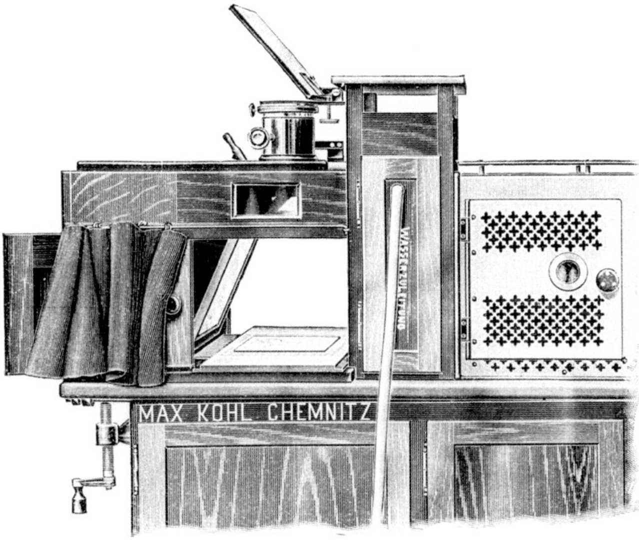

Einrichtung des „Megadiaskops" von Max Kohl in Chemnitz für Projektion im auffallenden Lichte.

Fig. 40. Projector for episcope and standard projection. "Megadiaskop" by Max Kohl, with water-filled glass tank installed in the light path.

Another constraint was the projection time. This was very limited since the Lippmann plates are very heat sensitive and the developed heat in the projection unit was substantial, even though cooling techniques were used. Overheating could cause delamination or "cooking" of the Canada balsam used to mount the prism, and the image material could also be destroyed. The Lehmann plate in the possession of the Preus Museum shows a very interesting "bubble pattern" that could point in the direction of overheating during projection (Figure 41). The episcope projectors and additional episcope equipment for standard and effective quality projection were manufactured by suppliers such as Lumière, Zeiss, Schmidt & Haensch, Liesegang, Fuess, Max Kohl, Krüss, and Reichert (Hassack and Rosenberg 1907, 96–107, Neuhauss 1908, 95–103), while typical light sources in these projectors include electric arc light, gaslight, hydrogen light (oxyhydrogen light), incandescent electric light bulbs, and also the less efficient petroleum light. The electric arc light was the most intense light source at the time and quite common if a large amount of light was needed, as when non-transparent objects were projected (Liesegang 1903, 45–144, Neuhauss 1906, 55–69).

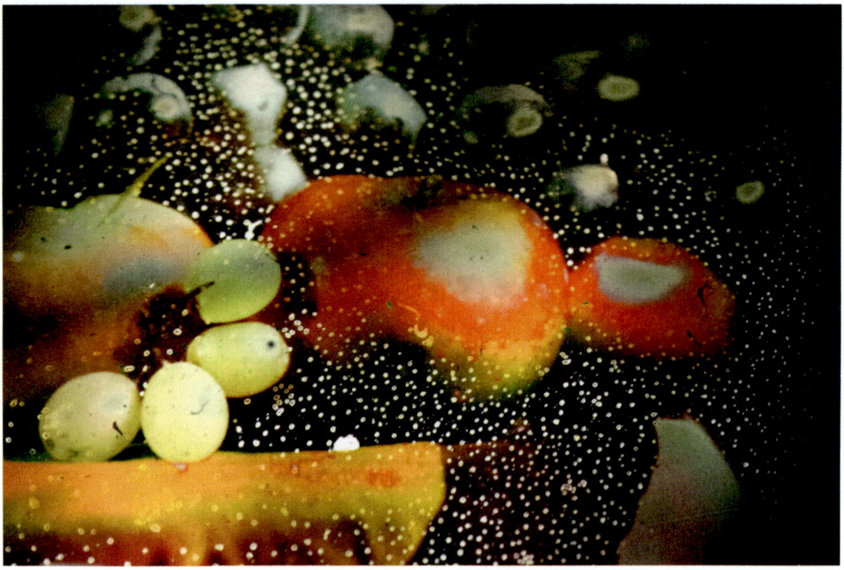

Fig. 41. Detail of the Lehmann plate in the possession of the Preus Museum showing an interesting "bubble pattern" that could point in the direction of overheating during projection.

Identification of Lippmann Colour Plates

The identification of Lippmann colour plates is usually, with a few exceptions, not too difficult. They seem to appear in four major groups with plate sizes from 22 mm x 80 mm up to 130 mm x 180 mm):

- Plates without any form of protective varnish, cemented cover glass, or prism
- Plates with protective varnish (not present in the Preus Museum collection)
- Plates with protective cover glass cemented on the image surface (not present in the Preus Museum collection)
- Plates with ten-degree prism cemented on the image surface.

Type I: Plates without any Form of Protective Varnish, Cover Glass, or Prism

Lippmann plates have a glass carrier with an emulsion side. Plates without any form of protective varnish, cemented cover glass, or prism have a visible emulsion. They appear in transmitted light as a rather flat and often yellowish black-and-white image. Colour is visible in reflected light but often not as easy and brilliant as described in the literature and above (Figures 26 & 43). An explanation could be that since the emulsion is exposed to the surrounding atmosphere, the image material deteriorates over time and so disturb or destroy the colour image recorded. Historical plates with image silver deterioration differ often significantly from new contemporary plates made by scholars or contemporary enthusiasts of nineteenth-century photography. Of all the plate types, these plates are the hardest to identify if discovered in a photographic collection without any context or provenance.

Type II: Plates with Protective Varnish

These plates show a clearer colour image than plates with uncoated and unprotected emulsion. A glossy varnish can be observed on the emulsion surface. The back of the plates is often covered with a black varnish. Typically, almost no image deterioration has taken place since the image silver is protected from air access.

Type III: Plates with Cemented Cover Glass or Mounted Cover Glass

A cover glass is cemented to the emulsion side, as may easily be seen from the side if it is not either sealed with paper or framed. Typically, almost no deterioration has occurred since the image material is protected from air access. Plates with glass covers show traces of the Canada balsam on the edges if they are not hidden with a paper seal. Air bubbles incapsulated in the sandwich or delamination on the edges are not unusual to see. The verso side of the whole package is often covered with a black lacquer or black glass mounted for example with Canada balsam. For protection and perhaps a nicer appearance, these plates are frequently sealed with paper tape.

Type IV: Plates with Ten-Degree Prism Cemented on the Image Surface

Plates with ten-degree glass prism cemented on the image surface are the most common ones when presented in the literature or in exhibitions (Figures 24,25,27-30,37). Their portion of the total amount of Lippmann plates worldwide needs to be recognised. Prisms were usually mounted on plates with a good image quality from the start. In terms of identification, they

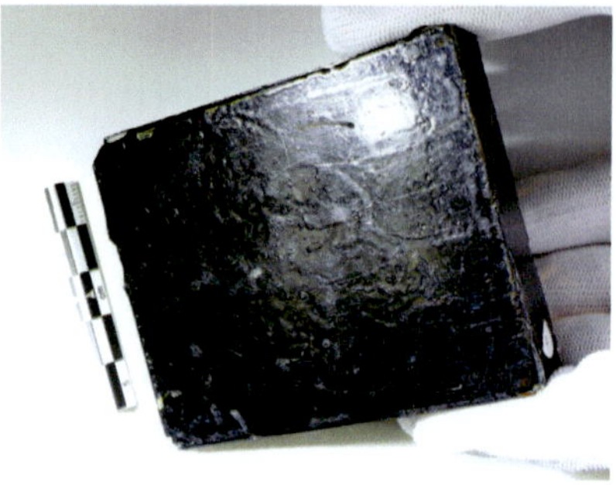

Fig. 42a. Verso side of the package covered with a black lacquer.

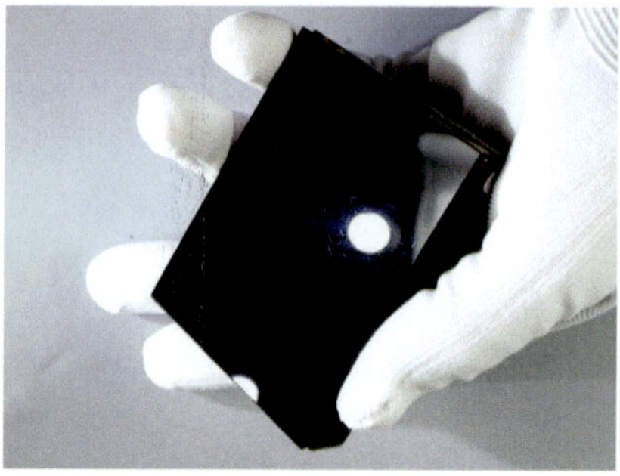

Fig. 42b. Verso side of the package covered with black glass.

represent the easiest to recognise and most beautiful objects of this kind. The glass carrier is mounted together with a ten-degree glass prism mostly with the aid of Canada balsam. As with type III plates, these plates may also show traces of mounting cement, delamination, air bubbles, and paper tape seals. Also, the verso side of the whole package is often covered with a black lacquer (Figure 42 a) or black glass (Figure 42 b) mounted for example with Canada balsam. For protection and perhaps a nicer appearance, these plates are frequently sealed with paper tape.

Labels with information about the images and maker are to be found on all of the above types of Lippmann plates (Figures 25,27,30). Many plates are frequently mounted on a second support for display and presentation purposes.

Typical Damages and Deterioration Signs: Deterioration of, and Damages to, the Photographic Image Silver

As said before, the photographic image of the Lippmann plates is made of microscopic silver particles which are located throughout the whole, but mostly close to the surface of the, photographic emulsion. The image silver is reactive in the presence of an aggressive atmosphere, but it is also partly sensitive to mechanical force. These two major factors can cause deterioration and damage to any photographic image material. One factor is the result of the use and handling of the photograph (Figure 43). The other factor is the effect of chemical reactions occurring with the presence of aggressive substances together with moisture, elevated temperatures, and light. The chemical reactions related to image silver deterioration are divided into two basic mechanisms which occur mostly parallel: the sulphide deterioration, also called silver sulphiding, and the oxidative-reductive deterioration of image silver. (Hendriks 1991, 57–61, Gold 2018, 83–86, Reilly 1986, 19–22).

Sulphide Deterioration / Silver Sulphiding

Silver, including image silver, is very susceptible to reacting with sulphur. When it does, it forms highly stable silver sulphide compounds. Potential sulphur sources include hydrogen sulphide from the direct environment, and to some degree air pollution. Another potential source is remnants of sodium thiosulphate (fixing baths) or exhausted fixing bath residues, together with the image silver forming silver thiosulphate as well as from an albumen emulsion containing sulphur by the materials nature. Characterised by their thin emulsion layer, Lippmann plates are unlikely to hold that many old process chemicals. Nevertheless, since these plates are produced under various circumstances, the stability of these materials certainly depends on many factors, for example the accuracy of the photographer's work. Sulphide deterioration can be seen on some unsealed examples in the Preus Museum collection. The deterioration products often have a brown or yellow colour, and the result of this process is yellowing and image fading. This kind of deterioration is very common with unvarnished, unsealed silver images on all kinds of carriers, including Lippmann plates (Hendriks 1991, 57–61; Gold 2018, 83–86; Reilly 1986, 19–22).

Oxidative-Reductive Deterioration

Another more common deterioration scenario for photographic silver is the oxidative-reductive deterioration. This kind of silver deterioration happens when silver with a large reactive surface, like photographic silver, is attacked by a variety of oxidants. These oxidants convert silver atoms to highly reactive, mobile silver ions. The silver ions may move from their original location, causing fading. But they can be reduced to colloidal silver and thereby represent a certain density in the image. This silver may form inside the photographic image layer, causing a yellow-brown, sometimes orange/red staining, but they may also form on the surface as a metallic sheen known as silver mirroring (Figure 43). Oxidative-reductive deterioration can be caused by a wide range of

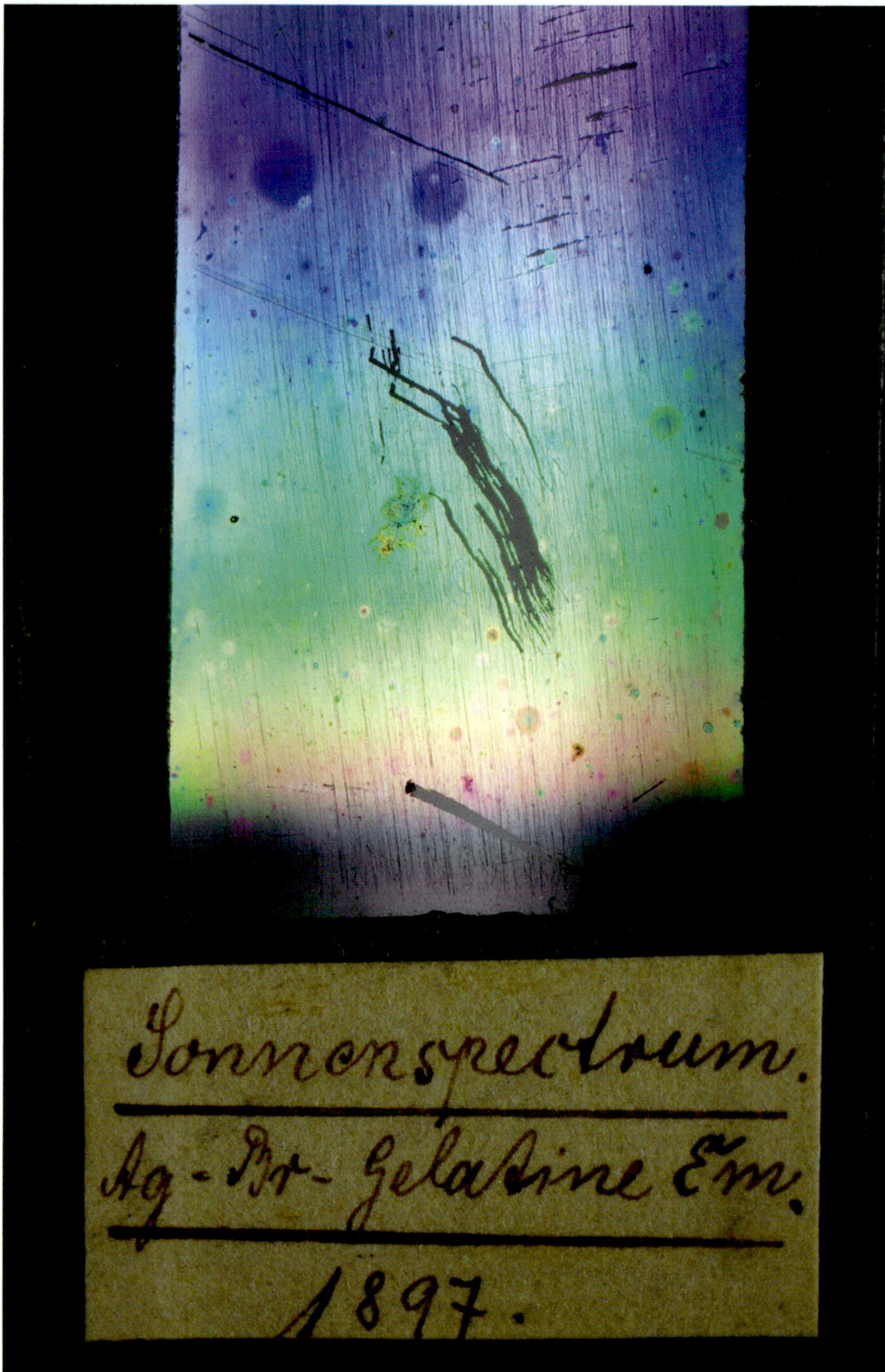

Fig. 43. Damages to the image surface.

complex reactions involving many different oxidising substances. Documented sources of these oxidants are for example atmospheric pollution by aggressive cleaning agents, old housings and inappropriate enclosure materials, paints and varnishes, solvents, rubber, oil, and organic materials like wood, leather, wool, and bone glue or hide glue. This type of deterioration is very common when the image silver is insufficiently protected. Lippmann images with the image silver protected by a protective varnish, glass sheet, or cement glass prism rarely display any sign of silver deterioration (Hendriks 1991, 57–61; Gold 2018, 83–86; Reilly 1986, 19–22).

Delamination and Glass Corrosion

Glass corrosion is an issue for any historical glass object, and it may impact the photograph in several ways, as glass may be found as a carrier of the image layer or as protection for the photographic emulsion. For photographic purposes, soda-lime glass was the glass of choice at the time of Lippmann plate production. For the prism, various high-quality glass types known from the optical industry were used. However, the composition of glass varies, as does its resistance to glass corrosion. Also, deterioration is influenced by the presence of aggressive gases, alkali, acids, salts, bacteria, fungi, light, temperature, and humidity fluctuation, among other factors. Still, there are two major factors concerning the stability of the glass. The first is composition. Because of better improved glass formulas, late nineteenth-century glass often had a lower alkali content, a circumstance that makes it less likely to develop glass corrosion than variations of earlier soda-lime glass. The other major factor is the presence of moisture in the glass environment. Water will cause alkali components to leak out from the glass, causing microscopic damage to the glass surface. This process is also called atmospheric corrosion or glass disease. With high humidity in the environment, or water condensation on the glass surface, more and more alkali leaks out of the glass surface. This brings about an alkaline environment and causes the pH on the glass surface to rise. The longer and more frequent the high humidity is, the more the alkali concentration will rise, causing the glass network to break down and the silica structure to dissolve. During this process of surface deterioration, alkaline glass corrosion products form on the glass surface (Gold 2018, 77–79; Davison 2003, 169–82 and 173–76; Koob 2006, 13–14; McCabe 1991, 41–73).

The humidity and temperature, but also the concentration of deterioration products on the plate, represents a range of factors that influence each other. The emulsion layer (both with and without protective coating and an existing black background varnish) and a mounted cover glass or prism offer some protection from humidity to the glass carrier of the Lippmann plates. But layers of different organic materials also absorb water and let it penetrate to some extent. Glass corrosion underneath the emulsion, especially on the edges and on the protective layers, is therefore not uncommon. The development of microscopic crystalline glass corrosion products makes the glass slowly separate from the layer it is attached to. This process slows down under low relative humidity but will eventually cause the complete separation from the glass support. Glass corrosion products can also cause both the silver image and the protective varnish layer to deteriorate (Gold 2018, 77–79; Barger, Smith, Deane, White, and William 1989, 1343–56).

Mechanical Damages

Old glass is prone to mechanical damage. Many Lippmann plates are also disfigured due to scratched glass, broken or missing glass parts, mechanically damaged varnish, and displaced sealing and emulsion parts. For photographic objects that have existed for about 130 years, this is far from ideal but expected. (Figure 44) (Davidson 2003, 169–71; Gold 2018, 80).

Deterioration and Delamination of Transparent Cement, Varnishes, and Binder Media

Delamination is an alarming issue for the Lippmann plates, especially those mounted to a prism or cover glass. To make an emulsion, resin, or varnish adhere to a smooth surface like glass over a long time is challenging, especially when the medium acts hygroscopically and the glass carrier is susceptible to glass corrosion. Additionally, binders such as gelatin, albumen, cellulose nitrate, and Canada balsam are prone to deteriorate. The binders will also change dimension and colour over time depending on the temperature and humidity conditions in their environment and as a consequence of their deterioration processes. In addition to high humidity and temperature and the fluctuation of these factors, examples of major

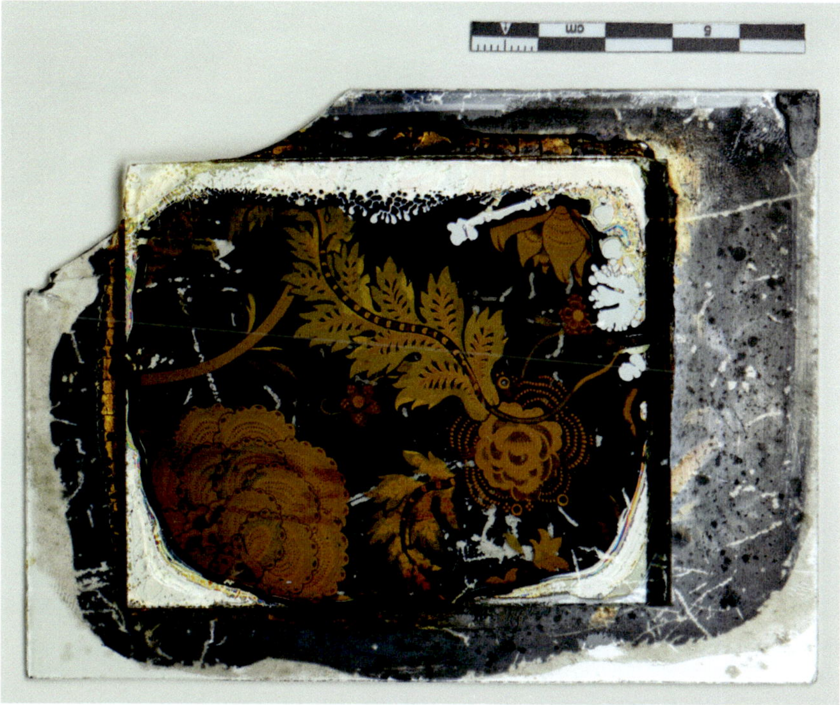

Fig. 44. Mechanical damages, delamination, and other damages.

causes of deterioration processes include bad binder combinations, air pollution, and UV light. The deterioration of the cements, binder, or varnish also leads to their hardening, which eventually contributes to a separation from the support (Figures 44 & 45) and the forming of cracks in the varnish layer or cement. The deterioration of the glass carrier that follows from the development of glass corrosion products also contributes to this varnish separation (Davison 2003, 337–44; Clark 1998, 231–33; Grinde 2005, 41–42). The types of deterioration mentioned above mostly emerge over time. But there are also damages related to delamination that may have been caused by the action of high temperature over a short period, for example under the influence of heat coming from a projection lamp. Damages of this kind are documented in the development of small bubbles in the cement between the photographic layer and prism (Canada balsam) or partly delamination of the prism from the photographic plate (Figures 27,29, 41,44,45,47). Of course, some of these "damages," like delamination and bubbles on the edges, could also be product from the mounting process of the glass prism.

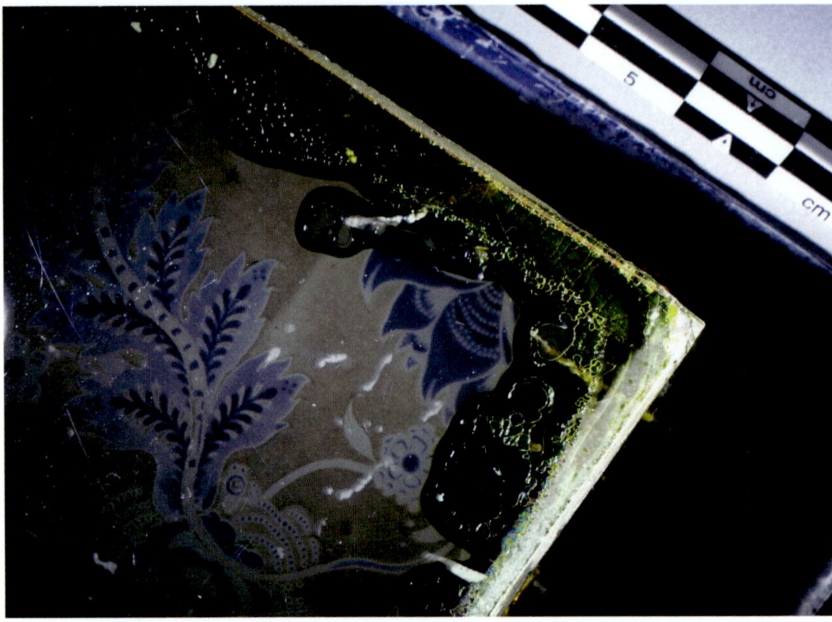

Fig. 45. Delamination damages and yellowing of the mounting cement.

Damages on Paper Seals, Paper Labels, and Board Material

The sealing tape, apart from its decorative purpose for housing, is one of the vital parts of the presentation design. It limits the access of air and moisture to the image material and cement, but it also protects the package from dust, insects, and larger particles as well as possible mechanical damage. It is often just an individual black sealing paper made by the materials available for the actual photographer. Broken or loose sealing tape is probably the most common damage on Lippmann plate packages (Figure 46). Such damage is not surprising considering that the paper materials over time have been exposed to the above-mentioned factors and every so often by unskilled or rough handling. Due to such factors, almost no original sealing paper is in good condition. The quality of the paper varies also, and when poor quality paper and board have been used, the risk of loss and decomposition is high. Such board and paper material can contribute to its own deterioration and may generate deterioration products like aggressive gases that can damage the silver image (Lavédrine 2005, 6–10; Ware 1994, 57–59; Gold 2014, 18–19; International Organization for Standardization, ISO 18916: 2007). At the end of the nineteenth century, several paper and board types were manufactured. While high-quality cotton/rag papers were still common, cheap paper and board made from wood pulp were

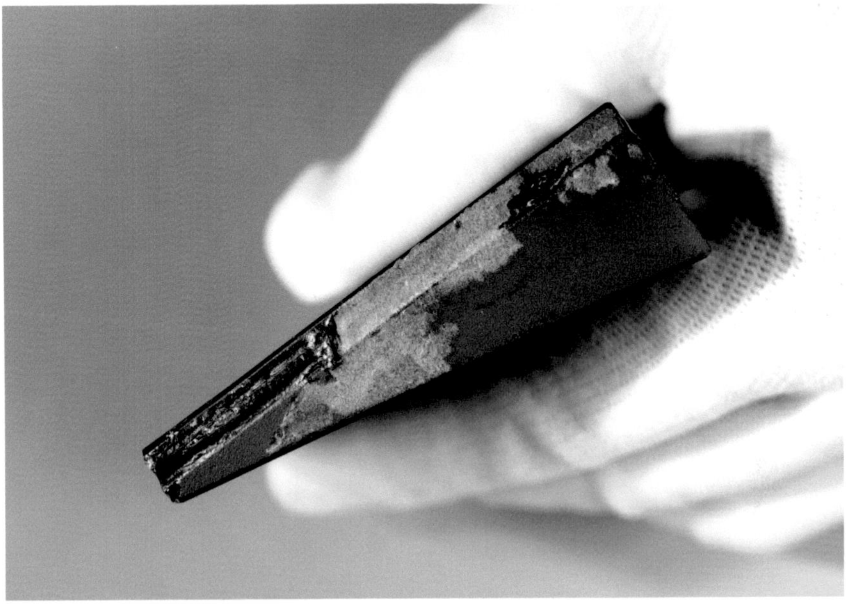

Fig. 46. Broken and missing sealing tape.

frequently used in the paper production. Wood pulp paper and board have a high content of lignin, which is a stabilising element in the cell walls of all cellulose containing plant material. The presence of lignin will cause a change in pH to the acetic over time, and the paper will therefore age faster. This is a major reason for the rapid change in the stability of these papers and their early deterioration (Whitmore 2015, 237–66 and 631). However, the main ageing process of any paper that causes the pH of the material to change into the acetic is called acid-catalysed hydrolysis. This is a change of the pH in the paper to an acid condition, causing the cellulose fibres to break and shorten, thereby weakening and destabilising the paper structure. Paper is also weakened by cellulose oxidation. This oxidation does not occur as fast as the acid hydrolyses but can be accelerated by factors like high humidity and temperature, an aggressive atmosphere (such as peroxides, ozone from for example electrical devices, solvent vapour, or hydrogen sulphur) and UV light. In total, indicators of paper deterioration are typically brittleness and an all-over yellow-brown colour change (Whitmore 2015, 237–66).

Papers and boards of varying qualities can be found as backing/covering/labelling paper. The used paper tape is often a simple piece of thin paper prepared with hide glue or bone glue. Over time the glue has hardened and changed dimension, eventually causing the paper tape to detach from the plate. The paper parts often show signs of deterioration, such as cracks and fine tears, weakening the material.

Preservation and Conservation

Recommendations for Storing and Displaying

When storing objects made of several materials whose properties differ greatly, determining the temperature and relative humidity must be a compromise between the requirements of the different materials. Since the photographic materials are the most vulnerable ones, however, the parameters for this material ought to guide this compromise. For Lippmann plates and any other photographic material, the guidelines of two ISO standards are important (Adelstein 2010):

- ISO 18918:2000 Imaging materials – Processed photographic plates – Storage practices
- ISO 18916:2007 Imaging materials – Processed imaging materials – Photographic activity test for enclosure materials (PAT)

The *recommended temperature and humidity* include a maximum storage temperature of 18 °C (with maximum fluctuation of 2 °C in 24 h) and a relative humidity of 30–50 per cent (with maximum fluctuation of 5 per cent in 24 h). But as these values are minimum recommendations, they may well be improved. Lower temperatures also slow down the rate of deterioration. A better temperature and humidity range, one that is used in many institutions with photographic collections, is around 10 °C (with maximum fluctuation of 2 °C in 24 h) and a relative humidity of 35–40 % (with maximum fluctuation of 5 per cent in 24 h). These values are on the lower range for materials like Canada balsam, hide glues, or paper, but still within the recommendations for these materials (Koob 2006, 141)·

When the recommended temperature and humidity levels are firmly followed, issues like *insect damage or mould* ought not be of concern, particularly not in the lower levels since neither mould nor insects will flourish in such conditions. However, storage areas or enclosures contaminated with any form of pests or mould must be avoided.

With regard to *air pollution*, it is recommended that the air in the storage and exhibition environment is free of aggressive gases (such as peroxides, hydrogen sulphur, solvent vapour from for example cleaning agents, or ozone from for example electrical devices) and particles (for instance from exhaust gases and dust) that can harm photographic materials.

The mentioned ISO standards for enclosure materials (ISO 18916:2007) are useful guidelines. For Lippmann plates and their housings, materials that have passed the Photographic Activity Test (PAT) are recommended. Materials used in direct contact with the photographic material, and glass, should have a neutral pH. Recommended are also materials, like paper and board, that to a certain degree can absorb aggressive gases and moisture coming from other materials in the environment or sudden changes in the conditions of relative humidity and temperature. Such materials also help prevent condensation damages.

Exhibition Recommendations of Neuhauss / Lehmann Plates at the Preus Museum

In regard to *exhibition lighting*, there are no ISO standards available for photographic materials, but recommendations given by international institutions and experts are documented in the conservation literature (Wagner, McCabe, and Lemmen 2010, 684–89, Lavédrine 2003, 163, 149–70, Gold 2014, 30). For exhibition purposes, light sources free of UV and IR levels should be preferred, for example LED light. Recommended light levels for rare and sensitive photographic materials like Lippmann plates are around 50 lux, with a maximum quantity of light from about 12,000 lux/h per year. However, these are recommendations for maximum values. In order to see colour images properly, the human eye needs a minimum light level often above these light levels. To find a compromise between light level and protection, appropriate exhibition installations and the use of electronic sensor technology is suggested. When it comes to the light setup, exhibition technicians may successfully use the recommendations from over 120 years ago mentioned above.

Permanence, Expected Conservation Work, and Conclusion

The long-term stability of Lippmann plates is good. They are comparable to dry plate and wet collodion negative and positive material. Plates with a cemented glass prism, varnish, or cover glass are expected to keep in good condition since the fine-grained silver image for the most part is sealed from aggressive air pollution or moisture. Experiences with even older

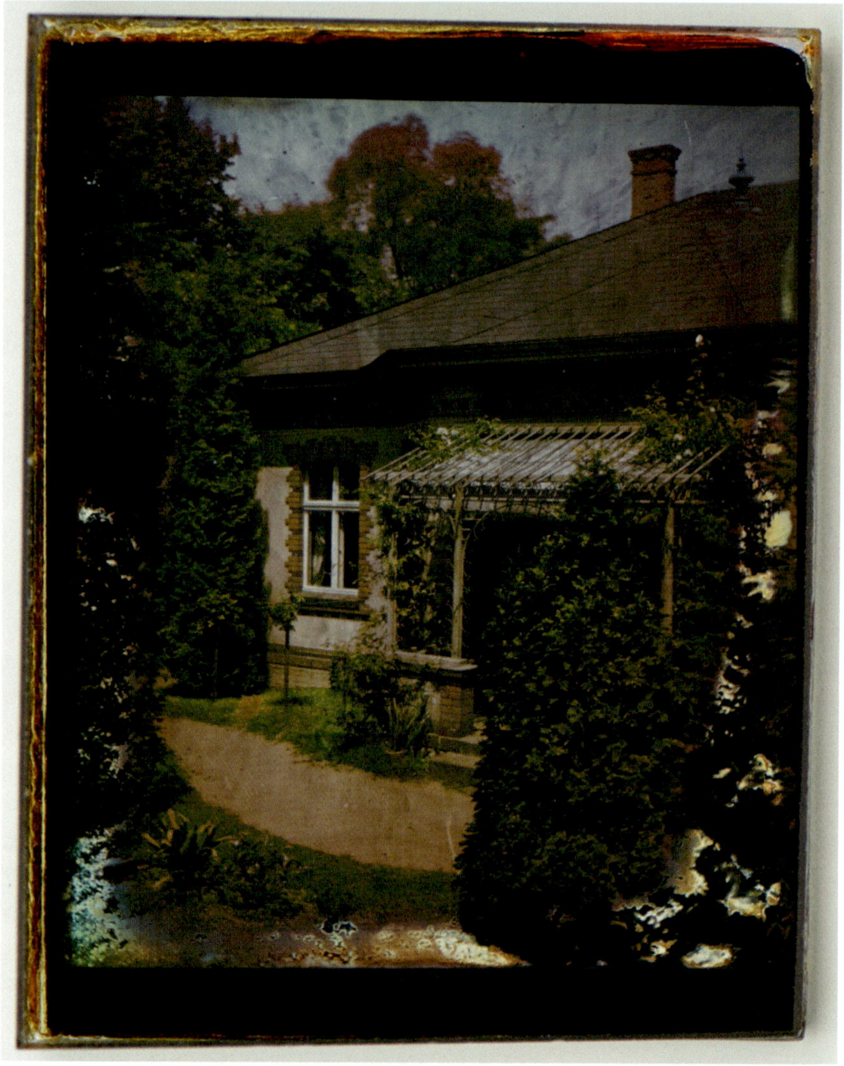

Fig. 47. View of the Neuhauss garden house in Berlin – Grosslichterfelde, plate made by R.G. Neuhauss. Delamination in the lower left and right image areas.

image materials sealed in similar ways have shown that this has been very beneficial for the long-term stability. Of course, long-term stability can be improved by following the above-mentioned preservation recommendations (Gold 2018, 33–35, 83–86).

Conservation work should be limited to a minimum, such as cleaning the surface of the glass materials, restoring papers seals and labels, and building suitable conservation housings and display materials. The major threat to Lippmann plates is any form of delamination and image silver deterioration. Such issues may largely be controlled by good environmental conditions. In the collection of Preus Museum, some plates have challenges with delamination of the optical prism (Figures 27, 29, 41, 44, 45, 47) – different stages of delamination). This is stable for now but is being closely monitored in the future, and consolidating the delaminated areas could be considered in order to secure the optical properties of the colour image.

After 130 years, Lippmann technology still reveals many interesting issues relating to these fascinating colour photographs. Of course, there are challenges regarding the preservation of this part of photographic culture heritage. But thanks also to its surprisingly outstanding permanence compared to many other early colour photographic materials, we will still have them with us for a long time.

Works Cited

Adelstein, Peter Z. 2010. *IPI Media Storage: Quick Reference.* 2nd edition. Rochester, NY: Image Permanence Institute, Rochester Institute of Technology

Barger, M. Susan, Deane K. Smith, and William B. White. 1989. "Characterization of corrosion products on old protective glass, especially daguerreotype cover glasses." *Journal of Materials Science* 24, no. 4: 1343–56.

Bjelkhagen, Hans I. 2008. "Gabriel Jonas Lippmann (1845–1921)". In *Encyclopedia of Nineteenth-Century Photography,* edited by John Hannavy, 862. New York: Routledge.

Bjelkhagen, Hans. 2014. "The History of Lippmann Photography (1891–1910)". Unpublished PowerPoint presentation, 2014. Available at http://www.hansholo.com/lippmann-photography-principle.html.

Bjelkhagen, Hans I., and Darran P.M. Green. "The True Colour of Photography". Paper, Available at http://www.hansholo.com/lippmann-photography-principle.html.

Clark, Susie. 1998. "The Conservation of Wet Collodion Positives". *Studies in Conservation.* 231–241.

Coote, Jack H. 1993. *The Illustrated History of Colour Photography.* Surbiton: Fountain.11–13.

Davison, Sara. 2003. *Conservation and Restoration of Glass.* Oxford: Butterworth-Heinemann. 169–171 & 337–344.

Eder, Josef Maria. 1932. *Geschichte der Photographie 2.* Halle (Saale): Knapp. 961–976.

Gold, Jens. 2018. "The Ambrotype / Wet Collodion Positives on Glass: Treatment Challenges on Complex Nineteenth-Century Photographic Objects". Master's thesis, University of Oslo. pp. 16–18, 33–35, 74–86 Available at http://urn.nb.no/URN:NBN:no-67385.

Gold, Jens. 2014. Skadet Fotografisk Materiale – En Veiledning i Håndtering av Skadet Fotografisk Materiale, Arts Council Norway 2014: 18–19, 30. Available at https://www.kulturradet.no/documents/10157/172383/skadet-fotografisk-materiale-web.pdf

Grebe, Leonhard, Arthur Hübl, E. J. Wall, and Alfred Hay. 1929. *Farbenphotographie.* Wien: J. Springer. 219–236.

Grinde, Lene. 2005. *Conservation of Stereo Daguerreotype: Examination and Documentation of the Characteristics.* George Eastman House – International Museum of Photography and Film and Image Performance Institute at Rochester Institute of Technology Rochester NY. 40–42.

Hassack, Karl, and Karl Rosenberg. 1907. *Die Projektionsapparate: Laternbilder und Projektionsversuche in ihren Verwendungen im Unterrichte.* Wien: A. Pichlers Witwe & Sohn. 13–16 & 96–107.

Hannouch, Hanin, Personal communication 2020, Motion Picture and Photo historian, Researcher, Curator for analog and digital media at the Weltmuseum Wien - Austria

Hendriks, Klaus B. 1991. *Fundamentals of Photograph Conservation: a Study Guide.* Toronto, Ont: Lugus. 57–61.

International Organization for Standardization. *ISO 18916: 2007: Imaging Materials – Processed Imaging Materials – Photographic Activity Test for Enclosure Materials.* Geneva: International Organization for Standardization, 2007. https://www.iso.org/standard/31940.html.

Koob, Stephen P. 2008. *Conservation and Care of Glass Objects.* London: Archetype Publications. 141.

Koshofer, Gert. 1981. "Geschichte der Farbphotographie in der Pionierzeit." In *Farbe im Photo: Die Geschichte der Farbphotographie von 1861 bis 1981*, 71–77. Cologne: Josef-Haubrich-Kunsthalle. 71–77.

Koshofer, Gert. 1981. *Farbfotografie, Band 1.* Laterna Magica: München 1981. 18–19.

Lavédrine, Bertrand, Jean-Paul Gandolfo, Sibylle Monod, and Sharon Grevet. 2003. *A Guide to the Preventive Conservation of Photograph Collections.* Los Angeles: The Getty Conservation Institute. 6–10, 149–170.

Liesegang, Paul. 1903. *Die Projektionskunst für Schulen, Familien und öffentliche Vorstellungen.* Verlag M. Eger: Leipzig 1903. 140–145.

Lippmann, Gabriel. "Nobel Lecture, December 14, 1908". In *Nobel Lectures: Physics 1901–1921.* Elsevier: Amsterdam. https://www.nobelprize.org/prizes/physics/1908/lippmann/lecture/.

MacCabe, Constance. 1991. "Preservation of 19th-century Negatives in the National Archives". *Journal of the American Institute for Conservation of Historic and Artistic Works*. 41–73.

Mutter, Edwin, Kurt Michel, and Joseph Stüper. 1967. *Die wissenschaftliche und angewandte Photographie: Erneuerung und Fortführung des Hay- v. Rohrschen Handbuchs der wissenschaftlichen und angewandten Photographie*. Wien: Springer. 111–118.

Neuhauss, Richard. 1898. *Die Farbenphotographie nach Lippmann's Verfahren*. Halle a.S: Knapp. 12–59.

Neuhauss, Richard. 1908. *Lehrbuch der Projektion*. Halle a.S.: Knapp. 55–69 & 95–103.

Pénichon, Sylvie. 2013. *Twentieth-Century Color Photographs. Identification and Care*, Los Angeles: Getty Conservation Institute. 17–18.

Peres, Michael. 2007. *The Focal Encyclopedia of Photography: Digital Imaging, Theory and Applications History and Science*. Oxford: Focal. 64.

Reilly, James M., and Constance McCabe. 2009. *Care and Identification of 19th-century Photographic Prints*. Rochester (NY): Eastman Kodak Company. 19–22.

Science Museum Group. "Emil Busch AG 1845". Available at https://collection.sciencemuseumgroup.org.uk/people/cp105648/emil-busch-ag.

Bolas, Thomas, Alexander A. K. Tallent, and Edgar Senior. 1900. *A Handbook of Photography in Colors*. New York: E. & H.T. Anthony & Co. 211–226.

Valenta, Eduard. 1912. *Die Photographie in Natürlichen Farben mit Besonderer Berücksichtigung des Lippmannschen Verfahrens sowie jener Methoden, welche bei Einmaliger Belichtung ein Bild in Farben liefern, Zweite Auflage*. Halle: Verlag von Wilhelm Knapp. 27–28, 64–65, 82–88, 105–107.

Wagner, Sarah S, Constance McCabe, and Barbara Lemmen. 2010.*"Guidelines for Exhibition Light Levels for Photographs"*. In *Issues in the Conservation of Photographs*, edited by Debra H. Norris and Jennifer J. Gutierrez. Getty Conservation Institute: Los Angeles. 684–689.

Ware, Mike. 1995. *Mechanisms of Image Deterioration in Early Photographs: The Sensitivity to Light of W H F Talbot's Halide-Fixed Images 1834–1844*. London: Dillons, Science Museum. 57–59.

Whitmore, Paul M. 2015. "Einfluss des Wassers auf die Papieralterung". *Mit Zusammenfassung (S. 266) Und Bibliografie (S. 267–273)*. 237–266 & 631.

Wiener, Otto. 1899. "Ursache und Beseitigung eines Fehlers bei der Lippmann'schen Farbenphotographie, Zugleich ein Beitrag zu ihrer Theorie", *Annalen der Physik* 69 (3), 488–530. 524–525.

Zenker, Wilhelm. 1900. *Lehrbuch der Photochromie – Photographie der Natürlichen Farben*. Braunschweig: F. Vieweg und Sohn. 149–157.

Photography:

Figure 24 – 33, 44 & 47: Ana Isabel Gonçalves - Preus Museum – Department of Photography

Figure 37, 41, 42, 43, 45, 46: Jens Gold – Preus Museum – Conservation Department

Reproductions, Jens Gold - Preus Museum – Conservation Department:

Sources of drawings & graphics from the Preus Museum Library

Figure 34: Valenta, Eduard. Die Photographie in natürlichen Farben. Halle an der Saale 1912: p. 49.

Figure 35: Valenta, Eduard. Die Photographie in natürlichen Farben. Halle an der Saale 1912: p. 65.

Figure 36: Mutter, Erwin. Die Wissenschaftlichen und angewandten Photographie. – vierter Band - Farbphotographie Theorie und Praxis, Wien & New York 1967: p. 111.

Figure 38: Mutter, Erwin. Die Wissenschaftlichen und angewandten Photographie. – vierter Band - Farbphotographie Theorie und Praxis, Wien & New York 1967: p. 117.

Figure 39: Neuhauss, Richard. Lehrbuch der Projektion. Halle an der Saale 1908: p. 100.

Figure 40: Hassack, Karl, and Karl Rosenberg. Die Projektionsapparate: Laternbilder und Projektionsversuche in ihren Verwendungen im Unterrichte. Wien und Leipzig1907: p. 106.

About the Author

Jens Gold has been a photograph conservator at Preus Museum: National Museum for Photography (Norway) since 2002. Currently, he is PhD Candidate at the University of Oslo – Institute for Archaeology, Conservation and History, with the research project: “Lippmann Interference Photography: History, Materiality and Treatment Challenges”. He completed his studies in photography conservation in 1998 from HTW Berlin- University of Applied Sciences, and he graduated from the University of Oslo in 2018 with an MA in object conservation. From 1999 to 2001, he was a fellow in the Mellon Advanced Residency Program in Photograph Conservation at the George Eastman House and the Image Permanence Institute in Rochester, New York.

10. Exhibiting Gabriel Lippmann: A Collaborative Challenge

Pauline Martin and Carole Sandrin

Abstract

This essay aims to show how the nature of cultural goods initiates specific collaborative approaches. Photo Elysée, Museum for Photography owns 138 Lippmann interferential plates, one of the largest collections in the world. Donated by descendants of Lippmann's in-laws in the 1990s, these plates are among the wonders of the collection, which counts over one million photographic objects. This article aims to explain what such a collection brings to the institution in terms of opening up to external disciplines and collaborations which, in turn, nourish the museum's work in a much broader way. At the intersection of the history of colour photography, science and contemporary art, Lippmann plates push the museum beyond its borders.

Keywords: Gabriel Lippmann, museum collections, museology, contemporary physics, exhibitions, preservation

With the largest collection of original interferential photochromes by Gabriel Lippmann in the world, Photo Elysée in Lausanne (Switzerland) has recently launched an ambitious research programme to deepen knowledge not only of Lippmann process, in which other researchers and institutions are also interested, but also of Lippmann's role in the history of photography. The complexity of interferential photochromy, Gabriel Lippmann's special position between the scientific, photographic and holographic worlds, and the diffusion and circulation of the process in Europe and throughout the world make it necessary to cross-reference fields of expertise to gain a better understanding of this subject. For this reason, Photo Elysée has embarked on an important interdisciplinary

Hannouch, H. (ed.), *Gabriel Lippmann's Colour Photography: Science, Media, Museums.* Taylor & Francis Group, 2022
DOI 10.5117/9789463728553_CH10

collaboration, which will open up new perspectives of collaborations for the institution for future projects.

An Atypical Collection

In 1985, when Pierre-Yves Suter learned that the Musée de l'Elysée (former name of Photo Elysée), a few kilometres from his home, was dedicated to preserving photography in all its forms, he persuaded his father, Gabriel Suter, to bequeath the Lippmann interferential photochromes he owned to this new institution. Indeed, Gabriel Suter and his sister, Antoinette Monition Suter, were the heirs of the physicist Gabriel Lippmann (1845–1921). Although they had never known their illustrious great-uncle, their mother, Charlotte Cherbuliez, had often spoken of him. She and her sister were very close to the childless Lippmann couple (Suter 1999). Their aunt, Laurence Cherbuliez (1857–1934), born in Geneva (Switzerland) and daughter of the academic Victor Cherbuliez (1829–99), had married Gabriel Lippmann in 1888. Involved with her husband, she accompanied him on his travels and helped him in his research. Widowed a few years ago, Laurence Lippmann decided to retire to Creuse (France) as a lady's companion where she lived out her days between reading and painting. Without a direct heir, she bequeathed her possessions to her nieces Charlotte and Laurence, including numerous interferential photochromes, among which are several portraits of the two young girls.

Gabriel Suter received about 150 Lippmann plates from his parents, which he bequeathed to Photo Elysée in two stages. The first set in 1989 was the most impressive: 126 photochromes, with and without prism, still lifes, landscapes of France and Switzerland, portraits attributed to Lippmann himself and a plate of the famous stuffed parrot, signed, dated and annotated by Richard Neuhauss, a German practitioner of the process between 1891 and 1910. A few years later, his son Pierre-Yves Suter completed this corpus by giving twelve other plates not as in good condition but not in the least less interesting.

From the very first donation, the institution realised that it holds one of the rarest treasures in the history of photography and will consider this ensemble as one of the jewels of its collection, out of the million other phototypes in its possession. Few institutions in the world possess Lippmann interferential photochromes in such quantities and in such good condition. Photo Elysée therefore appears to be a privileged place because of the large number of plates preserved (138 in total) and their proven provenance, directly linked to the inventor.

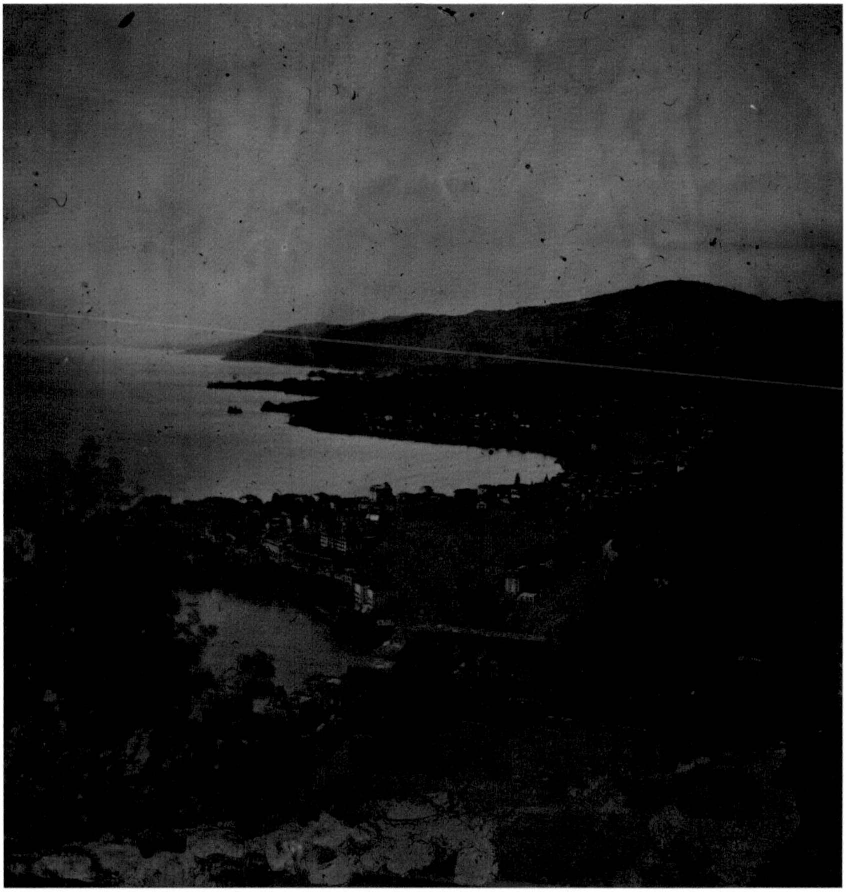

Figure 48 Gabriel Lippmann, [Clarens], 1892-1910, Lippmann interferential photochrome, 9,2 x 8,8 cm, Photo Elysée Collections, inventory number 009955

Catalogue Raisonné Project

The importance of this collection has recently given Photo Elysée the impetus to begin work on the *catalogue raisonné* of Gabriel Lippmann's photochromes. In the 1990s, William R. Alschuler had already carried out a very important worldwide census, which led him to count several hundred interferential plates. Today it seems important for a museum dedicated to the photographic medium to take up this question, to go beyond the physical science in which Lippmann was largely confined and to assess his place in the history of photography. The first step will be to list, as exhaustively as possible, all the interferential photographs made by Lippmann himself or attributed to him. Second, the identification of Lippmann's contemporary and non-contemporary practitioners who have produced plates by this same process will make it possible to assess how to include in this catalogue the interferential photographs made by others.

France, Germany, Austria and the United States are the main known centres where Lippmann and his disciples officiated, whether they were operators, photographers or distinguished physicists. To name but a few, Louis Lumière, Georges Goddé, Edmond Rothé, Auguste Ponsot in France, foreign colleagues such as Richard Neuhauss, Hermann Krone and Hans Lehmann in Germany, Eduard Valenta in Austria, Herbert Eugene Ives in the United States all contributed to perfecting and developing the Lippmann process between 1891 and the beginning of the 20th century. Most of their surviving plates are preserved in physics laboratories, institutions related to the history of technology or the origins of photography. There are, for example, fifteen plates by Auguste Ponsot at the University of Lille, as he was a teacher of physics there. Gabriel Lippmann left the results of his research in his physics laboratory at the Sorbonne (Sorbonne University, Paris) and at the French Society of Photography (Société française de photographie – SFP), of which he was elected president from 1897 to 1899. Possibly made with the help of an assistant, these plates were identified as "Lippmann Process" and attributed to him. However, doubts remain as to the identification of certain plates, particularly those kept at the SFP, which the work for the *catalogue raisonné* may resolve.

Initiated by a public institution,[1] the research aims first and foremost to make this important heritage accessible so that other researchers can inherit it and expand knowledge on the subject. The project would also like to be able to re-evaluate Lippmann's place in the history of photography, from the point of view of the history of techniques, but also by taking a fresh look at artistic exchanges and the particular photographic environments in which interferential plates played a role. The way Lippmann worked and his professional circles will also be better understood. As an example, several questions arise about the role of his wife, Laurence Cherbuliez, in the making of the plates, or even the aesthetic viewpoint that contributed to the fame of these photographs. It is known that Lippmann collaborated with his wife, but a more in-depth study will perhaps provide a better understanding of her importance, since, as several sources suggest, she seems to have played a more important role than that of sole assistant. In an article published in 1922, shortly after the physicist's death, the newspaper *Le Film* placed her in first

1 The project is piloted by Photo Elysée, Lausanne, and led by Nathalie Boulouch (University of Rennes 2, France), Pauline Martin and Carole Sandrin.

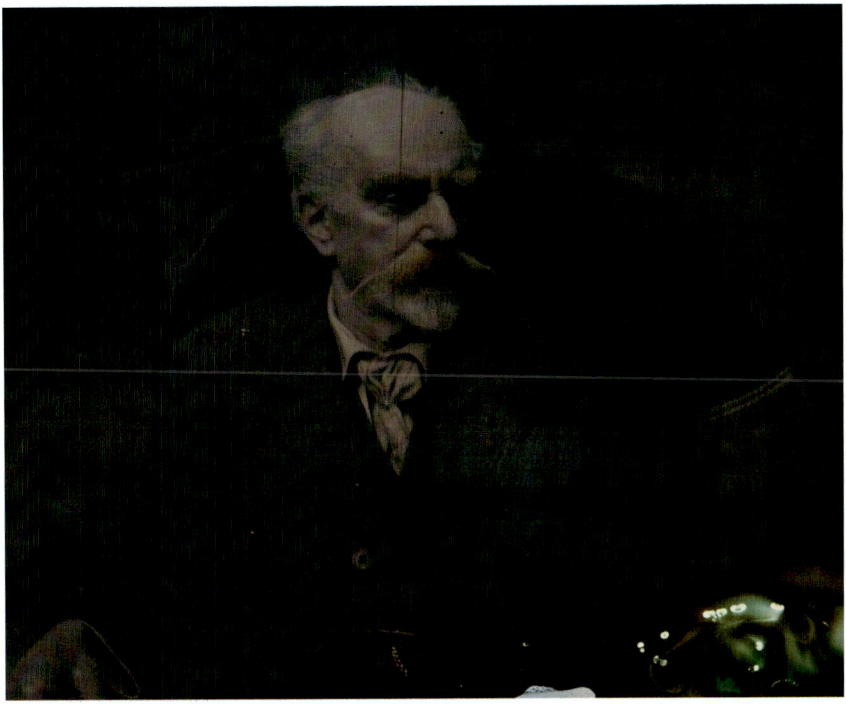

Figure 49 Laurence Lippmann (attributed to), [Portrait of Gabriel Lippmann], 1900-1910, Lippmann interferential photochrome, 6,4 x 8,5 cm, Photo Elysée Collections, inventory number 009079

place among Lippmann's contemporary practitioners: "During the last thirty years, the best operators of the Lippmann process have been Mrs. Lippmann, the scientist's operator, Messrs. Lumière, Valenta, Condamine and de Wattine [*sic*], who recently gave a brilliant lecture on the great missing person at the French Society of Photography.[2]" (Technic 1922) This long-term research will provide a better understanding of the importance of Lippmann's work, his influence on his contemporaries and his place in the history of photography.

Gabriel Lippmann through the Scientific Prism

A quick historical review of the exhibitions that presented Lippmann's photochromes indicates that it was mainly through one discipline – physics – that he gained public recognition. At the time of the invention of the

2 The person known as de Wattine is in fact C. de Watteville (see his lecture: C. de Watteville, "Séance du 16 décembre 1921", *Bulletin de la Société française de photographie*, 12 (December), 328–341).

process, very few plates were presented to the general public. They were shown rather confidentially in specialist circles, whether in the Académie des sciences – where Lippmann first officially presented his process on February 2, 1891 – or in various photographic associations and clubs. On several occasions, Gabriel Lippmann and the Lumière brothers, among others, presented the results of their work during sessions of the Photo-Club de Paris or the SFP. In August 1893 the Lumière brothers presented "to the Académie des sciences, the Photo-Club of Paris and the French Society of Photography chromophotographs [which] are true watercolours." (Forest 1893, 5–6) Here, the academics and the club members had the opportunity to see the plate of a bouquet of flowers, landscapes and reproductions of posters and various images. Projections of the plates were also regularly made, such as on May 12, 1894 at the session of the National Union of Photographic Societies of France (*Bulletin du Photo-Club de Paris*, 1894, 178–82) or on May 22 of the same year during a conference on colour photography given by A. Berget at the Versailles Society of Photography (*Bulletin du Photo-Club de Paris*, 1894, 188–89).

The first real exhibition to the public may have taken place in 1892, when six interferential photographs were presented at the International Exhibition of Photography in Paris. In *La Science illustrée*, Frédéric Dillaye expressed his amazement, placing the exceptional interest of this novelty on the same level as the invention of photography by Niépce and Daguerre fifty years earlier:

> There is, however, one thing that is unparalleled, one thing that stands out for its high importance and incalculable significance: it is the photography of colours by the interferential method so superbly demonstrated by Gabriel Lippmann [...] Today everyone can admire this wonderful discovery based on pure science [...] In my opinion, the technical part of the International Photography Exhibition is of the greatest interest to me. It alone would be enough to make it a success. It is to chromophotography what the discovery of Nicéphore Niépce and Daguerre was to photography itself (Dillaye 1892, 113–14).

Placed in the "technical part" of the exhibition, three plates represented the solar spectrum and were the result of the work of the Lumière brothers "who wanted to renew the master's experience on plates emulsified with gelatino-silver bromide." (Dillaye 1892, 113-14) The other three plates were made by Lippmann himself: two of them – a stained-glass window and a parrot – had been presented to the Académie des sciences a few months earlier, and the last one depicted a branch of holly with its fruits.

In 1900, interferential plates were offered to an even wider audience. Some of Gabriel Lippmann's productions were presented at the French Pavilion of the Paris World's Fair in what the exhibition guide called a "retrospective" of the history of photography: "This is of little importance, photography being a modern art. M. Vidal shows us an interesting collection of pictures by Daguerre; it is the beginning of the history of photography which is completed in the production of the colour pictures exhibited by M. Lippmann" (Lapauze, de Nansouty, da Cunha, Jarzuel, Vitoux, and Guillet 1900, 107). In addition to this presentation, there were also cameras and albumen prints by Hippolyte Bayard. While the authors of the guide readily admited that "the art photography that was unknown ten years ago now has an undeniable vitality" (Lapauze 1900, 106), it was nevertheless clear that interferential photography had not managed to gain recognition as a process available to artists, as its realisation seems so complex: "[T]he wonderful discoveries of Mr. Lippmann allow us to obtain images in the *laboratory,* but not in the *studio, so that* colour photography has yet to be put into practice." (Lapauze 1900, 106) The process was also presented abroad on a few occasions, for example in London between September and November of the same year, when a spectrum plate by Richard Neuhauss was shown at the Royal Photographic Society exhibition (*The Illustrated Catalogue of the Forty-Fifth Annual Exhibition of the Royal Photographic Society* 1900, 36).

Further research would make it possible to specify the number and type of Gabriel Lippmann's exhibitions organised in particular during the 20th century. In any case, it appears that it was only in the early 1990s that a real revival of interest in the process was shown in several exhibitions celebrating various anniversary dates linked to the presentation of the process (1891), the award of the Nobel Prize (1908) and Lippmann's death (1921). Although in May 1991 Photo Elysée included twelve interferential plates in an exhibition of works from its collection, we owe the most important initiatives in this field mainly to personalities from the world of science. In 1991, the physicist Jean-Marc Fournier organised an exhibition at the Palais des Congrès in Paris, then in Toulouse, in parallel with scientific congresses, entitled *Gabriel Lippmann, cent ans de Physique et de Technologie* (*Gabriel Lippmann, one hundred years of Physics and Technology*[3]). This exhibition took a global

3 Jean-Marc Fournier explains that, in parallel with the Opto'91 congress in March 1991, "an exhibition was held to underline the beauty of interferential colour photographs while at the same time enhancing the technical value of certain old and recent works, showing twelve photographs from the Lippmann collection of the Musée de l'Elysée de Lausanne, a few monochrome and colour holograms working according to the so-called Lippmann volume regime, as well as some embryos of integral photography and Bonnet photographs. The exhibition was taken up again,

approach to Lippmann's research, including integral photography and holograms of the Lippmann-Denisyuk type. Close to Photo Elysée, Jean-Marc Fournier has made a major contribution to increasing the profile of the process by presenting numerous conferences, both scientific and general public, and by initiating various events.[4] The most remarkable of these was undoubtedly the exhibition *Un monde en couleurs. De Gabriel Lippmann à la nanophotonique* (*A World in Colour. From Gabriel Lippmann to nanophotonics*), which he organised at the Palais de la découverte (Paris) from June 16, 2009 to May 2, 2010, and which he enriched with the help of the Gabriel Lippmann Public Research Centre in Luxembourg.[5]

As far as institutions and museums dedicated to photography are concerned, there have been no exhibitions entirely devoted to Lippmann. In 1993, an initiative was launched by the scientist William R. Alschuler, who contacted Robert Delpire, founder of the Centre National de la Photographie, to propose that he takes part in a major exhibition project on Gabriel Lippmann aimed at investigating both the process and its place in the history of photography (Alschuler 1993).[6] Unfortunately not successful, this attempt at a major exhibition on Lippmann, integrating both aspects related to the history of science and the history of photography, was not taken up again. Lippmann's photographs were thus only occasionally shown in thematic exhibitions in which two or three plates were integrated. In 1937, when he organised the first photography exhibition at MoMA, Beaumont Newhall included two: a reproduction of the 1892 spectrum and a 1900 plate showing a "garden scene" (Newhall 1937). More recently, three plates were shown in 2004 in the exhibition *L'Utopie photographique. Regard sur la collection de la Société française de photographie* (*Photographic Utopia. A look at the collection of the French Society of Photography*) at the Maison Européenne de la Photographie in Paris.[7] At Photo Elysée, Lippmann's portrait, recently

in a modified version, in Toulouse during the *Horizons de l'Optique* and *ICSO'91* congresses. "(Jean-Marc Fournier, "Interference Color Photography: One Hundred Years Later," *J. Optics*, 1991, vol. 22, (6), 259–266).

4 Jean-Marc Fournier organised an exhibition at the Physics Department of the University of Luxembourg in October 1995 and collaborated in the presentation of Lippmann photographs at Lake Forest College (United States) in 1997.

5 The exhibition was then presented at the École Normale Supérieure in Cachan (from August 25 to November 11, 2010) and at the Centre culturel de l'Abbaye de Neumünster in Luxembourg (from October 22 to November 20, 2011).

6 Letter from William R. Alschuler to Robert Delpire, dated June 28, 1993, Archives of Photo Elysée, Lausanne.

7 *L'Utopie photographique. Regard sur la collection de la Société française de photographie*, November 3, 2004–February 6, 2005, Maison européenne de la photographie, Paris (curators: Michel Poivert and Carole Troufléau Sandrin)

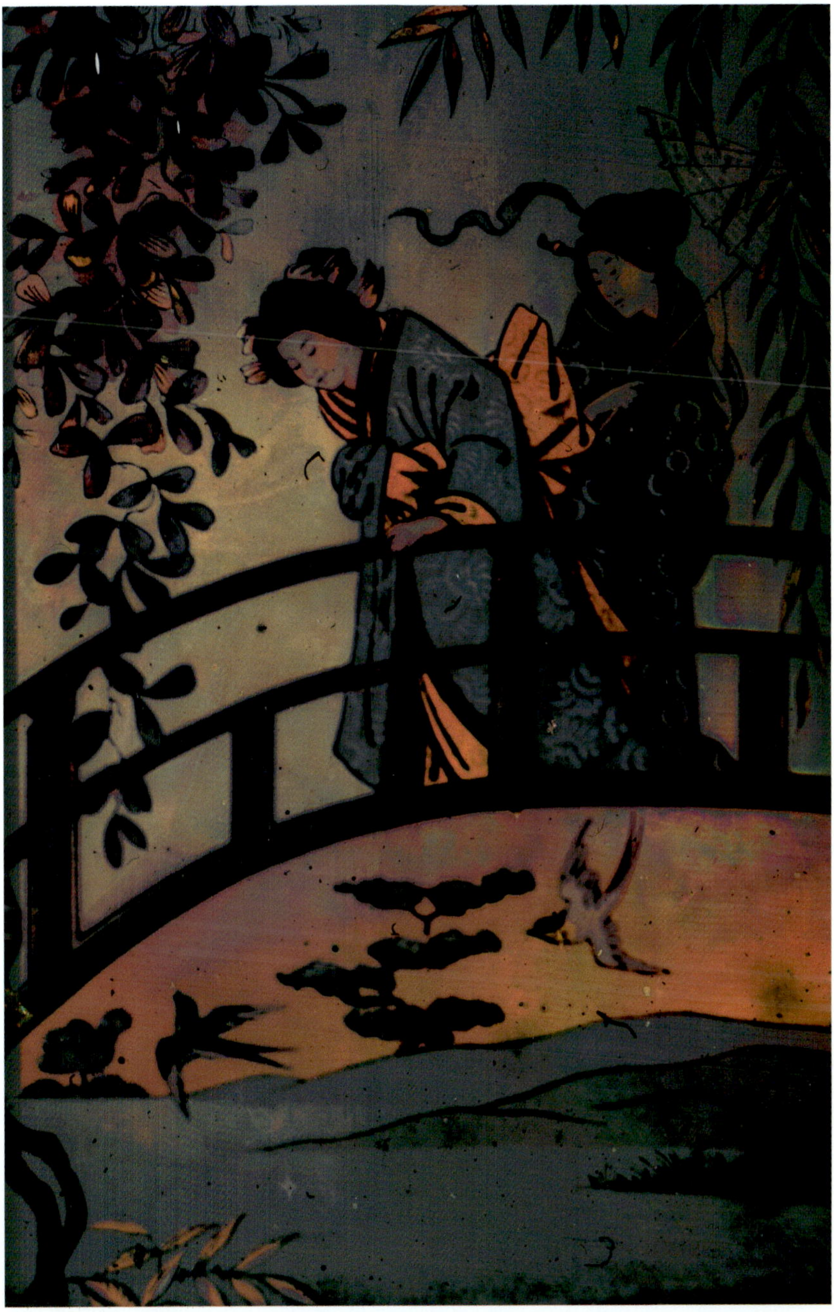

Figure 50 Gabriel Lippmann, untitled, 1892-1910, Lippmann interferential photochrome, 6,5 x 9 cm, Photo Elysée Collections, inventory number 009900

attributed to his wife Laurence Lippmann, was included in 2016 in the exhibition *The Memory of the Future* and three landscapes were also shown in 2017 in the exhibition *Vertical No Limit. Mountain Photography.*[8]

A Collaborative Museum Approach

This gap in the history of photography remains to be filled. Before finalising the *catalogue raisonné*, Photo Elysée is organising an exhibition that will take place in 2023 in the new building in which the institution has moved to on PLATEFORME 10[9] site. The programming of this exhibition demonstrates the importance given to the work of Gabriel Lippmann and the photography community's desire to showcase it from now on. It also indicates that research on this historical figure is part of a network of very contemporary interests, starting with the interdisciplinarity that is essential to the understanding of his work. The context in which PLATEFORME 10, a brand-new arts district in the heart of the city of Lausanne, emerges is particular and involves, in its very structure, interdisciplinary collaboration. Indeed, it brings together three museums with clearly identified disciplines in a simultaneous and completely innovative way, namely a museum of fine arts (MCBA), a museum of design (mudac) and a museum of photography (Photo Elysée). This new configuration is part of a broader societal movement that is now leading to the development of collaborations and synergies of all kinds. In this particular context, Gabriel Lippmann, a personality mostly inscribed in the history of science, nevertheless appears as an extremely rich figure to question the networks of collaborations which at his time were undoubtedly already fertile and complex.

In addition to showcasing this little known heritage, the ambition of this exhibition is also to be able to present to the public a complex photographic object that testifies to the multiplicity of issues at stake. From this perspective, interdisciplinary collaboration is essential in order to approach the Lippmann interferential process in three ways, which list the major fields of research in which it is involved: 1) the Lippmann plate as a scientific object, or the importance of the process, 2) the Lippmann plate as an art

8 *The Memory of the Future*, May 25 to August 28, 2016, Musée de l'Elysée Lausanne (curator: Tatyana Franck); *Vertical No Limit. Mountain Photography*, January 25 to April 30, 2017, Musée de l'Elysée Lausanne (curator: Daniel Girardin).

9 www.plateforme10.ch/en. The exhibition will take place from March 3 to May 29, 2023; curator: Pauline Martin; scientific committee: Nathalie Boulouch, Jean-Marc Fournier, Tatyana Franck, Manuel Sigrist, Martin Vetterli; production manager: Emilie Chenevard.

history object, or the importance of the image, 3) the Lippmann plate as a museographic object, or the importance of the visual experience.

Close Institutional Collaboration

For the first axis of the exhibition, which is also the one most often addressed in past exhibitions, Photo Elysée has partnered with the Swiss Federal Polytechnic School of Lausanne (École Polytechnique Fédérale de Lausanne – EPFL) to develop interactive experiences with the public to help them understand the process. The current team is developing several devices, including a vibrating string to demonstrate the principle of standing waves and an optical machine that diffracts light into spectral bands, thus showing what the decomposition of light consists of.

This partnership is part of the history of a much longer collaboration between the museum and the EPFL, dating back to 1988 when the Suter family made its first donation. Charles-Henri Favrod, the first director of Photo Elysée as a "museum for photography," was looking for funding to preserve these treasures and contacted Professor Land of the Rowland Institute at Harvard (Cambridge, Massachusetts, United States). Jean-Marc Fournier was then the Director of the Rowland Institute's Optical Structures Laboratory and was interested in Lippmann's work in the context of his research on holography (Lippmann type). Also a visiting professor at the Institute for Measurement and Analysis of Constraints (Institut de Mesure et d'Analyse des Contraintes – IMAC) of the EPFL, he contributed, with Professor Pierre Jacquot, to highlighting Lippmann's photographs at Photo Elysée, to exhibiting some of them, and to setting up panels explaining the principle of interferential photography. Fournier had also obtained several loans of Lippmann plates from Photo Elysée and from the University of Paris VI (Sorbonne University nowadays) to analyse them in his laboratory at Harvard by measuring their physical properties, particularly their spectroscopic aspect. He thus examined an extremely damaged photograph by cutting off two edges where the image was illegible and then separating the prism from the plate: this allowed direct access to the emulsion and observation of its microstructure under an electron microscope. In 1994, Fournier published the results of his research and recounted this very special, almost unique episode, which allowed him to confirm the validity of Lippmann's theories and to verify the colour rendering[10] (Fournier and Burnett 1994).

10 The colour image of the plate n°009915 preserved at Photo Elysée is still visible today.

In 2015, Lippmann returns to the heart of the discussions during the meeting between Martin Vetterli, Director of the Audiovisual Communications Laboratory (LCAV) at the EPFL, and Tatyana Franck, former Director of Photo Elysée. It was then a question of digitising the plates in 5D thanks to the technology developed by the startup Artmyn. Initial attempts were not sufficiently convincing, mainly because of the reflections generated by the prism of the plates. This was followed by 5D scans of works from the three PLATEFORME 10 museums, which were presented in 2016.[11] Currently, the team at the LCAV[12] are continuing their research on the process by learning how to redo it investigating holographic plates with a similar extremely fine-grain emulsion.[13] Also, student Gilles Baechler successfully completed his PhD thesis in 2018 demonstrating the mathematical formula for the wave nature of Lippmann plates (Baechler 2018).

Finally, the project for an exhibition planned for 2023 at Photo Elysée and supported by Agora, a programme of the Swiss National Science Foundation which aims to promote dialogue between science and society, was launched in 2019. This support seals the collaboration between the LCAV and Photo Elysée around the analysis of colours, their nature and its reproduction. By giving priority to an educational focus in the 2023 exhibition, this project aims to show the general public the difference between the light spectrum, whose electromagnetic waves are strictly speaking colourless, and the trichromatic representation of colours that our eyes perceive.

The Artist Faced with the Interferential Process

The second section devoted to the importance of the Lippmann plate in the history of photography will address the challenges of a research on its premises, culminating in the publication of the *catalogue raisonné*. With the desire to explore the iconographic subjects dealt with by Lippmann, Photo Elysée has joined forces with several museums, in particular the Musée des Beaux-Arts de Lausanne (MCBA), the mudac and the Museum of Zoology, with the aim of reproducing by the interferential process, with the help of the EPFL, objects that had already interested Lippmann in his time.

11 https://artmyn.com/explore/collections/elysee

12 The current EPFL LCAV team involved since 2017 up to now is mainly composed of Paolo Prandoni and Arnaud Latty. Jean-Marc Fournier joined the project in 2020 as an advisor.

13 Workshop conducted by Filipe Alves in September 2017: https://lippmannphotography.com/Lippmann-workshop-at-EPFL

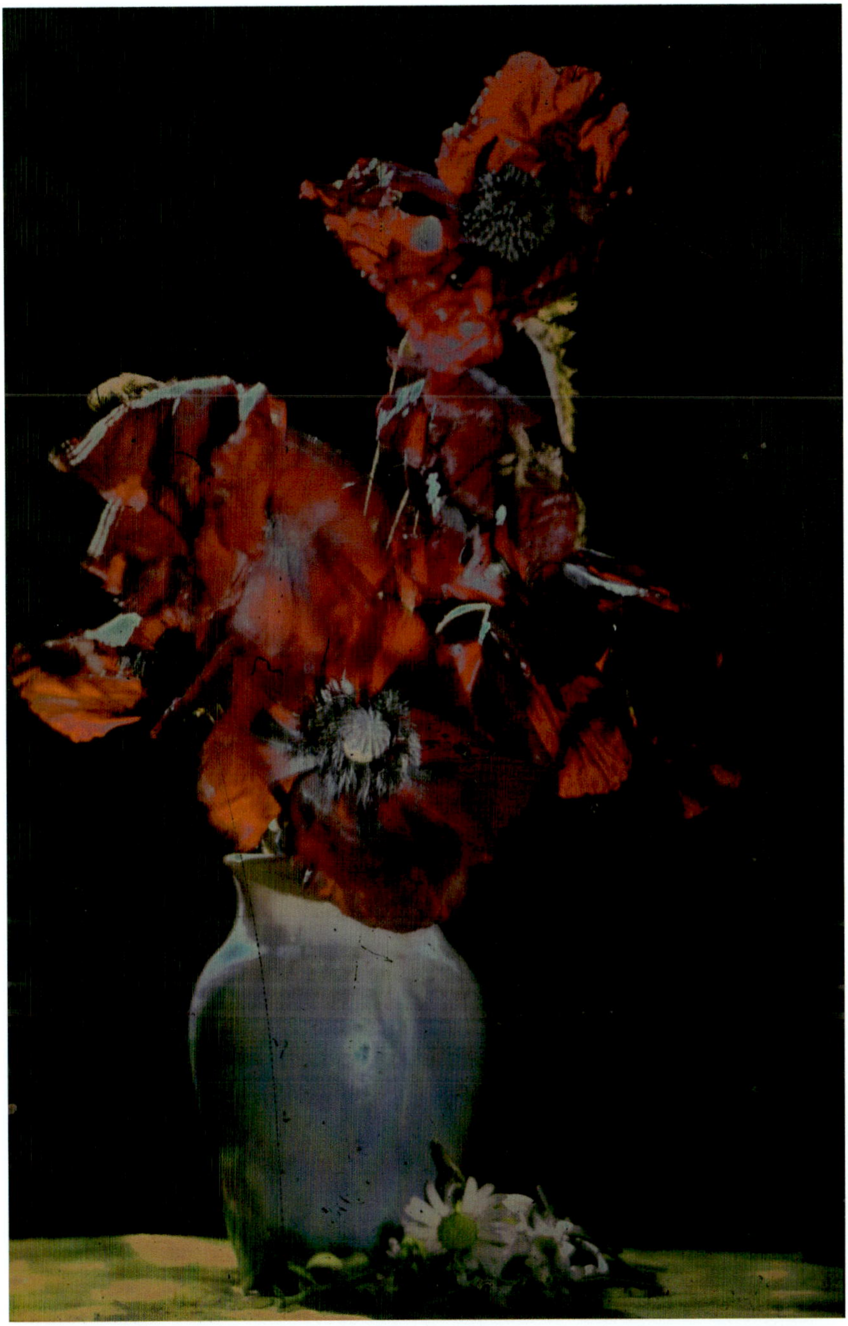

Figure 51 Gabriel Lippmann, untitled, 1892-1910, Lippmann interferential photochrome, 6,4 x 8,5 cm, Photo Elysée Collections, inventory number 009981

In addition, this section will also be an opportunity to question the contemporaneity of the process and its current vitality. Is it a historical process or is it still alive today? A look back at the history shows the great difficulties this technique has experienced in becoming part of the artistic history of photography. As soon as it was invented at the end of the nineteenth century, attempts were made to have it adopted by enlightened amateurs. Although it required considerable technical expertise, the process entered photographers' circles and was even the subject of competitions, just like other processes. In 1894, the SFP announced a competition for interferential photochromes for the following year. In 1895, the SFP included, for example, the interferential process in the range of "competitions of personal works" that it proposed to its members. In addition to the competition of oil print processes, which were highly prized by pictorialists, and the competition for stereoscopic prints, there was also a "competition for photographs using the Lippmann process," the aim of which was precisely to "encourage the popularisation of Mr. Lippmann's method" and at the same time to "enrich the collections of the Society with the works of its members" (*Bulletin de la Société française de photographie* 1895). This news encouraged research in an attempt to make the process more accessible to amateurs. F.-M. Richard, who was working to produce a "mercury frame designed to produce the interference and the dryer to dry the plates quickly" explains: "From the opening of the competition set up by the French Society of Photography to obtain colour photographs using the Lippmann method, I have been studying the equipment that amateurs may need" (Richard, 1895).

However, the efforts did not seem sufficient, as the process remained extremely difficult for amateurs to succeed. In 1900, the same competition was organised, with the similar aim of "encouraging the popularisation of Mr. Lippmann's method" (*Bulletin de la Société française de photographie* 1900). However, practically no entries were sent to the SFP, which was forced to award a prize to Georges Goddé, the only participant in the competition, whose plates are still in the SFP collection today. The commission admits to having had "to regret the lack of competitors; in fact, only one entry was submitted for its examination, and even this entry, which consisted of only three rounds, was not in conformity with the programme, which required six of different types"; and expresses "hopes that a greater number of members of the Society will set to work" (Vallot 1900).

Today, this process is still complicated to carry out, but some experienced practitioners – such as Filipe Alves, Nick Brandreth, Darran Green, Jerome

Klingenfus to name but a few – manage to obtain beautiful specimens.[14] The use of commercially available ultra-fine emulsion holographic plates and shooting without a mercury holder facilitates the process but does not guarantee results. Although holographists have long been the only ones to re-experiment the process, the revival of ancient photographic practices has won over other curious people eager to penetrate the secrets of this unique colour photography. Some purists try to prepare their own plates with albumen, collodion or gelatin emulsions. The fascination that this process exerts on aficionados of photographic know-how, whether they are physicochemists, holographists or just lovers of this alternative imagery, gives life to a new creativity.

A Visual Experience to Master

The third axis, which addresses the question of the visual experience offered by the Lippmann process, will cover several intertwined issues: museographic presentation devices, the conservation of the plates and their digital reproduction.

Like the daguerreotype or ambrotype, Lippmann's interferential photography can be seen in negative and positive. A low contrast negative placed on a black background reveals a positive image, a natural colour one in Lippmann's case, which makes it all the more difficult to see because it rarely appears at first glance. In order to find colour you have to tilt your head in one direction or the other. If one is lucky enough to hold it in one's hands, you can tilt the plate until the right light incidence is encountered and the colours appear. Used in university physics laboratories until the 1960s as a reliable spectroscopic recording device, it was put at the service of teaching by passing through the hands of the students in the amphitheatre (Gandolfo 2008, 223–25). But it is highly likely that anyone who is not prepared to experiment with this colourful standing wave phenomenon will consider the plate to be a dud or degraded plate, believing they are seeing only a poor negative. The handling is often made tedious due to inadequate artificial lighting. So how do you show these unique plates properly?

In 1991, Photo Elysée used displays designed by Pierre Jacquot and Jean-Marc Fournier. These wall-mounted showcases with integrated fluorescent tubes and fans allowed a visitor to easily see two to four plates. According

14 The works of these contemporary practitioners are notably visible on the Facebook group *Lippmann Photography*, a real platform for international exchanges on the process.

to Jean-Marc Fournier, the lighting sources, although corrected by coloured gelatin filters, did not do justice to the colour reproduction process, but had the advantage of not heating the plates. For the purposes of the exhibition at the Palais de la découverte in 2009, Gilles Pauliat at Sup'Optique Graduate School (Orsay) took inspiration from an old Zeiss patent to redesign and manufacture a catadioptric iconoscope, a replica of a period viewer that was originally placed in front of a window in diffuse daylight, the best possible light for observing Lippmann photochromes. This scenographic element was reserved for an original plate and a single viewer.

For the 2016 and 2017 exhibitions, Photo Elysée used a showcase with diffused zenithal lighting. This museographic choice was inspired by the one used for the exhibition *L'Utopie photographique* presented at the Maison Européenne de la Photographie (Paris) on the occasion of the 150th anniversary of the SFP in 2004, namely a table-top showcase containing several plates placed on fixed supports previously oriented and placed in front of brightly lit kakemonos. Visible as objects, the plates revealed their colourful image after a few genuflections by the spectator who had taken the time to stop and look for what there was to see.

Since 2018, the EPFL has been considering the development of a specific showcase for the 2022 Lippmann exhibition at PLATEFORME 10. A first device has been imagined, which could include a mobile support, allowing the rotation of the plate, and a modular LED lighting with diffusers arranged in an arc (Bergum and Skullstad 2018). Visitors could thus see the objects in their entirety, without receiving any specular reflection, orientate a few plates themselves and experiment with the changes in colour depending on the orientation of the object and the variations in the angle of light reflection. Modern interferential photographs are also to be made available to visitors so that they can experience how to manipulate them by themselves, while others would be projected. As mentioned above, this method of viewing was used by Lippmann himself and his fellow students. The German firm Zeiss even marketed a special projector in 1909 (Gandolfo, 2008, 223).[15] Jean-Marc Fournier frequently used projection during his lectures, mainly projecting plates of his own making. Anxious to show the visitor the reality of the colours reproduced by the Lippmann plate, he praised the qualities of projection, the only method that makes it possible to share the image simultaneously with several spectators while ensuring that the plate is

15 Under the direction of Jean-Paul Gandolfo at the Ecole nationale supérieure Louis Lumière in France, student Alban Chassagne had developed such an optical projection device. See Chassagne, 2010 (in French).

illuminated under optimal conditions, from the correct angle. However, one must take into account the high heat emission of the light source and the need to expose the plate for only a few seconds, as this may cause irreparable damage to the plate or even break the prism. Some of the original plates today show damages that could be linked to this abusive use.

The conservation of Lippmann photochromes remains a field to be explored more precisely and better publicised.[16] Although the plates at Photo Elysée have been preserved for more than seventy years in variable climatic conditions before reaching the stability of its storage, they seem to show a certain robustness.[17] The prism, whose optical function is undeniable, is also a real shield protecting the image layer from abrasion, pollutants, oxidation and climatic variations. This property is further enhanced by the sealing of gummed paper that borders the glass plate and the prism.

However, the inevitable ageing of the materials making up these photochromes (glass, Canada Balsam, varnish, silver emulsion) is sometimes coupled with manufacturing and handling defects (holes in the emulsion due to mercury or developer for example). The Canada Balsam, which ensures the adhesion of the prism to the coated plate, can thus present bubbles and cracks depending on the thermo-hygrometric conditions in which it was cast and dried (Lamotte 2009–2014). Canada Balsam yellowing, a flaky or lacunar black varnish on the back of the plate, a white veil on the emulsion can also hinder the reading of the image. But more important, the hygroscopic nature of the thin layer of gelatin emulsion is to be feared: a high ambient humidity may swell the gelatin causing a change in the spacing of the silver fringes that make up the coloured image and thus "shift" the colours (blue becomes green, etc.) to the point of losing their rendering. However, some plates that have remained bare (without prism or other emulsion protection) still offer an accurate colour image, such as spectra with a thicker and sometimes hardened image layer (made of bichromated gelatin emulsion).

16 In France, in addition to the generic works on the conservation of photographs by Bertrand Lavédrine and Anne Cartier-Bresson, which devote a chapter to interferential photochromy, the work of Marie Lamotte is one of the rare references devoted to the subject. See Lavédrine, Gandolfo, Monod, Frizot 2009, Gandolfo 2008 and Lamotte 2009-2014 (in French).

17 In parallel with the analyses and interventions of Jean-Marc Fournier mentioned above, the Lippmann photochromes of Photo Elysée have been the subject of various studies and treatments. The Swiss Institute for the Conservation of Photographs (ISCP) resealed the plates requiring it in 1991 (new borders or consolidation of existing ones). Marie Lamotte studied them for her dissertation at the INP-restauration. Marsha Sirven, conservator at the Atelier de restauration et de conservation des photographies de la ville de Paris (ARCP), was asked for advice before the 2016 exhibition. She inspected all the plates, sealed those that remained unmounted and consolidated those whose borders were still incomplete. See Sirven 2016 (in French).

The advantage of being able to identify the composition of the emulsion would help to better conserve these photochromes and prevent certain risks of degradation. Recreating the Lippmann process today is also a means of better distinguishing manufacturing artefacts from alterations due to the environmental conditions in which the objects have been stored.

The colours of interferential photochromes cannot fade due to the absence of pigments or dyes and are therefore considered to be physically stable over time. Nevertheless, there is still some doubt as to whether certain Lippmann process plates are not fixed (so as not to alter the colours obtained when they are photographed), which raises the question of their stability over time. Although the plate is said to be not very sensitive at the time of exposure, to what extent does repeated exposure of the developed plate to light (even excluding UV and infrared rays) and climate changes present a risk of modifying its nanostructure? The identification of the emulsion and previous chemical treatments could be determined by spectrometric and microscopic analyses.[18] Ageing tests of contemporary plates would also make it possible to better understand the "light capital" of these phototypes. This would measure more precisely the different risks taken if there was a change in environment, resulting in an accurate preventable preservation plan.

Digital reproduction is therefore an obvious way to help preserve these plates in the long term. The reproductions of Photo Elysée plates made by the EPFL photographer Alain Herzog in the 1990s for the publication of Luxembourg are still important visual references today (Pier and Massard 1997). Here again, the orientation of the lighting, its intensity, its diffusion and its spectral composition are all factors to be compiled to obtain the most faithful reproduction possible. The quality of the examination light must be close to natural light with a continuous spectrum so that the colours recorded by the plate are perceived correctly.[19]

The research project supported by Agora, conducted jointly by Photo Elysée and the EPFL, will make it possible to perfect the lighting system to be put in place in order to determine a reliable and reproducible digitisation protocol. By continuing research on the 5D digital acquisition of Lippmann plates, the ambition is to be able to offer a virtual experience of manipulating and viewing these photochromes, by varying the orientation of the object and the lighting.

18 A preliminary study on characterization and preservation of Lippmann plates is currently led by Marie-Angélique Languille at the Centre de Recherche sur la Conservation (CRC, Paris). Carole Sandrin has initiated this project in 2021 in collaboration with Bertrand Lavédrine.

19 The rendered colour is, theoretically, identical to the original colour of the recording only if the observation light is the same as that one used when taking the picture.

Artmyn has already realised this project with the Codex of Jean, the famous papyrus kept at the Martin Bodmer Foundation in Geneva, as well as with other works from Photo Elysée. This would give the observer a sense of the volume of the object and a privileged reading of the image and its surface by zooming into the emulsion. Finally, these digitisation campaigns should make Lippmann's photochromes widely accessible to the general public as well as to conservators and researchers from different backgrounds without endangering them.

Because of its complexity, interferential photochromy requires collaboration of Photo Elysée with many specialists from a variety of disciplines. It thus feeds the museum's network of skills and broadens its horizons.

Acknowledgements

The authors would especially like to thank Richard Benton, Nathalie Boulouch, Jean-Marc Fournier, Tania Passafiume, and Pierre-Yves Suter for their careful rereading. Unless otherwise stated, all French citations were translated by the authors.

Works Cited

Alschuler, William. 1993. Letter to Robert Delpire, 28 June 1993, Archives of Photo Elysée, Lausanne.

Baechler, Gilles. 2018. *Sampling the Multiple Facets of Light,* PhD thesis presented on 19 October 2018 at the Faculty of Computer Science and Communications – Audiovisual Communications Laboratory, EPFL. https://infoscience.epfl.ch/record/257308

Bergum, Kjell Sverre Hoyvik and Tarjei Nygaard Skulstad. 2018. *Design and development of a display device for Lippmann plates,* Mechanical Engineering Project I, EPFL,: https://infoscience.epfl.ch/record/255588

"Concours d'œuvres personnelles." 1895. *Bulletin de la Société française de photographie.* 2nd series, 11, 489–494.

"Concours d'œuvres personnelles." 1900. *Bulletin de la Société française de photographie.* 2nd series, 16, 472–474.

Chassagne, Alban. 2010. *Préliminaires d'étude sur la visualisation de plaques Lippmann par réflexion. Recherche d'un dispositif,* Saint Denis : Ecole Nationale Supérieure Louis-Lumière.

"Concours d'œuvres personnelles" 1895. *Bulletin de la Société française de photographie,* 2nd series, 11, 489–494.

"Concours d'œuvres personnelles" 1900. *Bulletin de la Société française de photographie,* 2nd series, 16, 472–474.
Dillaye, Frédéric. 1892. "L'exposition de photographie. La photographie des couleurs ", *La Science illustrée,* 242 (July) 113–114.
"Echos et Nouvelles," 1894. *Bulletin du Photo-Club de Paris,* 188–189.
Forest, Max. 1893. "La Photographie des Couleurs," *Gazette du Photographe Amateur,* (August), 5–6.
Fournier, Jean-Marc. 1991. "Interference Color Photography : One Hundred Years Later", *Journal of Optics,* 22(6), 259–266.
Fournier, Jean-Marc, and Paul L. Burnett. 1994. " Color Rendition and Archival Properties of Lippmann Photographs ", *Journal of Imaging Science and Technology,* 38 (6), 507–512.
Gandolfo, Jean-Paul. 2008. "Interferentiel (Procédé)", in *Le Vocabulaire Technique de la Photographie,* edited by Anne Cartier-Bresson, Paris : Marvel / Paris musées, 223–225.
The Illustrated Catalogue of the Forty-Fifth Annual Exhibition of the Royal Photographic Society, London: The Royal Photographic Society, 1900, 36.
Lamotte, Marie. 2009–2014. *Conservation et restauration d'un corpus de photochromies interférentielles obtenues par la méthode de Gabriel Lippmann provenant de la Société française de photographie et de l'Université Pierre et Marie Curie. Caractérisation de l'altération du baume du Canada employé comme matériau de collage dans le procédé interférentiel,* diplôme de restaurateur du patrimoine – spécialité Photographie, Institut National du Patrimoine.
Lapauze H., Max de Nansouty, A. da Cunha, H. Jarzuel, G. Vitoux, L. Guillet. 1900. *Le guide de l'Exposition de 1900,* Paris : Ernest Flammarion, 106-107.
Lavédrine, Bertrand, Jean-Paul Gandolfo, Sybille Monod, Michel Frizot (pref.). 2009. *(re)Connaître et Conserver les Photographies Anciennes,* Paris : Editions du CTHS, Collection Orientations et méthodes.
Pénichon, Sylvie. 2013. *Twentieth-Century Color Photographs. Identification and Care,* Los Angeles: Getty Conservation Institute.
Newhall, Beaumont. 1937. *Photography : 1839–1937,* exhibition catalogue, New York: The Museum of Modern Art.
Pier, J.P, and J. A. Massard [dir.]. 1997. *Gabriel Lippmann, Commémoration par la section des sciences naturelles, physiques et mathématiques de l'Institut Grand Ducal de Luxembourg du 150e anniversaire du savant né au Luxembourg lauréat du Prix Nobel en 1908,* Luxembourg : Jean-Paul Pier & Jos. A. Massard.
Richard, F.M. 1895. "Matériel pour la photographie des couleurs par la méthode Lippmann ", *Bulletin de la Société française de photographie,* 2 (11) 27–32.
Sirven, Marsha. 2016. *Rapport d'expertise / Rapport de traitement / Cahier des charges techniques de conservation préventive pour l'exposition temporaire d'une*

photochromie interférentielle attribuée à Gabriel Lippmann, Archives of Photo Elysée, Lausanne.
Suter, Gabriel and Antoinette. 1999. Letter sent from Biot to Professor J.P. Pier, Luxembourg, August 17. Archives Pierre-Yves Suter, Lausanne.
Technic, " Gabriel Lippmann ", *Le Film*, year 8, n°185, 1922, n.p.
"Troisième Session de l'Union Nationale des sociétés photographiques de France" 1894. *Bulletin du Photo-Club de Paris*, 178–182.
Vallot, Emile. 1900. "Rapport de la commission chargée de juger le concours de photographie des couleurs", 2nd series, 16, 537–538.

About the Authors

Curator and head of the exhibitions department at Photo Elysée, Museum for Photography, in Lausanne, **Pauline Martin** is an art historian and author of a PhD thesis about the history of blur in photography. She has published several books, including *L'oeil photographique de Daniel Arasse. Théories et pratiques d'un regard* (Fage éditions, 2012) and *L'Évidence, le vide, la vie. La photographie face à ses lacunes* (Ithaque, 2017).

Curator in charge of the photographic archives at the Institut pour la photographie des Hauts-de-France since 2021, **Carole Sandrin** has been curator head of the preservation department of Photo Elysée, Lausanne from 2011 to 2020. She was in charge of the Société française de photographie collection from 2005 to 2011 and graduated with a professional MA in the preservation of cultural goods (University Paris I Panthéon-Sorbonne) in 2010. Author of several articles on scientific photography, she has co-curated exhibitions on the SFP collection, the amateur photography around 1900, the history of projected photography, Charlie Chaplin, Agnès Varda and Bettina Rheims.

11. Modern Lippmann Photography

Hans I. Bjelkhagen

Abstract
Since the 1980s, there is a renewed interest in Gabriel Lippmann's colour photography technique and several people around the world recording Lippmann photographs, especially holographers. The development of colour holography shows similarity to interference photography. The same type of panchromatic ultra-fine-grain silver halide material is used both for holography and Lippmann photography. A short review of the history of the technique is provided with a presentation of how modern Lippmann colour photographs can be recorded nowadays. A current advantage is the possibility of recording them without using mercury, making it safer and easier to practice the technique today. Examples of recently recorded interference colour photographs are illustrated with digital photographs.

Keywords: interferential colour photography, colour holography, silver halide, emulsion, contemporary physics

Introduction

Few photographers today are familiar with the name Gabriel Lippmann (1845–1921), even fewer have seen a Lippmann colour photograph, but one hundred years ago, Lippmann was awarded the 1908 Nobel Prize in Physics for his invention of interference photography. This was for his early colour technique exploiting the phenomenon of optical standing waves. Lippmann's prize represents the only time this prestigious award has been given for a photographic invention. However, the process proved difficult and impractical, exposures ran into minutes, the colour were difficult to view and almost impossible to copy. Only a minority of photographers recorded successful Lippmann photographs. Those who did were skilled in making and coating their own emulsions and dedicated many years to the cultivation of the

Hannouch, H. (ed.), *Gabriel Lippmann's Colour Photography: Science, Media, Museums.*
Taylor & Francis Group, 2022
DOI 10.5117/9789463728553_CH11

technique. Original examples of this extremely rare, exotic and arcane medium number around five hundred specimens worldwide. Attempts at commercialisation floundered and by the time Lippmann was awarded his prize the more practical *autochrome* plates had become available. Despite the difficulties, Lippmann's photography remains, to this day, the only direct process of true colour photography known. It is a technique of exquisite beauty, both technically and aesthetically. The ultra-fine-grain plates, essential to the medium, display the highest photographic resolution ever achieved. The encoding of colour, as a periodic volume diffraction grating of colloidal silver offers excellent and unrivalled archival longevity.

Today there is a renewal interest in the technique and several people around the world record Lippmann photographs. It started in the late 1980s with several papers being published. The development of holography, in particular colour holography, has opened possibilities to study and record Lippmann photography. The same panchromatic ultra-fine-grain silver halide material is used both for holography and Lippmann photography.

History of Lippmann Photography

Gabriel Lippmann (1845–1921), professor of mathematical physics at the Sorbonne invented, demonstrated and mathematically formulated the process of *interference colour photography*, also known as *interferential photography*, or *Lippmann photography*, in the years 1891–1894. (Lippmann 1891, 1892a, 1892b). Lippmann's process was the first exploitation of optical standing waves for the direct recording of permanent colour realised after some twelve years of experimentation. Lippmann photography is the only imaging technique (theoretically) that can directly record the entire colour spectrum of an object or scene.

The process exploits the formation of standing light waves, by means of a mirror, and records these in a single-layer, panchromatic, ultra-fine-grain, *black-and-white* photographic emulsion. When illuminated with white light, the recorded interference fringes select and reflect the original colours registered in the image, and these construct, by Bragg diffraction, the colours of the original scene.

The principle of Lippmann photography is explained in this way: An ordinary plate camera is used to form an image on the photographic recording plate. But the plate is loaded backwards with the emulsion side placed away from the lens and in optical contact with a mirror of mercury. The incident light waves interfere with their own reflections forming a standing wave pattern within the volume of the emulsion, with a periodic spacing of $\lambda/(2n)$ that has to be resolved and recorded (λ is the wavelength of light in air and n is the refractive index of the emulsion). The colour information is stored locally in this way. The larger the separation between the fringes, the longer is the wavelength of the recorded light. When the developed photograph is viewed in white light, different parts of the recorded image produce different colours. This is due to the separation of the recorded fringes. In this way Lippmann's photographs reproduce all the monochromatic components registered on them. Since wavelength is related to colour of the spectra corresponding to the original scene, it can be reproduced with formidable accuracy, unlike any other colour photograph (Fig. 21, page 145). It may seem reasonable that pure spectral colours could be recorded this way, but what about most natural colours with complex and mixed spectra? Lippmann explained the polychromatic case using Fourier mathematics (Lippmann 1894).

In addition to Lippmann, in particular August and Louis Lumière, Richard Neuhauss and Hans Lehmann contributed extensively to progress in this field (Lumière 1893, 1897, Neuhauss 1898, and Lehmann, 1906). Richard Neuhauss, was a physician from Gross-Lichterfelde outside Berlin and one of the most experienced experimenters. Between 1894 and 1905 he made about 2500 Lippmann recordings and also examined the images microscopically which vindicated Lippmann's theory. Neuhauss' book Die Farbenphotographie nach Lippmann's Verfahren (1898) and his *Laboratory Notes* (1990) contain important information specific to making Lippmann emulsions.

The first Lippmann photographers had to make and coat their own emulsions. This was a very skilled craft, as the crystals in the emulsion had to be extremely fine-grain if any colours were to be recorded. In addition, the plates had to be not only panchromatic but also isochromatic. While eventually some dry plate manufacturers produced Lippmann plates, there is very little evidence of much public interest. The difficulty of recording and processing Lippmann photographs may have disappointed many photographers, and discouraged others from buying plates. Hans Lehmann's emulsion was used for manufacturing commercial plates at Richard Jahr's Dry Plate Factory in Dresden. Not until 1925, long after Lehmann's death in 1917, was his formula revealed by Richard Jahr (Jahr 1925).

Standard plate cameras of the time were suitable for recording Lippmann photographs. It was only the dark slides which had to be adapted to flow mercury behind the emulsion.

Carl Zeiss Kamerawerke in Jena manufactured mercury plate holders, filters, viewing and projection apparatus. Equipment was also made by other German companies: Stegemann, Braun, and the plate manufacturer Kranseder & Cie. In England dark slides were manufactured by Watson & Sons and Penrose & Co. Hans Lehmann described the special equipment from Zeiss Kamerawerke used in Lippmann photography (Lehmann 1909, 1910).

Old Lippmann Emulsions

Although albumen emulsions were initially used by Lippmann, the silver halide emulsion was the only "practical" recording material for Lippmann photography. They can offer extreme resolution, and the highest sensitivity due to the development-amplification factor. There were two basic ways Lippmann plates were prepared. The first technique was by diffusion precipitation of silver bromide in a pre-coated albumen or gelatin film by successive bathing in potassium bromide and silver nitrate solutions followed by optical sensitising. This was soon succeeded by precipitation of silver bromide in molten gelatin, with both salts in gelatin or halide only in gelatin with the silver added either in aqueous solution or dry and powdered.

Optical sensitisers were added before the emulsion was filtered and coated. Emulsions had to be coated rapidly after mixing to avoid crystal growth. The coatings were left to gel before washing to remove unwanted potassium nitrate. In all the old formulas except that formulated by Hans Lehmann, the silver was added to the halide at 35°C with gentle stirring over one to two minutes. However, good Lippmann images can be recorded on single-jet emulsions precipitated under silver excess, as performed by Lehmann, who kept his method secret as it was for a commercial product but was published later by Jahr (1925).

Processing Lippmann silver halide emulsions

The processing of the colour photographs was more or less done in the same way by most of the early photographers. They used developers based on *pyrogallol* and *ammonia*, which were formulated to suit the particular emulsion. The hardening effect on gelatin that pyrogallol provides is important. The *Lumière developer* (Table 1) was most often recommended.

Table 1. Lumière developer

Solution A		***Solution B***	
Pyrogallol	1 g	Potassium bromide	30 g
Distilled water	100 ml	Distilled water	100 ml

Working solution: Mix 20 ml Part A + 30 ml Part B + 140 ml distilled water.
Add 10 ml Ammonia (s.w. 0.960) at 18°C just before using. Development time: 60 s to 90 s.

Viewing Lippmann Photographs

There is one particular problem in viewing Lippmann photographs. The images have to be observed in parallel diffuse light and shielded from all side light. A good condition for viewing is a small window facing a brilliant white sky. To view, hold the plate at arm's length, emulsion side facing you, and stand with your back towards the window. Tilt the plate until it reflects a broad region of feature-less sky and you will see the colours "switch-on." Important to improve image contrast is to separate the specular surface reflection from the gelatin surface from the recorded image. This can be done by adding a ten-degree glass wedge (*Wiener Prism*) which will improve viewing conditions. Index-matching glue, for example, Canada balsam was used to attach the prism to the emulsion side of the plate. The back of the plate was painted black covered with black paper and the edges sealed. The Lippmann plates could also be viewed enlarged by projecting them. However, the image could not be projected like a modern slide. A special projector, an *aphengescope*, had to be used. Carl Zeiss in Jena, Germany, manufactured equipment for viewing and projecting Lippmann photographs (Lehmann 1909, Lehmann 1910).

Early Investigations of Lippmann Emulsions

It was of interest to investigate the emulsion of Lippmann photographs to confirm that interference fringes actually existed inside the emulsion. However, using optical microscopy to study the periodic silver grain structure in the emulsion is pushing this technique to its very limit. Nevertheless, Neuhauss (1898) succeeded in microscopically imaging the interference structure of red light recorded in a Lippmann emulsion. He used a red-wavelength recording where the interference layers are most widely separated. Flatau from the Anatomic Institute in Berlin helped him to prepare thin emulsion layers. The quality of the microscope lens was considered also very important, which is why the oil-immersion technique combined with a high-quality apochromat lens (numerical aperture 1.40) from Zeiss was used. Neuhauss succeeded in recording the microphotographs of the emulsion where the separation of interference layers was clearly visible and the distance between them could be measured. The results of the experiment were in good correlation with the theory. Thereafter, optical microscopy was used by Lippmann photography scientists to study the interference fringes looked like for different recorded colour images, for example, investigations by Lehmann (1906).

Medicine and Physiology Nobel laureate Ramón y Cajal, Madrid, was also a pioneer in colour photography in Spain. Cajal who was an expert on histological preparations was well prepared to investigate the structure of Lippmann emulsions (Cajal, 1907). Like Lehmann; he studied and recorded microphotographs of different interference patterns as a result of monochrome, two-colour, white light and natural scenes. Under- and over-exposure and different processing techniques were also investigated. More recently the emulsion of old Lippmann plates has been investigated using modern techniques. One scientist involved is Jean-Marc Fournier (1994) at Rowland Institute for Science in Cambridge, MA, United States, and later at Swiss Institute of Technology in Lausanne, Switzerland. At both institutes he devoted a lot of time on Lippmann photography, its history and performed micro structure analysis with electron microscopy on old Lippmann plates. In Lausanne, at the Musée de l'Elysée, he got access to the Lippmann plates collection there. In addition to electron microscopy, he measured spectra from 'white snow' and 'pink rock' in old Lippmann plates. Very detailed electron microscopic images of the recorded interference fringes in the emulsion could be obtained and analysed. He also recorded Lippmann photographs using holographic photopolymer and silver halide materials as well as prepared ultra-fine-grain silver halide emulsions. He was able to record high-quality Lippmann plates on his prepared plates. More recently the emulsion of old Lippmann plates has been investigated

Modern Lippmann Photography

The recent history of Lippmann's photography has become almost inextricably woven into the modern development of holography. The analysis of Lippmann's technique in relation to holography is comprehensible to holographers but must seem highly esoteric to photographers. It is in the recording of an interference pattern and the replay of colour by diffraction from that pattern that holography resembles Lippmann's photography. The Lippmann colour image is a distinctly a 2D photographic one and recreates the recorded continuous visible electromagnetic spectrum, while a colour hologram usually sample only three red, green, and blue (RGB) wavelengths. The fundamental difference is that, in the Lippmann case, there is *no phase recording* involved; the recorded interference structure is a result of *phase-locking* the light by the reflecting mirror. In holography, the *phase information is actually recorded*, being encoded as an interference pattern created between the light reflected from the object and a coherent reference beam. Modern examples of interference colour photography have been made using improved panchromatic ultra-fine-grain recording materials combined with special processing techniques. The revival of interest in Lippmann's technique began in the mid-1980s, and manifested in several publications. The most important ones are: Phillips et al. (1984), Connes (1987), Naried (1988), Fournier (1991, 1994), Hariharan (1998), and Alschuler (1998, 2002, 2003). The author's first experiments on Lippmann photography were carried out on photopolymer materials Bjelkhagen (1997). The colour holography panchromatic photopolymer material HRF-700X from E. I. du Pont de Nemours & Co. was selected since it was easy to test. No wet processing needed, only UV-curing and baking upon exposure. However, it is much less sensitive than silver halide emulsions. The recording of a Lippmann photograph on photopolymer is simple. Before the recording takes place, a reflecting mirror foil was laminated to the polymer film. A silver-sputtered (800 Å) polyester film without the standard anti-dust oxide (InO) top layer was used as a mirror. The reflecting foil was specially produced by Courtaulds Performance Films. The polymer is slightly sticky and easy to laminate to the reflecting foil and load in the dark slide. After exposure (typically many minutes) is finished, the dark slide can be opened in full daylight, the reflecting foil detached and the polymer film is exposed to direct sunlight (or UV light – DuPont recommends about 100 mJ/cm^2 exposure at 350–380 nm). After that it is put in an oven at a temperature of 100°C for two hours in order to increase the brightness of the image. The first test was a spectrum projected on the photopolymer film with the reflecting foil. Another test was a bowl of fruit recorded in a camera. Two 500 W halogen lights illuminated the scene at a distance of half a metre. The exposure time was about ten minutes at aperture F:4. A 1996 publication on a recorded Lippmann photograph was reported by Dickerson and Rich (1996). Bjelkhagen (1999a) made a review of the history of Lippmann photography and reported that it was possible to record such photographs in panchromatic holographic silver halide materials without using the mercury reflector. Bjelkhagen (1999b) introduced Lippmann photographs on photopolymer film as a new Optical Variable Device (OVD) which could be used on security documents, such as passports and ID-cards. Chapter 2 in the *Ultra-Realistic Imaging* book by Bjelkhagen and Brotherton Ratcliffe (2013) is devoted to Lippmann photography, and the theory and mathematical treatment of Lippmann photography are included.

The Recording Material

Ultra-fine-grain (extreme resolution) and panchromatic materials are essential for recording Lippmann photographs. These must exhibit extremely low light scattering, be practically transparent, and be able to resolve about 10,000 lines/mm or more. It is very difficult to make ultra-fine-grain emulsions and only a few people have been able to do that since the time Lippmann's photography was practised. However, the Russian company *Slavich* (address) manufactures colour holographic plates. *PFG-03C* which is a panchromatic ultra-fine-grain silver halide emulsion (grain size about 10 nm) suitable for Lippmann photographs. The company can provide different size glass plates, for example, 4 x 5 inches, coated on 1.5 mm thick glass (Table 2). In 1995 Bjelkhagen (1999a) recorded his first Lippmann photographs using this emulsion. Even if this emulsion is panchromatic, it is not isochromatic, which means that colour correction filters in front of the camera lens were required.

Table 2. Characteristics of the *Slavich* emulsion

Silver halide material	PFG-03C
Emulsion thickness	7 µm
Grain size	12 – 20 nm
Resolution	~10000 lp/mm
Blue sensitivity	~1.0 – 1.5µ10^{-3} J/cm^2
Green sensitivity	~1.2 – 1.6µ10^{-3} J/cm^2
Red sensitivity	~0.8 – 1.2µ10^{-3} J/cm/^{2}
Colour sensitivity peaked at:	633 nm and 530 nm

Even if this emulsion is panchromatic, it is not isochromatic, which means that colour correction filters in front of the camera lens may be required.

Recording Lippmann Photographs without Mercury

When Lippmann introduced his new colour photographic technique he wanted strong reflected light exerting the emulsion to create the standing waves within it. In addition, since the coherence lengths of white light is extremely short, he wanted to make sure that the reflector is as close as possible to the emulsion. Mercury was a reflector which could accomplish this. However, it had to be perfectly clean; it is heavy and has some safety hazards. In contact with the emulsion it can cause chemical interaction causing streaks and fog. You need a specially made plate holder containing mercury. The use of mercury also makes it difficult to record large plates. The early scientists and photographers who wanted to record Lippmann photographs assumed that they also needed mercury to be able to record colour photographs. However, Edmond Rothé (1904a, 1904b) at Nancy University in France, instead of mercury, he wanted to test if *air* in contact with the emulsion (*Fresnel reflection*) could work. If the emulsion side of the plate is not facing the camera lens, but turned around it could be tested in any camera for glass plates. The difference between the standing wave pattern for the *mercury reflector* as compared with *the gelatin/air reflection* can be described in this way: A node is located at the mercury reflector, an optically thicker medium than air, and coincides with the gelatin surface. The phase shift there is $\lambda/2$. On the contrary, an anti-node (wave crest) is located at the emulsion surface in the gelatin/air interface, an optically thinner medium than gelatin, which means, since no phase shift occurs in this case, a silver layer will be created at the emulsion surface after development. In the mercury case the first silver layer is located $\lambda/4$ inside the emulsion. When using air reflection, the exposure must be increased to bring the recording up on the linear part of the Hurter-Driffield curve. The weaker fringe modulation caused by Fresnel reflection at the gelatin/air interface can be amplified in the developing process. The problem, pointed out by Wiener, about the surface reflection being out of phase with the image when viewing a Lippmann photograph only applies to the mercury case. When using the air reflector, the first fringe is in phase at the emulsion surface. Correct exposure time is very important in Lippmann photography. In the air reflection spectrum case, by increasing the exposure time and combined it with a different developing technique which can enhance the contrast of the recording interference fringes it is possible to obtain the same spectrum as with mercury. The principle of recording Lippmann photographs without mercury is very simple. Almost any plate camera can be used. The ground glass of same

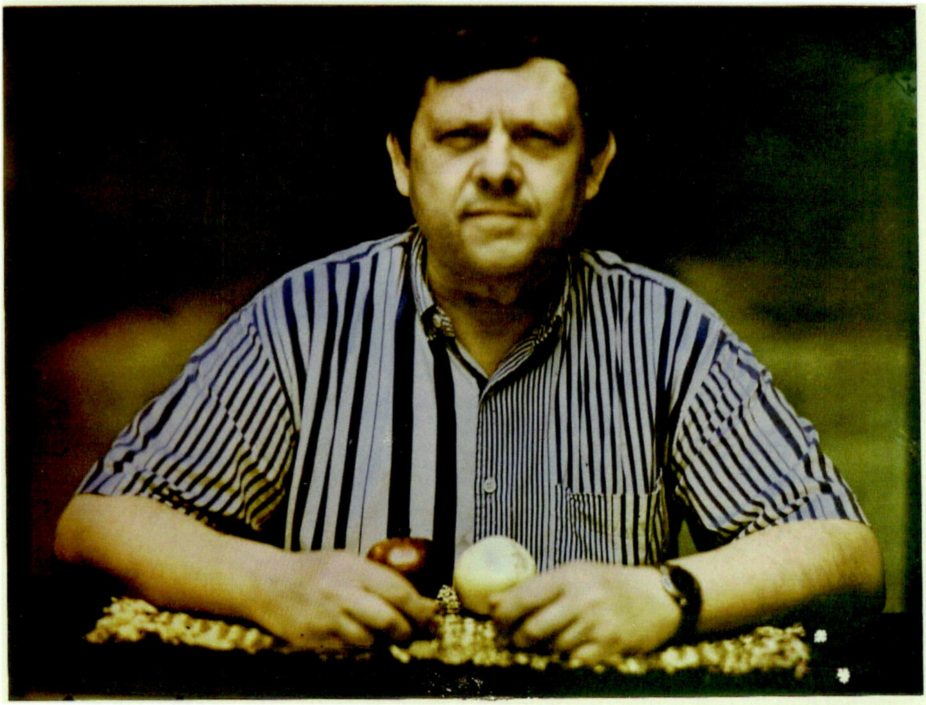

Fig. 52. Lippmann portrait of the author (4 x 5 inch glass plate).

thickness as the plates has to be positioned so that focusing can be made at the emulsion side. The camera and modified dark slides the author used was an old Eastman Kodak Co. (Folmer & Schwing Div.) Auto Graflex 4 x 5 inch camera equipped with a Kodak Aero-Ektar f/2.5, 178 mm lens. Graphic Film Pack Adapters were modified for the plates. They were loaded with the emulsion side facing away from the lens, so that he light reflected at the interface between the emulsion and air enters the emulsion to create the interference pattern. Any other light in the dark slide has to be absorbed using black velvet inside the dark slide. The processing of the Lippmann photographs is very critical. Emulsion shrinkage and other emulsion distortions caused by the developer must be avoided. Among the old Lippmann developers, the *Lumière pyrogallol-ammonia* developer gives good results (Table 1). To avoid shrinkage the plates are not fixed, only washed after development. This is not a problem since developed colloidal silver is not light sensitive.

The plates for recording Lippmann photographs must be transparent with no or very low light scatter. The ideal crystal size for Lippmann emulsions is assumed to be less than 10 nm. Inevitably Lippmann emulsions are spectacularly light insensitive, and are rated at around 0.001 ASA. However, the resolution is a formidable 10,000 lp/mm outstripping the sharpest lens and is able to resolve half-wavelengths of violet light (200 nm). Exposure times are very long, even by the standards of traditional black-and-white plates in Lippmann's time. Processing is an extremely sensitive procedure. Solution-physical developers are the only type of developers that give satisfactory results. In the period, pyrogallol-ammonnia was a favourite. Today developers used in holography, such as GP2 also perform very well for Lippmann photography. The solvent content of the developer has to be balanced for every emulsion, and very small changes in solvent concentration give drastic changes in visual efficiency. Lower levels give "pure" but dim colours, and higher levels result in impressive efficiency but muddy colours. There is a trade-off between colour resolution and efficiency. After being processed, the back of the Lippmann plate is painted black. For better viewing of the image it is recommended that a wedged glass prism (*Wiener prism*) is cemented to the emulsion side of each plate. A modern 4 x 5 inch Lippmann photograph is a portrait of the author reproduced in Fig. 52. The photo was recorded in bright sunlight, softened with a diffuser held above the head. Exposure time was two minutes at aperture f/4. Skin tones are remarkably realistic looking in Lippmann photographs.

The Slavich plates were processed with the Lumière developer (Table 3) in the following steps:

Table 3. *Slavich* emulsion processing

1 Pre-hardening step (6 min) before development in:

Distilled water	750 ml
Formaldehyde 37% (Formalin)	10 ml (10.2g)
Potassium bromide	2 g
Sodium carbonate (anhydrous)	5 g
Add distilled water to make	1 l

2 Development (1½ min) in *Lumière developer* at 15 degree C.
3 (No fixing) – Wash (10 min) + Dip in distilled with wetting agent
4 Dry (slowly) in room temperature.

Darran Green, Bury St Edmunds, UK, is a very experienced Lippmann photographer who has spent many years to perfect silver halide emulsions, recording and processing Lippmann plates. When not using mercury it is much easier to record large Lippmann photographs, provided one has access to large photographic plates. An example is an 8 x 10 inch plate recorded by the author and Darran Green on May 1, 2000 in the *Abbey Gardens*, Bury St Edmunds, UK, Fig. 53. Recording data: Iston view camera with a Kodak Aero-Ektar f/2.5 178 mm lens., aperture f/11, exposure time three minutes on a Slavich PFG-03C glass plate. The plate was developed in the holographic GP8 developer. The very high image resolution of the recorded Lippmann plate can be demonstrated by enlarging details of the recorded scene. The image resolution is actually limited by the quality of the camera lens and not by the resolving power of the recording material. In the United States the art photographer Jesús López (2003) has built a large Lippmann plate camera named Gabriela for the recording of glass plates up to 60 x 60 cm. This camera is equipped with a 1940s Bausch & Lomb Aero Tessar lens, f/6.0, focal length 610 mm. He is using the recording materials from Slavich (PFG 03C, most often 30 x 40 cm glass plates). The exposure time varies between five to twenty minutes in his Michigan studio equipped with a very strong lighting system. He got excited about Lippmann's technique when he saw some recently recorded Lippmann photographs by the author in the mid-1990s. In particular he was impressed how the colour of skin could be reproduced in such a photograph. López' interest in art has always been the beauty of the female figure and he found that Lippmann photography would be the most suitable media for this. Since he wanted to create large-format Lippmann photographs he needed an equally large camera. When mercury is no longer needed; making large-format Lippmanns is no longer an impossible task, but still a lot a work went into the fabrication of the Gabriela camera.

Conclusion

The first Lippmann colour photographs recorded by the Lumière brothers were inspected by members of the Royal Photographic Society in 1893, and settled any lingering doubts over the veracity of the reports – true colour photography had been realised. In 1897, Lippmann himself addressed the same society, demonstrating his process, apparently speaking in "faultless English." For a period spanning two decades, photographic journals carried numerous reports on progress in Lippmann photography. Whole books were devoted to the technique detailing every aspect of the process, from emulsion making and coating, to processing and display. With the exception of the known early Lippmann photographers, it is not known how extensive its practice was among photographers generally. There is little doubt that technical difficulties inhibited its widespread use. The

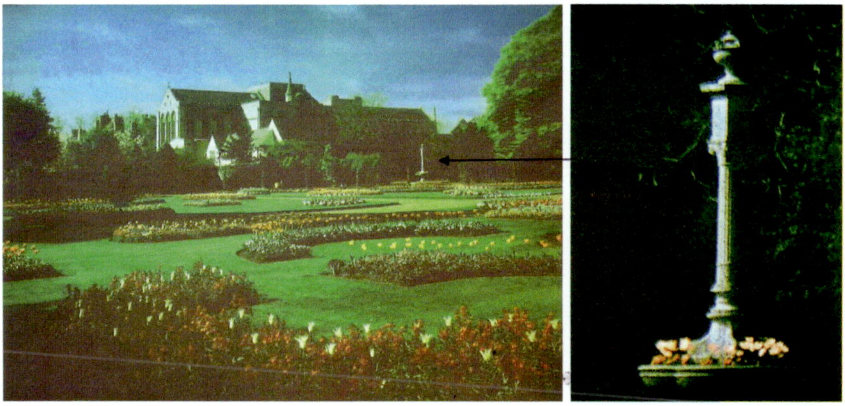

Fig. 53. *Abbey Gardens*, Bury St. Edmunds, 8 x 10 inch Lippmann plate (left.) Enlarged detail (right).

impression Lippmann's images made on photographers of the day is clearly expressed by at least one master of monochrome – Edward Steichen who wrote to Stieglitz in 1908:

> Professor Lippmann has shown me slides of still-life subjects by projection, that were as perfect in color as in an ordinary glass positive in the rendering of the image in monochrome. The rendering of white tones was astonishing, and a slide made by one of the Lumière brothers, at a time when they were trying to make the process commercially possible, a slide of a girl in a plaid dress on a brilliant sunlit lawn, was simply dazzling, and one would have to go to a good Renoir to find its equal in color luminosity. (Newhall 1981, 272)

The impact of Lippmann's process on the scientific community is evidenced by his nomination for the Physics Nobel Prize in 1908. His process had vindicated Maxwell's electromagnetic wave theory of light, and his mathematical analysis, using Fourier transforms, was considered to be of extraordinary elegance. Finally the concept of the spectral registration of colours in a volume interference pattern and replay by diffraction combined with the extreme resolution plates necessary to realise this, contributed directly to the development of holography.

Today, Lippmann photography has limited applications in photography and colour imaging, but may well appeal to artists and art photographers. The Lippmann photograph is virtually impossible to copy, which makes each image a unique, one-of-a-kind photographic recording, combined with extreme archival longevity. There is a growing interest in Lippmann photography today with several people practising this type of photography with discussions and photos of their plates posted at the Lippmann Photography Facebook group.

Note: (Since conventional photography, analogue or digital, cannot record colours recorded in a Lippmann photograph, only by viewing real Lippmann photographs, one can appreciate the quality of them. It is also difficult to obtain the very high image resolution of the Lippmann images when reproducing them.)

Works Cited

Alschuler, W. R. 1998. "On the Physical and Visual state of 100 year old Lippmann Color Photographs".In *Sixth International Symposium on Display Holography SPIE 3358*, edited by T. H. Jeong and H. I. Bjelkhagen. Proc: 84–94.

Alschuler, William R. 2002. "Lippmann Photography and the Glory of Frozen Light: Eternal Photographic Color Real and False". *IEEE 31st Applied Imagery Pattern Recocnition Workshop* (AIPR'02).

Alschuler, William R. 2003. "Lippmann Color Process, the Experience of Amateurs in England." *The PhotoHistorian, Journal of the Historical Group of the Royal Photographic Society, APIS 2002 Special double edition*: 41–142.44–51

Bjelkhagen, H. I. 1997. "Lippmann Photographs Recorded in DuPont Color Photopolymer Material [3011–35]". *Proceedings- SPIE The International Society for Optical Engineering* (3011): 358–366.

Bjelkhagen, Hans I. 1999. "Lippmann Photography: Reviving an Early Colour Process." *History of Photography* 23, no. 3: pp. 274–280.

Bjelkhagen, Hans I. 1999b. "New Optical Security Device Based on One-Hundred-Year-Old Photographic Technique". *Optical Engineering.* 38 (1): 55.

Cajal, S.R.y.1906. "Estructura de las Imágenes Fotocrómicas de G. Lippmann." Revista de la Real Academia de Ciencias Exactas, Físicas y Naturales de Madrid 4 (4) [German transl.] 1907."Die Structur der Lippmannschen Photochromien." *Zeitschrift für wissenschaftliche Photographie* 5: 213–245. [English transl.] 1907. "The Structure of Lippmann Heliochromes." *British Journal of Photography*: 59–79.

Connes, P. 1987. Silver salts and standing waves: the history of interference color photography. *J. Optics* (*Paris*) 18, 147–166.

Dickerson, Lydia, and Chris C. Rich. 1996. "Lippmann Photographic Process put to Practice with Available Materials". *Proceedings of SPIE – The International Society for Optical Engineering.* 2688: 88–95.

Fournier, J.- M. 1994. "An Investigation on Lippmann Photographs: Materials, Processes, and Color Rendition". In *Practical Holography VII*, edited by S.A. Benton. Proc. SPIE 2176: 144–152.

Fournier, J.-M., and P. L. Burnett. 1994. "Color Rendition and Archival Properties of Lippmann Photographs". *Journal of Imaging Science and Technology*. 38 (6): 507.

Hans Bjelkhagen, and David Brotherton-Ratcliffe. 2013. *Ultra-Realistic Imaging: Advanced Techniques in Analogue and Digital Colour Holography*. Abingdon: CRC Press.

Hariharan, P. 1998. Lippmann Photography or Lippmann Holography? *Journal of Modern Optics* 45:1759 – 1762.

J.- M. Fournier. 1991. "Le Photographie en Couleur de type Lippmann: Cent ans de Physique et de Technologie." *Journal of Optics*. 22: 259–266.

Jahr, Richard. 1925. "Über die Herstellung Sogenannter "Kornloser" Platten für das Interferenz-Farben-Verfahren von Professor Lippmann nach Dr. Hans Lehmann." *Photographische Industrie*. 23: 1013–1014

Lehmann, Hans. 1906. *Beiträge zur Theorie und Praxis der Direkten Farbenphotographie mittels stehender Lichtwellen nach Lippmanns Methode*. Freiburg i. Br: Trömer.

Lehmann, Hans. 1909c. "Die Praxis der Interferenzfarbenphotographie", *Photographische Rundschau* 23, 125–134.

Lehmann, Hans. 1910. "Practical Application of Interference Colour Photography. II. *British Journal of Photography Colour Supplement*:92–95.

Lippmann, Gabriel. 1891. "La Photographie des Couleurs." *Comptes Rendus Hebdomadaires des Séances de l'Académie des Sciences* 112: 274–275.

Lippmann, Gabriel. 1892. "La Photographie des Couleurs.[deuxième note]". *Comptes Rendus Hebdomadaires des Séances de l'Académie des Sciences* 114: 961–962.

Lippmann, Gabriel. 1892. "Photographies Colorées du Spectre, sur Albumine et sur Gélatine Bichromatées." *Comptes Rendus Hebdomadaires des Séances de l'Académie des Sciences* 115:575.

Lippmann, Gabriel. 1894. "Sur la Théorie de la Photographie des Couleurs Simples et Composées par la Méthode Interférentielle." *Journal Physique* 3 (3): 97–107.

López, J. 2003. *After all These Years, Modern Lippmann Photography*. Chicago: López's Photonics Studio. Wicklander Printing Co.

Lumière, A. and L. 1897. "Untersuchungen über die Herstellung einer Lichtempfindlichen Kornlosen Schicht." *Eder's Jahrbuch für Photographie und Reproductionstechnik*. 11: 27–30.

Lumière, A. and L. 1893. "Sur les procédés pour la photographie des couleurs d'après la méthode de M. Lippmann." *Bulletin de la Société française de Photographie* (20 série) 9: 249–251.

Nareid, H.1988. "A Review of the Lippmann Color Process." *Journal of the Photographic Science*. 36: 140–147.
Neuhauss, Richard. 1898. "Nachweis der dünnen Zenkerschen Blättchen in den nach Lippmann's Verfahren aufgenommenen Farbildern." *Annalen der Physik und Chemie* 65:164–172.
Neuhauss, Richard. 1898a. *Die Farbenphotographie nach Lippmann's Verfahren neue Untersuchungen und Ergebnisse*. Halle a. S: Knapp
Richard Neuhauss. Journals 1894–1900. The Lippmann Color Process. Prof. Dr. Albert Narath Archives, Preus Museum: The National Museum of Photography, Norway. https://holowiki.org/data/lippmann/Neuhauss_Journals.pdf
Newhall, Beaumont. 1982. *The History of Photography*. Boston: Little, Brown. 272.
Phillips, N. J., H. Heyworth, and T. Hare. 1984. "On Lippmann's Photography." *The Journal of Photographic Science* 32: 158–169.
Rothé, E. 1904. "Photographies en Couleurs Obtenues par la Méthode Interférentielle sans Miroir de Mercure." *Comptes Rendus Hebdomadaires des Séances de l'Académie des Sciences* 139: 565–567.
SLAVICH Joint Stock Co., Micron Branch Co., 2 p1. Mendeleeva, 152140 Pereslavl-Zalessky, Russia.

About the Author

Dr. **Hans I Bjelkhagen**, a consultant at Hansholo Consulting Ltd, was awarded his doctoral degree in 1978 by the Royal Institute of Technology in Stockholm, Sweden. Over the last fifteen years, Bjelkhagen has received much international recognition for his work in the field of colour holography. In 1983, Bjelkhagen joined CERN in Geneva, Switzerland and, in 1984, Fermilab in Batavia, Illinois. Between 1985 and 1991, he was at Northwestern University, working on medical applications of holography.

12. "It's not just a picture, it's a magical object": Interview with Contemporary Lippmann Photographer Filipe Alves

Hanin Hannouch

Abstract
This interview with Lisbon-based artist, scientist, and contemporary Lippmann photographer Filipe Alves (*1979) explores his committent to producing interferential colour photographs today. Alves's practice of analogue colour photography has reshaped his studio into a site of media archaeological re-enactment of the photographic process. It has also lead to the adaptation of the chemistry behind it for current use, as well as to nuanced reflections about its history. In this conversation, Alves sheds light on his colourful endeavour so far, the enduring lure of the analogue, and the future trajectory of the Lippmann plate as a creative medium.

Keywords: contemporary photography, interferential colour photography, Filipe Alves, media archaeology, artist studio

Filipe Alves (*1979) is an artist and scientist based in Lisbon, where he studied architecture. With his associates, he co-founded "18:25," a studio for virtual photography and architectural visualisation as well as the photography studio "Silverbox" where he practices historical analogue photography, such as the wetplate collodion process.

He has been using interferential colour photography for over a decade and has exhibited his work in various museums across Europe, including at the exhibition *Slow Color Photography: Richard Neuhauss and Hans Lehmann's Images from the Preus Museum's Collection* (2020) at Preus Museum: Norway's National Museum Photography, curated by Dr. Hanin Hannouch and Jens Gold, which was dedicated to the centenary of Lippmann's death (Preus Museum

Hannouch, H. (ed.), *Gabriel Lippmann's Colour Photography: Science, Media, Museums.* Taylor & Francis Group, 2022
DOI 10.5117/9789463728553_CH12

2021). As a key figure in the growing artistic use of the Lippmann process, Alves's work is crucial to any comprehensive understanding of this vivid medium because it reveals one of its most exciting future trajectories. Adapting the chemistry needed for the interferential photograph to contemporary use, managing the repeated experimental failures to produce it, and creating photographs that resonate with current audiences, constitute both a challenge and a pleasure to Alves. In this interview, which has been edited and condensed for clarity, Alves sheds light on the intricacies of his endeavour thus far.

Hanin Hannouch: Out of all possible colour photographies, both old and new, why did you choose Lippmann photography?
Filipe Alves: I discovered it around 2010 when I started practising alternative processes such as the gum bichromate process and the cyanotype in my free time. I love to take pictures and to develop them in a darkroom, so I began buying chemicals to make some of the processes and then I found out about dichromated gelatin being used for holography. I don't remember very well why, but I just decided in that moment that I would do this because it looked compelling and easy enough, so I began practising holography. I bought some lasers and I made an optical table using a slab of marble. Through working with holography, I discovered Lippmann photography, but I left Lippmann photography aside for a while until we [my partner and I] moved to Lisbon and opened a photography studio where we produced wet plate collodion images [Silverbox studio]. In Lisbon, I stopped doing holography. You need hard floors and an optical table for holography so that everything around you remains stable for the duration of the exposure. You cannot practice it on a wooden floor because if anything in the room moves for even a few hundred nanometres, the interference patterns shifts and you cannot register anything. The hologram will be gone. With Silverbox studio, I had space and time to explore Lippmann photography which for me is a visible manifestation or proof of the interferential nature of light, like a proof that the earth is round. I read Hans Bjelkhagen's papers and found out that you can use an emulsion-air interface instead of liquid mercury to make a successful emulsion. It seemed easy to do, so I bought panchromatic holographic plates from GEOLA and used the Lumière brothers' recipe for development. I exposed one image, the gelatin emulsion took a lot of time to dry and after waiting for a while I started seeing the image appear in colour. It was really amazing for me to see a Lippmann plate emerge for the first time, I got hooked!
HH: What did your first photograph depict and why did you choose it?
FA: It was a photo of a painting of mine. It has a suitable dynamic range and is quite colourful. I did this painting as a present for my girlfriend, so

there is a special meaning attached to it. It was the first one I tried. After that, I did several, you can see some of them in the videos on my youtube channel (Alves 2014). It was really amazing, but right now, I hardly see any colour in them. This is weird because the brain adapts to the circumstances and no longer sees the hues in the same way as when it first observed them.
HH: Is it because now you are looking for more colour saturation?
FA: Yes, but maybe I'm not necessarily looking for more saturation, the hardest thing to achieve is panchromaticity, and good colour balance. I like images that are not very saturated, but what I want is to be able to control the saturation, to be able to choose if I want to saturate the image or not and to choose the mood of the picture. So with the GEOLA plates I was using, I had twenty of them back then, and I could undertake several experiments and take many photographs. But I wanted to keep on going and increase the complexity of my work. If I can just buy plates and shoot the photos, then the process is easy-peasy. I can do that. But then I would get bored very quickly because I would not have the challenge of being able to make the plates and of doing the entire process, both from scratch, which is often compelling and immersive.
HH: You use a really interesting word: control. When reading the historical literature of Neuhauss, Lippmann, and others, they were really complaining about not being able to control interferential colour photography. How do you deal with certain steps in the photographic process going wrong on a regular basis?
FA: I don't manage! Maybe I manage a little bit, but it is somewhat of a mystery to me. The amount of variables that the process has, known and unknown, are too many for a somewhat chaotic or disorganised investigation. What I have learned in these few years is how to identify the kinds of problems I encounter since they relate to the workflow of interferential colour photography. For example, sometimes if I use some eggs and the plates [made with albumen emulsion that is based on egg-whites] come out in a specific way, as in sensitive to some colours more than others, I can do adjustments to maximise what I can do with them or just throw them away and try again some other time. Sometimes I get lucky, I have good eggs and all the stars align and I think this is wonderful, and have the feeling that I am finally controlling the process, but then, something happens and the emulsion fails for the next three months. But I keep trying again and again. It's an endless roller-coaster ride of excitement and frustration.
HH: What is your favourite step of the Lippmann process?
FA: My favourite step is drying the plate after developing, because that is where it all happens. When you develop, you develop with a backlight,

so you can see the negative forming and it forms differently each time: sometimes it forms very rapidly and builds up. Sometimes, it forms slower, and the colours change and become darker. The parts that are exposed the most are the darkest, so they start getting lighter and more yellow, and finally the brightest parts of the image start turning positive (Alvez, "Lippmann Photography"). Just looking at the negative and knowing when to stop developing the image also is a key factor in the outcome of the photograph. After developing and washing, during the drying process, you can see the colours appearing and blueshifting as the plate dries and shrinks the albumen to its original size. There, finally, you can see how bright and saturated the plate came out.

HH: When scientists first tried Lippmann photography starting 1891, they did not have many other options to generate colour images mechanically. Hans Lehmann who tried Lippmann photography ca. 1910 while working for Zeiss could have used three-colour photography which was somewhat easier, but chose not to. What motivates you to keep trying, knowing that other and more forgiving colour processes are available? Why put yourself through this?

FA: I think it's similar to people today who decide to use analogue photography instead of just digital. Right now everyone has several cameras, in the phone, on the computer, everywhere, all digital. What is special [in comparison to digital image-making] are the difficult processes, handmade processes which have a different allure to them. That's why people keep making daguerreotypes, wet plate collodion, and other old processes, because you can actually do everything from scratch. I think that the drive to do something by hand will never disappear. It is both a challenge and a vice that I am addicted to because it entails so many emotional changes. When you fail, it is incredibly frustrating. But when you are drying the plate and see it emerging, going from red to yellow, to green, and then it starts to dry, it goes from infrared to ultraviolet, then you see all the shades coming up gradually and at last you can see the image, then it is incredible and beautiful. It is amazing to see when they [the plates] are mounted [with a glass wedge prism] and are bright and there is no other way of making them, except through the Lippmann process. The result [of the photographic process] is not just a picture, it's a magical object that can only be seen by one person at a time. You cannot have two people looking at it simultaneously and if you do, it is not very comfortable for them. You have to actively engage with the image to be able to see it and sometimes it peeks through at you. It is as if you are looking at a hole in the wall and get to observe what is on the

other side. Moreover, it is a process that was not explored artistically. It was just investigated for a short time, and most of the pictures that were made using it were of test objects.
HH: True. The Lippmann process's most iconic photograph is of Richard Neuhauss's colourful stuffed parrot, which was his favourite test object.
FA: This is interesting because the experiments of yesterday, over time, become the art of today. And being able to engage in dialogue directly with that short epoch is special. We are here but sometimes I feel like I am re-living something done over a hundred year ago. I imagine that Neuhauss and others were obsessed, making plate after plate and taking notes like crazy. They were struggling with the things that I am struggling with as well.

HH: For the exhibition about interferential colour photography which I guest-curated at Preus Museum: National Museum of Photography in Norway titled *Slow Color Photography: Richard Neuhauss and Hans Lehmann's Images from the Preus Collection* (co-curator is Jens Gold), the museum displayed one of your plates "Body and Soul," made in 2020. It is a still life featuring a clothes steamer and a small antique bust. How do you create the visual compositions for your Lippmann plates?
FA: When I get good plates and emulsion, I get a terrible sense of dread: what am I going to photograph? Because if I am not prepared and I have no idea what to photograph, I would have to search around the house frantically to find objects to make a composition. And I think this is the hardest part of the process. It is not about making the Lippmann plate [as in the photosensitive emulsion] as such but about making the image [what is depicted] which is much more demanding. It is harder because I need to find a meaning for the objects that I place in front of the camera in order to have an appealing and significant composition. I also must accommodate the plate, sometimes it does not register green, sometimes the whites can be solarised, so I have to carefully choose what I photograph depending on the quality of the plates. I think in the case of "Body and Soul," I was doing a full set of images that confront the classical still life with modern objects made of plastic, or with something that is recognisably modern, but whose aesthetic preserves the overall mood. For me, the hardest part is finding the best composition, the right objects. This is also what I do for a living, creating computer-generated images of architecture, so I am always thinking of the composition.
HH: What advice would you give to those starting to try the Lippmann process?

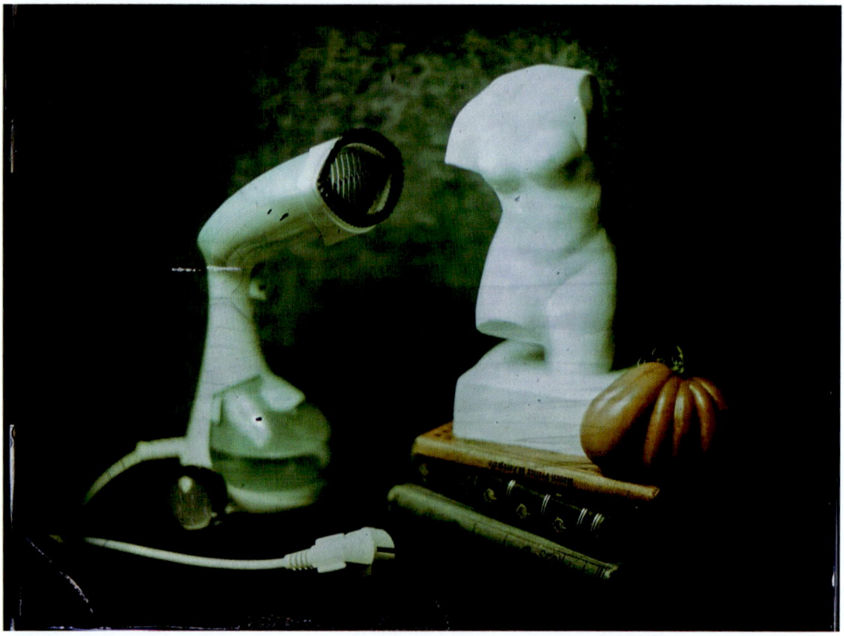

Fig. 54. Filipe Alves (1979–). "Body and Soul" [original title]. Lippmann process. 9.7 cm x 8 cm. Glass plate with wedge prism at 10 degree angle. 2020. Courtesy of the artist.

FA: The first advice I would give them is to buy some pre-made panchromatic holographic plates and try the Lippmann process using them because this is the best way to start. Put aside the production of the emulsions yourself and focus on taking the picture and developing it as a first step, because after that you will get hooked, it is so amazing that you will become fascinated. The second piece of advice is not to expect repeatability. Or learn to live with the fact that the process is very unpredictable. It is also very sensitive, maybe to ghosts. Who knows? A third piece of advice is learn to deal with your frustration. I remember going to several dinners celebrating the success of a Lippmann plate but the next day, the plates I produced were a disaster.
HH: What do you think is the future of Lippmann photography?
FA: I think there is a mini-boom of Lippmann photography right now because people see that results actually can be achieved. I think it is a process reserved for dedicated or curious people who want to try it because it is so obscure and weird and beautiful. I also think one of the biggest contributions and improvements that will allow the process to flourish is the fact that using liquid mercury is no longer needed and can be replaced by air. Still, on a variety of levels, there is still a learning curve which interested photographers and scholars can look forward to.

About the Author

Dr. **Hanin Hannouch** is curator for analog and digital media at the Weltmuseum Wien (Vienna). She has guest-curated the exhibition "Slow Color Photography" about interferential colour photography at Preus Museum: Norway's National Museum for Photography and is the guest-editor of the journal PhotoResearcher Nr. 37 "Three-Colour Photography around 1900: Technologies, Expeditions, Empires". She was post-doctoral researcher at the Kunsthistorisches Institut in Florenz, Max-Planck Institut and at the Ethnologisches Museum in Berlin. She is currently writing her monograph about the history of colour photography in Imperial Germany.

Afterword: Building Bridges over Standing Waves

Nathalie Boulouch

Abstract
This afterword examines how Gabriel Lippmann's work can generate interdisciplinary viewpoints. Starting with Lippmann's cooperation with the Lumière brothers and their reception in the artistic press, it then examines interferential photography's position in the history of the medium and asks questions on how it can be renewed through interpretations from a cultural or *media archaeology* point of view. Last, it demonstrates how research on the Lippmann process is being renewed thanks to recent interest in museum collections, digitisation projects and upcoming exhibitions.

Keywords: interferential colour photography, Lumière brothers, aesthetics, art history, museum collections

On Monday February 2, 1891, Gabriel Lippmann, a member of the Physics section since 1886, presented a communication to his colleagues at the Académie des Sciences de Paris that could be seen as a wager: "I have decided to obtain a coloured image of the spectrum on a photographic plate in such a way that the image will remain permanently fixed and can be observed in full daylight without fading over time." (Lippmann 1891, 274).

While Frederik Eugene Ives's composite heliochromies[1] drew attention to the reproduction of colours by additive colour synthesis, the French physicist sought to capture a colour image "directly." Immediately and unanimously acclaimed for its elegance, interferential photography seemed to accomplish the desire to "paint with light" (Becquerel 1868, 218). Following work begun

1 Ives took out an American patent on July 22, 1890 (number 432.530) called Composite Heliochromy. Mentioned in *Le Moniteur de la photographie*, Sept 1 (1890): 130–31.

Hannouch, H. (ed.), *Gabriel Lippmann's Colour Photography: Science, Media, Museums.*
Taylor & Francis Group, 2022
DOI 10.5117/9789463728553_AFTER

in 1848 by Edmond Becquerel then Niépce de Saint-Victor and the colour synthesis "solutions" presented by Louis Ducos du Hauron and Charles Cros in 1869 (Cros 1869, du Hauron 1869),[2] Gabriel Lippmann proposed an entirely new approach, drawing on the wave-based nature of light. He broke with existing paradigms, opening an entirely new field of research into colour photography that was mainly physical.

Unfortunately, the remarkable scientific value of his research could not compensate for the complexity of his method and the irregular nature of its results. The process never really got beyond the experimental stage and was only ever employed by a handful of experienced users around the world. While the process undoubtedly solved the hitherto unachievable goal of colour permanence, interferential photography recorded colour but "could only create it on a single image" (du Hauron 1897, 474). Comparing his brother's three-colour method to Lippmann's, Alcide Ducos du Hauron stated: "While interferential colour photography is admirable, should we criticise the fact that it doesn't have the gift to reproduce?" (du Hauron 1897, 87).

If the Lippmann process, like the daguerreotype, was sterile from this point of view, where is its "fertility" today? The texts assembled by Hannin Hannouch in this book invite us to consider the exemplary nature of Lippmann's work for its ability to generate interdisciplinary viewpoints. In the same way that the 1991 celebration of the centenary of the presentation to the Académie des Sciences re-ignited interest in Lippmann's process, 2021 offers a fresh occasion to take stock of work currently underway.

As William Alschuler said, Lippmann "froze light and passed it around." The process never had widespread popular appeal, but it was used by experts on an international scale and particularly in Germany. Georges Goddé, Herbert Eugene Ives, Herman Krone, Louis and Auguste Lumière, Hans Lehmann, Richard Neuhauss, Auguste Ponsot, Santiago Ramón y Cajal, Edmond Rothé, I. F. Usagin, Eduard Valenta, and a number of others experimented and attempted to improve it. From 1892, the Lumière brothers made an essential contribution with the preparation of very thin gelatin emulsion containing extremely fine grains of silver halide and reduced exposure times significantly (Lumière and Lumière 1893, 249–51). They rapidly obtained excellent results that were applauded by both the scientific and photographic communities.

2 By coincidence, Cros and Ducos du Hauron presented their two papers, which reached almost identical conclusions, on the same day at the meeting of the Société française de photographie on Mai 7th 1869. Ducos du Hauron's paper followed a patent (number 83.061) with an identical title registered on November 23, 1868. Cros meanwhile had deposited a sealed letter with the Académie des sciences on December 2, 1867 : "Procédé d'enregistrement et de reproduction des couleurs, des formes et des mouvements."

In 1895, several projections of the Lumière brothers' interferential photographs even accompanied the first demonstrations of their Cinématographe. The colour images were shown after projections of the film *La Sortie des usines Lumière*, adding to the experience of viewing images that aimed to reproduce reality: "When seen at a certain angle by reflected light, a beautiful, iridescent image becomes apparent, which when shown by reflection in a projecting lantern, has all the startling realism that the image in a mirror conveys" (Steichen 1908, 14). In a single evening the viewing public could have a double experience of "sensations of real life": through the moving pictures of the film and the colours of the interferential photographs. In the dark of the auditorium, the appearance on the screen of the enlarged images, intensified by the bright projector beam, reinforced the striking effect of the colours. The naturalness of the images vied for attention with the perceptual and spectacular seduction of the chromatic iridescences.

This fascinating realism was regularly referred to in the dedicated press. Noting the particularly faithful reproduction of colours, one commentator said, "all of these colours are reproduced and *sing,* as painters say, with their respective intensities [...] Amidst this orgy of complex tonalities, the whites remain pure. As I have explained, faithfully reproducing white is the great challenge of chromophotography" (Dillaye 1894, 153)[3]. But while the subjects of the images (portraits, landscapes, still life) corresponded to the criteria of the Beaux Arts, how would it be possible to evaluate the artistic value of this process, which through its use of colour quickly invited comparisons with painting? In 1898, colour gum prints presented by the Austrian pictorialists at the *Salon photographique de Paris* provoked a heated debate among members of the artistic photographic community (Boulouch 2006). Constant Puyo, highly critical of practitioners of colour photography, referred to Lippmann's interferential photographs to defend his view. If the method was, in his opinion, "the most perfect, because it is mathematical," he was categorical about its absence of any aesthetic interest:

> The dazzling bouquets that Mr. Lippmann projects on the screen have precisely the same effect on me as illuminated fountains or the Loïe Fuller; it is a visual spectacle, a physical caress of the retina, an exquisite sensation, but it is a pure sensation that refuses to develop into a feeling [...]these images are utterly impersonal; they are scientific products and, despite the fact that they are beautiful, they have no soul. (Puyo 1898, 224)

3 This astonishing reproduction of whites also struck Steichen, "Color Photography," 14.

This label of a scientific process devoid of any artistic merit stuck doggedly to the Lippmann process. In 1907, when the commercialisation of the autochrome led Alfred Stieglitz to proclaim, "Color photography is an accomplished fact," he simply mentioned Lippmann to note, "his invention is only of scientific value "(Stieglitz 1907, 20).

If the Lumière brothers contributed brilliantly to the improvement of the interferential process, they subsequently contributed to its confinement to a few laboratories run by, "a handful of lapidiaries lovingly crafting dazzling stones for the jewel boxes of aristocratic collectors" (du Hauron 1897, 89). In 1895, the Lyon-based industrialists decided that it would be impossible to commercialise the Lippmann process on a large scale. They abandoned the scientific beauty of interferential colours and redirected their efforts towards the more empirical principle of the "indirect" method (Lumière and Lumière 1895, 1034–1038). At the turn of the century in France, the popularity of polychrome projections promoted by Léon Vidal and the commercialisation of several devices inspired by Ives's Kromscop, such as Ducos du Hauron's Mélanochromoscope,[4] assured the supremacy of the "indirect" method. The technique was honoured in the "retrospective museum" of photography at the Paris World's Fair of 1900. Lumière three-colour transparencies were presented alongside other three-colour processes and cameras, while Lippman's images were relegated to a corridor dealing with the question of education and training (Vidal 1901). In 1908, Lippman's Nobel Prize for Physics made no change to the destiny of the interferential process, whose fate had by now been well and truly sealed by the industrial production of autochromes. Later, the links with the invention of holography, would definitively tie interferential photography to the world of remarkable scientific inventions.

As a consequence, the historiography of the Lippmann process has been marked by an overwhelming emphasis on its scientific importance to the detriment of its place in the history of photography, where discussions are generally limited to its early technical role in the development of colour photographs. Studies of the intrinsic value of the images themselves, their iconographic status and their cultural reception, have often been overlooked. It is true that there are only a few hundred examples of Lippmann photographs known to exist in collections today. Photographic curiosities, they have remained invisible, undervalued and often poorly identified. They have become victims of their status as photographic objects. Their dark, prismatic

4 *Mélanochromoscope* (I.N.P.I. brevet n° 288.870 du 16 mai 1899; addition du 4 sept. 1899) following a first *système d'appareil chromographique et chromoscopique* manufactured by Ducos du Hauron et Mackenstein (I.N.P.I. brevet n ° 271.704 du 26 oct. 1897).

forms that only reveal their coloured images if viewed at a certain angle have relegated Lippmann plates to museum reserves or university collections, where they have been used as scientific specimens or teaching aids. The museographical challenge posed by these photographs is set out by Rolf Sachsse: "Around the time of the exhibition *Farbe im Photo* in the 1980s, the Lippmann plate of a stuffed parrot produced by Neuhauss was an unlikely candidate as an exhibit." Today, institutions such as the Photo Elysée in Lausanne and the Preus Museum – National Museum for Photography in Horten (Norway), but also universities such as the Sorbonne in Paris or University of Lille are re-appraising their valued collections. They are working alongside scientists, historians, curators and conservators as Jens Gold, Pauline Martin and Carole Sandrin explain in this book. The exhibitions currently being conceived will present new opportunities for displaying Lippmann images, which will hopefully allow the general public to better appreciate these remarkable works while ensuring optimal conservation conditions.

The question of visualisation methods is also at the heart of the history of the process, as Jens Schröter and Hanin Hannouch explain in their texts, using a media studies approach. Whether the method for viewing images is for individual viewers or by projection, interferential images need to be presented using specific equipment. In this regard, the Lippmann process is not fundamentally different to other colour processes, including photographic slides such as autochromes or Kodachromes, to mention the two most widely used processes of the twentieth century.

Studying the history of the way the process was used, the ways it was presented and distributed (projections, exhibitions), studying the iconography: all of these areas of research need to be opened up. Richard Neuhauss's image of the stuffed parrot, which has often been used to illustrate the Lippmann process in photographic history books, is an excellent example. Chosen as an object for study by Hanin Hannouch, this bird, photographed on several occasions and therefore available in several different collections, allows us to explore how images were shared between the scientific and photographic domains. The body of interferential photographs as a whole allows us to examine the dividing lines between scientific and amateur photography.

The body of interferential photographs as a whole allows us to examine the dividing lines between scientific and amateur photography. The aesthetic quality of images made by Gabriel Lippmann and his wife Laurence Cherbuliez, but also by other practitioners such as Auguste Ponsot[5] show

5 Fifteen plates made by the physicist Auguste Ponsot belong to Lille university's heritage collection of scientific instruments.

that for a few very technically gifted photographers – in this regard their scientific training was indispensable – the process became a means of artistic expression in its own right. This leads naturally to an examination of the aesthetic reception of interferential photographs. Elizabeth Cronin sketches the initial outlines: "the realms of art and science overlapped in a variety of ways.'. But research of this kind can only really advance when the images themselves are readily accessible. The project to create an online catalogue raisonné, initiated by the Photo Elysée in 2020, aims to fulfil this need by virtually assembling the different collections and making them visible and consultable. Beyond the desire to better understand Gabriel Lippmann's place within the scientific, photographic and intellectual networks of his day, this project will allow, we hope, to re-place these images in a cultural, scientific and aesthetic history of photography.

During his lifetime, Lippmann was described as, "a physician who has not remained completely confined to theory and whose name has become famous [...] in the photographic world" (Rameau 1894, 10). His process remains a bridge between science and photography, as the diversity in the summary of this book shows. Today, Lippmann photographs are proving to be promising historical objects for visual studies of science.

"Life is short and progress is slow" (Lippmann 1908), It was with these words that Gabriel Lippmann concluded his acceptance speech for the Nobel prize on December 14, 1908. Now seems a perfect time to re-assess the role interferential photography played in the history of colour photography processes. A time to bring together researchers from highly diverse backgrounds, collection curators and conservators concerned with questions of museography. Finally, many photographers who are passionate about old processes, including Filipe Alves, Nick Brandreth, and Darran Green, whose works can already be found in museum collections, are making very high-quality contemporary interferential photographs. Surely all of that is testament to a profound gift of fertility.

Translated by Simon Coss.

Works Cited

Becquerel, Edmond. *La Lumière, ses Causes et ses Effets*, Paris: Didot, 1868.

Boulouch, Nathalie. 2006. "Autochromes and Pictorialism: An Element of Color in a Monochrome Universe." In *Impressionist Camera. Pictorial Photography in Europe, 1888–1918*, edited by Francis Ribemont and Patrick Daum, 268– 283. London, New-York: Merrell.

Cros, Charles. 1869. *Solution Générale du Problème de la Photographie des Couleurs.* Paris: Gauthier-Villars.

Ducos du Hauron, Alcide. 1897. *La Triplice Photographique des Couleurs et L'imprimerie. Système de Photochromographie Louis Ducos du Hauron. Nouvelles Descriptions, etc.* Paris: Gauthier-Villars et fils.

Ducos du Hauron, Louis. 1869. *Les Couleurs en Photographie: Solution du Problème.* Paris: A. Marion Éditeur.

Frédéric Dillaye. 1894. "La Chromophotographie." *Les Nouveautés Photographiques.* Paris: A la Librairie Illustrée.

Lippmann, Gabriel. 1891. "La Photographie des Couleurs", *Comptes Rendus des séances hebdomadaires de l'Académie des Sciences* 112 (5), 274–275.

Lippmann Gabriel. 1908. "Nobel lecture", Dec. 14, 1908. Online at https://www.nobelprize.org/prizes/physics/1908/lippmann/lecture/

Lumière, Auguste and Lumière, Louis. 1893. "Sur les Procédés Pour la Photographie des Couleurs d'après la Methode de M. Lippmann." *Bulletin de la Société Française de Photographie* (2) 9: 249–251.

Lumière, Auguste and Lumière, Louis. 1895 " La photographie des couleurs, ses méthodes et ses résultats", *Revue générale des sciences pures et appliquées* 23:1034-1038.

Puyo, Constant. 1898. "À Propos des Envois de Vienne et de Hambourg au Salon de Paris", *Bulletin du Photo-club de Paris.* no.90 (July): 224.

Rameau, Alexandre.. 1894. "Les Savants Contemporains. M. Lippmann", *La Science illustrée .* (February): 10.

Steichen, Edward. 1908 "Color Photography" *Camera Work* 22 (April): 14.

Stieglitz, Alfred. 1907. "The New Color Photography-A Bit of History", *Camera Work*, n° 20 (Oct): 20.

Vidal, Léon. 1901. *Exposition Universelle Internationale de 1900 à Paris. Rapports du Jury International. Classe 12: photographie. Rapport de M. Léon Vidal.* Paris: Impr. nationale.

About the Author

Nathalie Boulouch is an associate professor at Rennes 2 University. She teaches history of contemporary art and history of photography. Between 2014 and 2019, she was director of the Archives de la critique d'art and coordinated research programmes exploring the archives of art critics on different topics. Since completing her PhD on autochrome photography in France (1903–1931), her main research has focused on the history of colour photography (19th–20th centuries) from a technical, cultural, artistic, and critical point of view.

Index

Page numbers in *italics* refer to illustrations.

For Product Safety Concerns and Information please contact our EU representative GPSR@taylorandfrancis.com Taylor & Francis Verlag GmbH, Kaufingerstraße 24, 80331 München, Germany

Batch number: 10722100

Printed by Printforce, the Netherlands